Counterculture
Green

CULTUREAMERICA
Karal Ann Marling
Erika Doss
Series Editors

COUNTERCULTURE GREEN

The *Whole Earth Catalog* and American Environmentalism

Andrew G. Kirk

University Press of Kansas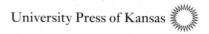

© 2007 by the University Press of Kansas

Published by the University Press of Kansas (Lawrence, Kansas 66045), which was organized by the Kansas Board of Regents and is operated and funded by Emporia State University, Fort Hays State University, Kansas State University, Pittsburg State University, the University of Kansas, and Wichita State University
Library of Congress Cataloging-in-Publication Data
British Library Cataloguing-in-Publication Data is available.
Printed in the United States of America

Library of Congress Cataloging-in-Publication Data
Kirk, Andrew G., 1964-
  Counterculture green : the Whole earth catalog and American environmentalism / Andrew G. Kirk.
      p.   cm.
  Includes bibliographical references and index.
  ISBN 978–0–7006–1545–2 (cloth : alk. paper)
  1. Environmentalism—United States—History—20th century.
  2. Counterculture—United States—History—20th century.
  3. Counterculture—California—San Francisco—History—20th century.
  4. Technology—Environmental aspects—United States.
  5. Appropriate technology—Catalogs—History. 6. Brand, Stewart.
  7. Whole earth catalog. 8. Coevolution quarterly.
  I. Title. II. Title: Whole earth catalog and American environmentalism.
  GE197.K58 2007
  333.720973'0904—dc22   2007028130

10 9 8 7 6 5 4 3 2 1

The paper used in this publication is acid-free and 100 percent recycled from postconsumer waste. It meets the minimum requirements of the American National Standard for Permanence of Paper for Printed Library Materials Z39.48–1992. ♾

For Quinn

Contents

Preface and Acknowledgments

Counterculture historians Michael William Doyle and Peter Braunstein describe the American counterculture as "an inherently unstable collection of attitudes, tendencies, postures, gestures, 'lifestyles,' ideals, visions, hedonistic pleasures, moralisms, negations, and affirmations." They use the term only after thoughtfully historicizing its path from analytical concept to near-useless catchphrase. First coined by sociologist Talcott Parsons in the 1950s and later popularized by Theodore Roszak's *The Making of a Counter Culture* (1969), by the early 1970s the term had been reduced to a catchall for all "1960s era political, social, or cultural dissent, encompassing any action from smoking pot at a rock concert to offing a cop."[1] Braunstein and Doyle conclude that perhaps the only way to fit the "anti-linear, anti-teleological . . . countercultural mode" into history is to avoid the conflation of counterculture with subculture and instead focus sharply on the specific activities of individuals and organizations who worked to create alternative ways of thinking and living during the 1960s and 1970s.

Many of the protagonists of this book did not like the term *counterculture* and never considered themselves hippies. Some of the central contributors to *Whole Earth* and the Point Foundation, like environmentalist Huey Johnson, were not countercultural in any sense other than that they were exceptionally innovative. *Whole Earth* succeeded as well as it did because its contributors were as eclectic as its subject matter. Still, *Whole Earth* was clearly a product of the counterculture moment and the vital "countercultural mode" that helped shape American history far beyond fashion, music, drugs, and sex. One of the ways to see beyond the stereotypes the term *counterculture* conjures is to look at the pragmatic environmental sensibility, a counterculture green, captured best on the pages of *Whole Earth*. Reconciling nature and culture toward a sustainable future was a theme that ran through the counterculture in many forms from music and literature to design and marketing, and was a central theme of *Whole Earth*.

Culture/Nature debates, particularly those focused on whether technology is good or bad, have to a certain extent limited the appeal of American environmentalism. Despite the remarkable environmental revolution of the twentieth century, many Americans are not actively engaged with environmental politics, in part because the challenges seem so daunting and the solutions so often out of sync with the way we really live day to day. Environmental and cultural historians have explored this problem for the past twenty years—in fact, it has been the focal point of debates within the evolving field of environmental history. Over the past five years, academic discussions have moved into the popular media because of growing concerns about global warming and renewed scarcity—fears that have sparked the most significant environmental debate since the early 1970s. Commenting about *World Changing: A User's Guide for the 21st Century*, the latest and best *Whole Earth* spin-off, *Whole Earth's* creator Stewart Brand praised the hefty guide to green living for signaling and accelerating "a crucial shift toward tech-friendly green thinking and action." The same can be said for *Whole Earth*, which "signaled and accelerated" a critical reevaluation about the relationship between nature and postindustrial culture and helped build the intellectual foundation for the pragmatic everyday environmentalism presented in *World Changing*. At times Brand took his enthusiasm for technology in directions that alienated environmentalists, including long-term supporters and contributors to *Whole Earth* and the Point Foundation. His enthusiasm for space colonies in the 1970s and his recent advocacy for nuclear power and genetic engineering caused much controversy. Still, historically it was more often the case that Brand brought together individuals who supported subtler heresies and a broader environmental sensibility that bolstered the efforts of the more traditional environmental groups while offering a forum for maverick environmentalists working to reconcile their technological enthusiasm and human-centered pragmatism with their love of nature.

This book explores an important trajectory of the environmental movement with an eye toward telling the stories of the many individuals and organizations who worked between the 1950s and the 1970s to construct a philosophy of pragmatic environmentalism that celebrated human ingenuity directed toward sustainable living in many forms. *Whole Earth* became a critical forum for environmental alternatives and a model for how complicated ecological ideas could be presented in a hopeful and even humorous way. The ecological thinking woven throughout the catalogs was sometimes so different for its day that many would not even recognize the content as environmentalist at all. Yet the research fostered and presented through *Whole Earth* enabled later environmental advocates like Al Gore to explain our current "inconvenient truth," and the actions of *Whole Earth's* Point Foundation showed how the epistemology of the catalogs could be put into action in meaningful ways that might foster an environmental optimism distinctly different from the jeremiads that became the stock in trade of American environmentalism.

Whole Earth is a fruitful subject for research, and many people are exploring its history from a variety of perspectives. The two most significant books, Fred Turner's *From Counterculture to Cyberculture* and John Markoff's *What the*

Dormouse Said, focus insightfully on the way that *Whole Earth* was a focal point for early thinking and research in personal computing and the information revolution it spawned. What most distinguishes this book from this recent literature is a sharper focus on the environmental thinking for which *Whole Earth* was most celebrated in its day. Also, I have tried to place *Whole Earth* within a western regional context and explore the ways that the publication tapped into deep regional traditions (libertarian-leaning individualism and distrust of centralization, for instance) and refashioned them for a new generation. The focus on regional history raises questions about the extent to which utopian communitarianism was key to understanding *Whole Earth.* In fact, I argue, it was its enlightened pragmatic individualism that explains the catalog's wide appeal.

While western individualism is a particular theme of this book, the history of *Whole Earth* is also a story of collaboration and mentors. Ralph Waldo Emerson said, "My chief want in life is someone who shall make me do what I can." Most of the protagonists of this story had fortuitous meetings with remarkable mentors, meetings that so inspired them they spent the rest of their lives mentoring others. At its best *Whole Earth* was a forum for mentors who enabled an inventive generation to do what they could. Likewise, as the teller of this version of the *Whole Earth* story, I am indebted to many mentors who led me to *Whole Earth* as a research topic and helped me during the research and writing of this book. I saw my first *Whole Earth* in my junior high friend Mike Reed's basement. I don't remember much about it except that I thought it was weird and cool. Thanks, Mike. The greatest debt I owe is to the historical actors who were uniformly kind and generous during the research of this book.

Because many of the protagonists of this history were quite young when they began making significant contributions to American culture, they are mostly alive and very actively engaged in their work. They are the focus of this study because they were exceptionally talented, remarkably productive, and genuinely innovative—and still are. I hope that the many contributors to the *Whole Earth* publications will appreciate this version of their story, although I realize some may not. Stewart Brand has been the subject of journalists, photographers, and writers for most of his adult life. Anyone who has endured the kind of media scrutiny that Brand has might be understandably reluctant to share his time with a new group of pesky scholars writing in an academic forum he famously rejected. But Brand was gracious and patient with my questions over the four years I spent researching this book. I greatly appreciated his willingness always to take my calls, reply to my correspondence, and carefully answer my questions. Three other key figures in the environmental history of *Whole Earth* were equally generous with their time. Jay Baldwin provided a critical overview of his own life, his career with *Whole Earth,* and his work as an ecological designer. I owe him special thanks for a tour of the tool truck, and for thoughtful help with aspects of my manuscript. Likewise, Peter Warshall endured my questions with humor and patience. His insights were particularly invaluable and helped shape this book in many ways—some obvious, some not. I especially thank Peter and Diana for the warm welcome for my family and the wonderful dinner underneath the Arizona stars. Finally, Huey Johnson went out of his way on several occasions to accommodate my requests for meetings and images and graciously let me use the offices of the Resource Renewal Institute for copying and the study of his personal papers. Internationally recognized as an innovative environmentalist, Huey is also a wonderful storyteller and keen observer of human nature.

I also had the opportunity to speak with John Perry Barlow, Steve Baer, Lloyd Kahn, Bill Bryan, Ilka Hartmann, Jay Kinney, Michael Phillips, and Dick Raymond. Stewart Brand, John Perry Barlow, Bill Bryan, Ilka Hartmann, and Michael Phillips also contributed images from their collections. Dick Raymond and Stewart Brand deserve special thanks for donating their wonderful collections of *Whole Earth,* Point, and Portola materials to Stanford. Mauna of the Auroville community helped secure images from India, and Tom Frost and Glen Denny were generous with their remarkable collections again. Many of my images and most of my research materials came from the Department of Special Collections at Stanford University. Director Roberto Trujillo was especially helpful with issues of access and use. Likewise, Polly Armstrong, Mattie Taormina, and all of the staff helped maximize my time on many hectic research trips.

Thanks once more to Virginia Scharff for being a great mentor and for introducing me to Michael Doyle and Peter Braunstein in 1998, who along with several reviewers provided critical early comments and published my first *Whole Earth* research efforts in their book *Imagine Nation.* Friend and mentor Tim Moy taught me more than he knew and died tragically before I could say a proper thank you. Anyone interested in technology should read his brilliant work. I was lucky to know him. The editorial staff and a series of anonymous readers for *Environmental History* also gave early support and excellent critical suggestions for revision. Likewise Mark Carnes and reviewers for Columbia University Press provided feedback on post–World War II context. Conference panel participants and audience members too numerous to name helped with gentle shoves as I moved along the process. Of these, Jeff Roach, David Weber, and Sherry Smith of Southern Methodist University's Clements Institute deserve special thanks for organizing and sponsoring the Political Legacies seminars in Taos and Dallas. Sincere thanks to Jennifer Price, Mark Fiege, Amy Scott, John Herron, Char Miller, CultureAmerica editors Erika Doss and Karal Ann Marling, and anonymous reviewers who provided insightful criticism during this project. Special thanks to Erika Doss for critical language about American Publishing and *Life* magazine. Generous colleagues and fellow *Whole Earth* researchers Fred Turner and Simon Sadler shared their significant insights, as did the remarkable participants of the Technocultural Studies *Whole Earth* symposium, at University of California–Davis, organized by Simon.

My longtime University Press of Kansas editor, Nancy Jackson, helped instigate this project but left the press before I finished. She deserves credit and thanks for pushing me to expand an essay into this book and for steadfast enthusiasm for my ability to do so. Kalyani Fernando bravely took over for Nancy and brought the project home in good style and with great enthusiasm. Also at UPK, director Fred Woodward, Michael Briggs, Susan Schott, Sara Henderson White, and Susan McRory reminded me through their professionalism and patience why Kansas is such a great press.

At the University of Nevada–Las Vegas, many friends, colleagues, and graduate students contributed to this project. When I arrived, Andy Fry was the chair of my department and also my assigned faculty mentor. A better mentor I could not have hoped for. Through the years, Andy has been an excellent and generous adviser and a good friend to me and my family. He read all of this book in raw form and as usual provided insight and excellent advice for revision. Hal Rothman was chair when I applied to the UNLV Faculty Senate for the sabbatical leave that provided the time to finish writing this book when public history demands had sidelined the project. Hal was a force of nature and one of my strongest supporters over the years. Few people have had such confidence in my abilities, and Hal always made sure that he shared his abundant confidence and legendary energy with me when mine was flagging. I don't expect to ever meet anyone like him again in this life—he was a true original. I owe David Wrobel special thanks for careful readings of the entire manuscript in several forms while he had very little time to spare and two babies on the way—a true friend. Elizabeth Nelson also read parts of the manuscript and shared her cultural history insights and good cheer. Gene Moehring, Kathy Adkins, Lynette Weber, Elizabeth Fraterrigo, Michelle Tusan, Todd Jones, Bret Birdsong, Robert Futrell, Marc Langia, and Mary Palevsky all shared their enthusiasm for my project. Laurie Boetcher did a fantastic job with oral history transcription. Thanks also to graduate student researchers Mary Wammack, Michelle Lettieri, Mike Childers, Chip Palmer, Leisl Carr Childers, and Chris Johnson for discussions and help along the way.

Finally, thanks to Art, Stephanie, Gordon, and Gus Evans for a base camp in Marin, vital encouragement, great food, and surfing during the research of this book. Lisa, Harrison, and Quinn traveled along to meet interesting people and see unusual sites with their usual enthusiasm. Quinn derailed my research schedule with her ill-timed arrival, but more than made up for it by brightening our lives every single day; this book is dedicated to her.

Counterculture
Green

INTRODUCTION
One Highly Evolved Tool Box

We *are* as gods and might as well get good at it. So far, remotely done power and glory—as via government, big business, formal education, church—has succeeded to the point where gross obscure actual gains. In response to this dilemma and to these gains a realm of intimate, personal power is developing—power of the individual to conduct his own education, find his own inspiration, shape his own environment, and share his adventure with whoever is interested. Tools that aid this process are sought and promoted by the *Whole Earth Catalog.*

Stewart Brand, Whole Earth Catalog, *1968*

In March 1968, while flying over Nebraska returning from the funeral of his father, countercultural innovator Stewart Brand concocted an idea and scrawled it over the end pages of Barbara Ward's environmental book *Spaceship Earth.*[1] Trying to think of a practical means to save "spaceship Earth" and "help my friends who were starting their own civilization hither and yon in the sticks," Brand envisioned a blueprint for a new kind of information delivery service.[2] He wanted a system modeled on the L.L. Bean catalog, which he viewed as a priceless and practical "service to humanity." Brand's counterculture version of L.L. Bean would be a "catalog of goods that owed nothing to the suppliers and everything to the users."[3] He hoped to create a service that would blend the liberal social values and technological enthusiasm of the counterculture with the emerging ecological worldview he cultivated as a Stanford University biology student. Further, he wanted to provide information on emerging ideas about "appropriate technologies" and commonsense advice for individuals who wanted to participate in the process of invention he hoped might lead to a new environmental culture in sync with the technological enthusiasm of one wing of the counterculture. "So many of the problems I could identify came down to a matter of access," he remembered. "Where to buy a windmill. Where to get good information on beekeeping. Where to lay hands on a computer."[4] Having spent several years traveling the American West, visiting Indian reservations and the network of intentional communities that were springing up all over the region, Brand wanted to come up with an information system that could connect these dispersed like-minded folks. "Shortly, I was fantasizing access service," an "Access Mobile . . . with all manner of access materials and advice for sale cheap."[5] He settled on a "truck store, maybe, traveling around with information and samples of what was worth getting

1

and information on where to get it" or a "catalog, continuously updated, in part by the users."[6]

In almost any used-book store in America, one may find, slumped in a corner, a couple of unwieldy, and slowly disintegrating, *Whole Earth Catalogs*. Too big to fit on any normal shelf, about the size of a road atlas but five times as thick, they are relegated to the oversized area or just dumped in a convenient corner. Into which corner to do the dumping has caused a generation of store owners, distributors, and librarians considerable problems. Unless you are willing to give the catalogs some serious time and thought, you will be hard pressed to decide in which section they belong or how to categorize them. They could go in history, culture, landscape architecture, survival, home improvement, outdoor adventure, religion, agriculture, technology, lifestyle, craft, tools, information, architecture, or environment and nature.

But that was the point. Like equally oversized *Life* and *Rolling Stone*, *Whole Earth* appealed to readers because of its unusual size and creative text-image format. Its size, scale, and general initial incomprehensibility suggested that the catalog was almost too full of ideas and subjects to come to terms with an overall synthetic view of them all. For some readers, *Whole Earth* remained a mystery that they were content never to unravel. Trying to puzzle out the linkages between the disparate ideas was actually part of the appeal, even if it was also a source of frustration. Brand and his collaborators wanted to generate a holistic, synthetic guide to modern life that defied reductive categorization and promised all readers a return to personal, individual agency and autonomy. The ecological sensibility conveyed through *Whole Earth* captured a new alchemy of environmental concern, small-scale technological enthusiasm, design research, alternative lifestyles, and business savvy that created a model of environmental advocacy that bears closer study. The thesis of this book is that the genuinely holistic and human-centered environmental pragmatism expressed through *Whole Earth* represented a more viable path toward a sustainable world in the twenty-first century than the guardian path of Progressive regulation alone could ever provide.

The *Whole Earth Catalogs* were large and ephemeral at first. Self-published on low-quality newsprint and stapled together, those copies that survive are fragile relics of a time before computers and desktop publishing. Even new, they smelled of ink and earthy pulp. The early covers were smooth, heavy, black paper with startling images of the bright blue and white ball of the Earth floating in an endless sea of black space and black ink. No one could see this remarkable publication and fail to pick it up and flip through its massive pages; the format was captivating. The first edition featured fifty-one carefully handcrafted black and white pages profusely illustrated with photos of books, designs, art, people, and machines, all accompanied by wonderful tiny essays and reviews exploring complicated topics like appropriate technology, whole systems, and cybernetics.[7] The reviews, "written as if a letter to a friend," were Brand's singular contribution to American literature.[8] As the catalogs aged, the photos and text took on a sepia tone as sun and air turned the pages a deeper yellow with browned edges that only accentuated the convergence of technology and nature, old and new, that the catalogs conveyed through their content. Over time, the catalogs grew in size, reaching maximum density with the huge 1980 *Next Whole Earth Catalog,* running to an amazing 608 pages and three-plus pounds. Production quality also increased, with sophisticated color covers and higher-quality interior design features. The catalog sold millions, and in 1972 won the National Book Award.

The catalogs appeared at irregular intervals over a period of thirty-plus years, sometimes quarterly, sometimes with several years between versions. Even the "last" catalog had many editions and remained in print for years. Two quarterly magazines emerged to provide more regular coverage, the *CoEvolution Quarterly* and later the *Whole Earth Review;* a semiregular series of "supplements," thematic books, and even a briefly nationally syndicated newspaper version of *Whole Earth* rounded out the remarkable run of this unique publishing venture. Counting the magazine-format versions and supplements, there were more than 150 different editions of the publication during its long run.[9] Additionally, virtually all of the primary contributors to *Whole Earth* published at least one book that built on their work with the catalogs, and Brand himself authored several critically acclaimed books. The pathbreaking format also inspired a host of imitators who created a subgenre of "whole" something-or-other catalogs. All this activity resulted in a remarkable library of alternative intellectual activity that benefited several generations of readers and provides a significant archive for scholars.

Brand's creation capitalized on a critical cultural moment by offering a tantalizing burst of optimism in one of the bleakest periods in American history. In a year marred by the assassinations of Martin Luther King Jr. and Robert Kennedy, the debacle at the Democratic National Convention, and the alarming Tet Offensive in South Vietnam, *Whole Earth* was a breath of fresh air for a generation suffocating on relentless bad news. Above all, *Whole Earth* presented an optimistic view of the future from the now iconic, but then very new, image of the Earth floating through space on the front cover, to the hopeful back-cover message, "We Can't Put it together, it is together." *Whole Earth* offered a simple proposition to readers: Empower yourself in an increasingly homogenized modern culture through access to creative information about alternate paths and good tools to get the job done.

Stewart Brand famously said, "Information wants to be free."[10] The catalog gave the information its freedom with only concise, insightful, and subtle signposts to lead readers gently into a new way of looking at the world. One of the

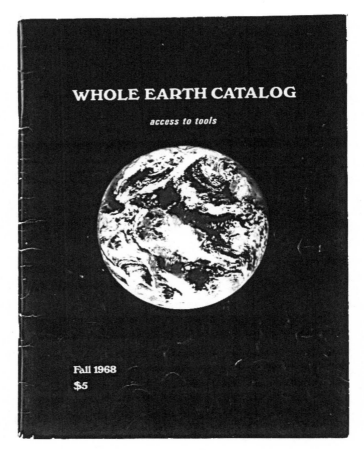

The first *Whole Earth Catalog* from fall 1968. Reflecting upon the thirtieth anniversary of the catalog, editor Peter Warshall said that the catalogs "fed a deep hunger in America—a hunger to know new stuff not taught in schools." Restoring agency to learning became a guiding principle of the catalog and the constellation of enterprises it spawned. (Courtesy Stewart Brand)

guiding principles of the *Whole Earth* was the desire to present information in as unmediated a fashion as possible. "Except for occasional self-dubious asides," Brand wrote, "I think we should deny our contemporaries and ourselves all this fucking interpretation. Such indulgence leads rapidly to advertising language such as 'lifestyle,' where we are the commodity. We are far better employed as journalists or field anthropologists, immersed in story. Let other poor drabs do the critiquing."[11] The pages of the catalog brimmed with positive book reviews—negative reviews were never printed—and thoughtfully culled information that forty years later still seems fresh and different.

Whole Earth was one of the best examples of the changing world of magazine publishing and journalism in the 1960s and 1970s. Magazines like *Sunset, Popular Mechanics*, and *Life* set an early standard for popular general information magazines, with a twist that *Whole Earth* followed. *Popular Mechanics* featured the kind of how-to essays and articles that would become a central feature of *Whole Earth*, and, like Brand's later creation, it was archived and shared by a do-it-yourself–obsessed generation, giving it a readership that exceeded its actual print runs.[12] Starting in 1898, *Sunset* became the leading voice for an emerging western regionalism and eventually an influential forum for regional design and nature appreciation aimed at middle-class readers who were immigrating to the region in large numbers during the twentieth century.[13] Although intended for very different audiences, *Whole Earth* and *Sunset* shared a fundamental desire to link regional traditions to modern design while celebrating the authenticity of western nature and its seemingly limitless recreational possibilities. Like *Sunset*'s most dynamic regionalist, Laurence W. Lane, Stewart Brand provided countercultural immigrants to the West with regionally specific design and environmental information at a time when the publishing industry still had a decidedly East Coast bias.

The magazine that *Whole Earth* most physically resembled was floppy, large-format *Life*. *Life*, especially in the 1950s, was an eclectic mix of politics, culture, and business. In 1968, when Brand started working on the first *Whole Earth*, *Life* was on its way out and *Whole Earth* stepped in to fill the void for a new generation. Not unlike *Life*'s dynamic publisher, Henry Luce, Brand wanted to manage information to shape and direct American attitudes on subjects he deemed most important for a new generation. Throughout his professional life, Brand's talent for promoting himself and his ideas generated hundreds of articles, an appearance on the *Dick Cavett Show*, and extensive media coverage, mostly positive, for almost everything he ever did. Brand's media savvy ensured that *Whole Earth* became the most successful and enduring example of the underground media that profoundly influenced American publishing in the last third of the twentieth century.

The emergence of a new underground media in the late 1960s converted some of the diverse components of the messy countercultural phenomenon into significant trends that both reflected and influenced the broader American culture. By "disseminating knowledge and validating shared understandings," underground publications like *Whole Earth* were able, according to historian Beth Bailey, to "create community by establishing the central elements of that specific community's countercultural identity."[14] More specifically, the underground press, Robert Gottlieb has observed, "became champions of a new ecological consciousness and related way of life."[15]

The environmental pragmatism promoted by *Whole Earth* is only now, in the first decade of the twenty-first century, taking hold. It has taken so long because the spectacular success of environmentalism as a collective ideological revolution obscured efforts to work toward solutions aimed at individuals and because the organized environmental movement succeeded in part by fostering a techno-phobic declensionist narrative. By pitting humankind against nature, environmentalists unwittingly contributed to what historian James Livingston has called the "tragic frame of acceptance" that became a common narrative form in American literature and an unfortunate characteristic of the politicized environmental movement as it entered the 1970s. According to Livingston, the tragic frame can blind us "to the possibilities of our own postmodern time" and our ability as individuals to affect positive change. Further, he suggests that the tragic frame of acceptance exiles us "from the present" and reduces our ability to "acknowledge the good faith and political capacity of our fellow citizens."[16] A declensionist narrative and focus on preserving areas where humans weren't lent a pessimistic tone to the environmental movement even during its high point in the late 1960s and early 1970s. Human ingenuity and inventiveness did not seem to have a place in the environmental movement. *Whole Earth* and the appropriate technology movement it promoted offered a potential solution to this profound conundrum.

Whole Earth was a cynosure for the emerging appropriate technology movement that found its voice in the underground press before moving slowly toward the mainstream. The key insight of the appropriate technology movement was the idea that individuals working within specific local environments could make everyday choices to use small-scale technology, enabling, if multiplied across a nation, a sustainable economy. Appropriate technologists celebrated human ingenuity at a time when environmental advocates tended to draw a clear line between people and nature, with preference given to the latter. Many of the ecological arguments made through the choice of material presented in the catalogs, or explicitly by editors and contributors, were so far outside the mainstream of the environmental thought of the day that they were considered heresies. The environmental views of Brand and his publication alienated some who might have been allies and who would not remember *Whole Earth* as a voice of environmentalism, while creating a very strong bond with techno-ecological readers and contributors who found a welcome forum for their views in Brand's catalogs.[17]

In a letter to Stewart Brand, fan and documentary filmmaker Mark Levenson wrote, "Since the initial publication of the first *Whole Earth Catalog* . . . there has been an evolution in the American style of living." "The cause and effect relationship," he continued, "between this change and the catalog's existence is

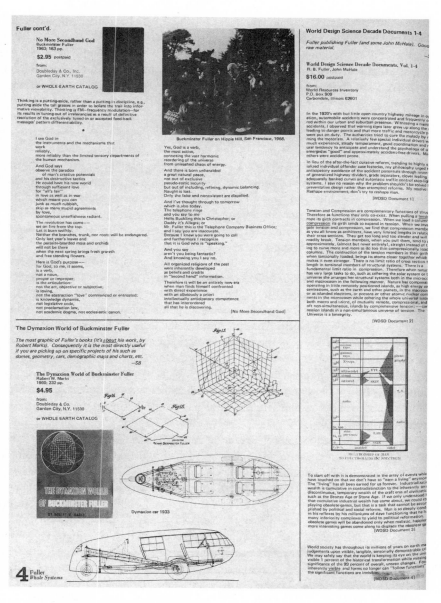

The first pages of the first *Whole Earth Catalog* through the *Updated Last Whole Earth* led off with a Whole Systems section prominently featuring the work of Buckminster Fuller and his "Dymaxion" design philosophy. (Courtesy of Department of Special Collections and University Libraries, Stanford University Libraries)

not absolutely clear. But we do know that for those who have made the basic decision to alter the living patterns as they were in the 'sixties' the Catalog has been an essential guide book—access to tools, physical and mental."[18] Levenson echoed what a generation of Americans learned from firsthand experience and extensive popular media coverage of the catalog—that somehow this strange book was a key ingredient to a period of significant cultural reinvention. In the past ten years, a growing number of scholars have searched for the cause-and-effect relationship that Levenson knew was there but could not quite identify in 1975.[19] I am studying the *Whole Earth Catalog* because, like Levenson, I think that there is a direct link between the catalogs and the rise of a particularly western pragmatic environmental sensibility, with a special emphasis on the union of technology and nature. This union is of great importance to our understanding of the emergence of the environmental movement in the post–World War II years and to our ability to craft thoughtful realistic responses to the ecological problems of the twenty-first century.

There is a rich literature on the nexus of technology and nature in American culture, with Leo Marx's classic study *The Machine in the Garden* (1964) the most notable. Marx elegantly depicted one of the central tensions in American culture—the desire to reconcile our technological prowess with our love of the extraordinary nature that characterizes North America—and shaped our early conception of democracy and history. He pays special attention to the goal of several generations of American writers and thinkers who fostered hope for a "middle landscape" where tools and nature harmoniously united.[20] Historians of technology have long reminded us of the ways that Americans have linked their appreciation of nature with a love of technical innovation.[21] Technology has been a dominant theme in American history partly because "the deep implications and lasting effects of our technological innovations almost always elude us."[22] The hope for a middle landscape faded for many in the years following the technological devastation of World War I. Still, the majority of Americans, as historian of technology Thomas Hughes has observed, retained a steadfast faith in American ingenuity and the ability of technological progress to solve future problems until the decades following World War II, "when the temper of the times began to change markedly."[23]

The postwar period was a time of unprecedented technological growth spawning much ambivalence about the power of our tools to remake the world and resolve concerns about resource scarcity. For a generation coming of age in the 1960s, *Whole Earth* became a forum for reevaluations of the tangled and shifting relationship among design, science, consumption, and ecology in postwar America.[24] Historian of technology Timothy Moy compared the thoughtful reevaluation of science and technology in *Whole Earth* to Thomas Kuhn's notion of a "paradigm shift." Kuhn's handy nomenclature for describing fundamental redirections of thinking and research has been much overused since its appearance in 1962. But his thesis about the structure of scientific revolutions grew out of the same 1960s intellectual ferment that inspired Brand and his collaborators to present a new perspective on technology and science for their peers. Kuhn's formulation "provided further ammunition for the counterculture critique of technocracy by highlighting the value-laden character of scientific revolutions."[25] One of *Whole Earth*'s central revelations concerned the nature of scientific and technological authority. By reclaiming an amateur tradition of invention and technological development and celebrating an ecological focus to technological research, *Whole Earth* provided moral support for young optimists working to map a brighter future free from flaws of technocratic thinking but not free from technology. These appropriate technologists believed a survivable future was still a possibility if technological development could be wedded to insights emerging from ecology and environmentalism while avoiding the political entanglements of Right/Left ideologies.[26] In a manner so subtle it is easy to miss, Stewart Brand and his collaborators incubated an environmental paradigm shift of critical importance to the goal of a sustainable future.

A closer look at the environmental thinking conveyed through *Whole Earth* does not provide an easy road map to a technologically enthusiastic environmental pragmatism or a program for a postindustrial environmentalism, but it does demonstrate some of the best efforts in this direction—efforts that deserve closer scrutiny and acknowledgment. Stewart Brand's intense interest in ecological living but deep ambivalence about environmentalism as an ideology enabled him to see the potential of alternative environmental paths more clearly than most of his generation.

Moreover, a history of the environmental sensibility conveyed through *Whole Earth* offers an opportunity to explore the social and cultural construction of notions of appropriate technology and the pragmatic society that might emerge if only Americans worked to unite their technological enthusiasm and desires for a healthy environment. *Whole Earth* provided a publishing vehicle and an audience for a new generation of ecologically minded "outlaw" designers who were exploring the possibilities of ecologically sensitive housing, industrial design, energy production, and farming. The catalog figures very prominently in the history of the ecological design movement, and by presenting the work of promising ecological designers decades before most Americans had heard of the idea, *Whole Earth* helped create a discipline that had good practical answers to some of the nagging questions of environmentalism. How do you reconcile consumption and materialism with the principles of ecology? How can you help shape a meaningful everyday environmentalism that empowers average people to make a difference in the world? How can you market those ideas without selling out?

While reading *Whole Earth*, it's tempting to focus on the utopian episodes that erupted periodically on the pages and spawned some of the most interesting debates among readers and contributors. Aspects of *Whole Earth* over the years were certainly utopian, with a spirited debate about space colonies perhaps the most interesting. As a new century dawns, though, far more intriguing are the ways that behind this veneer of utopian enthusiasm lay a surprisingly pragmatic vision of environmentalism that offered an alternative set of potential solutions to deeply rooted problems and the opportunity to dramatically expand the constituency of the environmental movement.[27] Trends that were initially utopian often get tempered by time, evolving into more practical versions of the revolutionary thinking that spawned a period of great creativity. Insights and products from the tempered revolution then work their way into culture, politics, and markets in ways that are sometimes obscured by assumptions about the perceived failure of the initial utopian programs. At first glance, the Ecotopian dreams that graced the pages of *Whole Earth* might appear as failed as the broken-down geodesic domes that litter the landscape of the rural West, but the alternative technology research and design that the counterculture spawned worked its way into American culture to a greater degree than most might assume. To see how, one must look beyond the utopian aspects of *Whole Earth* and explore the pragmatic dialog about the nature/culture relationship that sustained the publication beyond its initial burst of popularity and fostered an intellectually diverse community that wanted to, in the words of geographers Bruce Braun and Noel Castree, "retie rather than endlessly attempt to untangle, the Gordian knot between nature and society."[28]

In the early part of the twentieth century, pragmatists such as John Dewey, with his emphasis on instrumentalism as opposed to the spectacular theory of knowledge, laid the intellectual groundwork for future generations of optimists like Brand. The foundation of pragmatism rests on the belief that ideas are tools—tools, as Louis Menand reminds us, like "forks and knives," that should be used to accomplish a purpose.[29] Dewey spoke of ideas as plans of action, means to ends, tools for effecting social change. It is that symbiosis of the idea and the social environment that made pragmatism so influential and perhaps the only distinctly American contribution to philosophy. American pragmatists, like Dewey and William James, pointed to a practical, problem-solving, tool-using, environment-transforming strain of American culture that, historian David Wrobel reminds us, is what many foreign observers have commented on as the most distinctive element of the American mind and the foundation of evolving conceptions of American exceptionalism.[30]

American pragmatism was characterized by an enormous optimism in our ability as a people to effect change when we put our minds to it. That optimism made a household name out of John Dewey precisely because Dewey and the movement for which he was the most visible and longest-lived popular representative offered a hopeful vision of the future during a period of astounding change. The ecological message expressed through the millions of copies of *Whole Earth* sold from 1968 through the 1990s provided a hip new spin on this pragmatic tradition. Dewey never presented his readers with images of nude women modeling lightweight backpacks, but Dewey and Brand shared some fundamental ideals, nonetheless. Like the pragmatists who preceded him, Stewart Brand provided readers with a captivatingly optimistic vision of the future and sources and suggestions on how to travel a different road toward sustainability and ecological equilibrium.[31]

The loose coalition of readers and contributors to *Whole Earth* used the publication as a forum for ideas about alternatives to the legislation- and regulation-driven environmental advocacy that achieved great gains during the twentieth century, but by the end of the 1960s was already showing its weaknesses: polarization on the Left and Right, grinding slowness, and emphasis on national scale programs that could ignore local issues and foster racial and class-based animosities. The model of individualistic ecological living represented in the catalogs offered a corrective to the impersonal top-down activism institutionalized by the end of the 1960s. *Whole Earth* presented an early model for reconciling environmentalism, technology, and consumerism, and an opportunity to help create design guru Buckminster Fuller's ideal of an apolitical design science revolution, where thoughtful attention to the front end of production could preemptively solve potentially detrimental environmental consequences.[32] This constructive human-centered environmental philosophy tapped deep into conservative impulses like thrift, ingenuity, technical know-how, tinkering, and individual responsibility and agency.

Frederick Jackson Turner identified some of these same impulses as key intellectual traits of the American pioneer and talked about the practical ingenuity required to transform nature. Turner is more a precursor of the counterculture tool guys than one might initially presume. All talked extensively about adopting Indian ways, crafting canoes from raw materials, and using whatever tools were most appropriate to effect a transformation of self and place. Like Turner's pioneers, counterculture environmentalists were obsessed with new frontiers and the tools required for conquest. Unlike earlier pioneers, however, their stage was wider than wilderness alone, and they found their frontiers in the countryside, the city, and even space.

Whole Earth reached its zenith of cultural influence in 1972 after Brand won the National Book Award and *The Last Whole Earth* reached the best-seller lists, but it continued to exert influence disproportionate to its sales in the coming decades.[33] The catalog's environmental vision implied a clear understanding on the part of its creators of the important role technological progress and human ingenuity could play in an ecologically viable future and intuitively rejected what Cindi Katz has called "apocalyptic environmentalism."[34] Stewart Brand was never a fan of environmental "calamity callers" or jeremiads, and he consistently presented authors and innovators who focused on solutions instead of problems.[35] Although the first editions of *Whole Earth* became legendary as rural communard bibles, the enthusiasm for radical communal disconnects was short lived at the publication. As the catalog evolved, pragmatic environmentalism without a hint of wilderness romanticism became the dominant theme even as the catalog continued to feature tools and articles that celebrated rural life and the wilderness recreation experience.

Forty years later, the model of a holistic human-centered pragmatic environmental culture predicated on individual agency and technological enthusiasm captured by Brand, his *Whole Earth* collaborators, and readers still feels fresh and urgent. By the turn of the millennium, a human-centered everyday environmentalism seemed quite natural, and the environmental heresies of the 1960s have become orthodoxy by the early 2000s.[36] It is hard to remember how radical Brand's "we are as gods" ideal, which melded technological enthusiasm with an ecological sensibility, must have seemed to fellow environmental advocates in 1968, when public health and quality-of-life issues were subservient to wilderness in the mainstream movement. But, while these ideas may have seemed outrageous to many deep-green advocates in 1968, they actually struck a harmonious chord with the larger American public by tapping into a deep-rooted affinity in American culture for the ideal of making a place for the machine in the garden.

CHAPTER ONE

Environmental Heresies

> Tools are the revolution . . . A tool—any tool—is possibility at one end
> and a handle at the other. There are many ways to change the world,
> but I think the most direct way . . . is to adopt new tools.
> *Kevin Kelly,* Whole Earth Review, *Winter 2000*

"We *are* as gods and might as well get good at it," the opening lines of the first *Whole Earth Catalog* read.[1] When Stewart Brand issued his clarion call for technological acceptance in the opening lines of his first catalog, the America environmental movement was in the midst of a period of significant ideological and political reorientation. Until the mid-1960s, most environmental advocacy was aimed at preserving the American wilderness from industrial development and urban encroachment. Environmental activists, from John Muir to Howard Zahniser, focused the environmental debate on the problems of industrial technology and constructed a sharp dichotomy between nature and civilization. In this ideological tradition, wilderness became the ultimate symbol of environmental purity and abundance, with the polluted modern technological city its antithesis. This bipolar, often antimodernist, framework served conservation and preservation activists well in early fights to convince the American public of the reality of scarcity and the necessity for preservation of some forestlands and remote natural treasures.[2] This simple dichotomy was less effective, however, when applied to increasingly complex environmental and social politics after the mid-1960s. Further, it failed to take into account that American relations with nature during the twentieth century were extensively mediated by popular activities like automobile tourism, hiking, camping, and a wide variety of forms of outdoor recreation. People involved in these activities have long reconciled the contradictory connections between consumption, leisure, technology, and nature. The move to protect wilderness and the rhetoric of the wilderness movement only gave the appearance that environmentalism was a fight between technology and nature.[3]

Following the 1964 passage of the Wilderness Act, environmental activism was enmeshed in the social struggles, political upheaval, and cultural tensions of the 1960s and 1970s. A new generation of environmentally minded countercultural innovators struggled to resolve long-standing tensions between the modernist faith in progressive reform and the antimodernist distrust of technology and desire to return to a simpler time. Ironically, the success of the wilderness movement created an ideological crisis for environmentalists who found it increasingly difficult to define their movement in terms of progress in contrast to preservation. The spirit of cooperation that united a diverse coalition of environmental advocates behind the banner of wilderness disintegrated in the years following the passage of the Wilderness Act, and environmental politics became increasingly complicated and the boundaries of the debate harder to define.⁴ Almost immediately after the successful passage of the Wilderness Act, wilderness ceased to be a defining environmental issue as quality-of-life issues emanating from urban areas gained national attention. As the 1960s progressed, Americans increasingly focused less on preserving a pristine nature and more on preserving the whole environment and particularly the places where they lived and worked.

In particular, a new generation of counterculture environmentalists attempted to move beyond the progress-versus-preservation debate and redefine the parameters of the environmental movement. This countercultural environmental sensibility embraced the seemingly contradictory notion that the antimodernist desire to return to a simpler time when people were more closely tied to nature could only be achieved through further technological innovation. In looking to technology and innovation to solve environmental problems, counterculture environmental advocates tapped into a long tradition of pragmatic environmental thinking going back to Theodore Roosevelt and the wise-use conservation movement. Roosevelt and his fellow conservationists had a strong faith in progress and the ability of human ingenuity to resolve environmental problems. What differentiated the counterculture technology enthusiasts from this earlier American tradition was their rejection of traditional notions of progress. Cultural historian Peter Braunstein has argued that one of the central characteristics of the youth driven counterculture was "a deep ambivalence about teleology . . . the notion that ideas, phenomena, people should be tending *toward* something, heading in some direction."⁵ This distrust of long-term thinking was central to the psychedelic culture of LSD (lysergic acid diethylamide) that encouraged seekers to live for the moment or, even better, regress to a childlike state of perpetual wonderment. The focus on the *now* sometimes undermined the ability of counterculturalists to achieve lasting social changes and made living-in-the-moment proponents like Timothy Leary easy targets for critics who scoffed at ideas like dumping LSD into the water supply so the nation could move toward blissful utopia.

For those in the counterculture who were thinking about the environment and wondering how to restore balance between people and nature, LSD could provide useful inspiration as it did for Stewart Brand at a key moment in his life when he realized the importance of the image of the whole Earth and, with less practical, but more poetic, results, for environmental prophet Gary Snyder, who mused

Stewart Brand, in white jumpsuit (far right), hangs on to the back of Ken Kesey's Merry Pranksters bus as they ride through the streets of San Francisco in preparation for the upcoming Acid Test Graduation, October 1966. (Ted Streshinsky/CORBIS)

about a techno-utopian future in 1967: "So what I visualize is a very complex and sophisticated cybernetic technology surrounded by thick hedges of trees . . . and the rest of the nation a buffalo pasture."[6] Snyder's early drug-fueled musing illustrates a tendency that set counterculture environmentalists apart from peers mainly concerned with the popular culture aspects of the era. Environmental proponents like Snyder, even at their most psychedelic, *were* interested in progress toward something, something that united human ingenuity with the natural environment toward a more sustainable future.

Counterculture environmentalism simultaneously encompassed both anti-modernism *and* modernism. Inside the covers of the *Whole Earth Catalog,* the seemingly neat bipolar world of twentieth-century environmental politics became a messy mélange of apparently incongruous philosophies and goals united under the banner of whole systems, cybernetics, and alternative technology. It is important to stress that prior to the rise of the counterculture environmentalists, twentieth-century environmental politics only *appeared* to be neatly bipolar. In fact, the jarring juxtapositions on the pages of *Whole Earth* only exaggerated old and deep tensions in American environmental politics. Henry David Thoreau, for example, was a pencil designer and entrepreneur. John Muir began his adult life as an inventor locally renowned for his mechanical genius, and Aldo Leopold was a scientific forester. All of these men struggled to reconcile their modernist epistemology and technological enthusiasm with their desire to preserve a pristine nature.

One of the popular misconceptions about environmental advocacy in American history stems from the desire to celebrate the few individuals who advocated the preservation of a nature where humans weren't, while often ignoring those who worked to use their technological enthusiasm to benefit nature. Historical actors in the drama of twentieth-century environmental advocacy are often rated on a sliding scale according to the purity of their wilderness vision. Thoreau, Muir, and Leopold rank high based on their early, and seemingly complete, conversions to the wilderness ethic. Those who fail to make the full conversion are, for the most part, forgotten if they were ever known to the public.[7] The wilderness purity test tends to force public environmental debates toward the areas where environmental politics appear black and white and the actors in the drama are easier to pigeonhole. This overenthusiasm for wilderness prodigals in American culture fosters a misleading sense of ideological purity in environmental politics that is not supported by the historical record. Historian William Cronon has argued that "the time has come to rethink wilderness." Cronon's "trouble with wilderness"[8] stems from his belief that by venerating a mythically pure wilderness we cede ground in the rest of the environment where most of us live.[9] This scholarly insight reflects the ambivalence with the wilderness movement that drove Stewart Brand and a generation of design enthusiasts to provide a very different model of environmentalism in the pages of *Whole Earth*. The environmental ethic that emerged on the pages of the catalog and the counterculture milieu it grew from complicates the story of American environmentalism and forces one to reconsider what constitutes and constituted environmentalism.

While it is important to understand *Whole Earth*'s role in capturing an alternative to the wilderness-based and technophobic environmentalism, it is equally important to recognize that *Whole Earth* was an expression of popular culture advocating something really different from environmentalism—really not an "ism" at all, but a nexus of ideas about nature, New Urbanism, technology, and quality of life that offered a different path toward ecological harmony that tapped into the widespread acceptance of technologically mediated nature recreation, urban preservation, environmental justice, and the search for appropriate technologies to solve problems with modern life. Although *Whole Earth* was an intentionally iconoclastic publication, it was, throughout its many incarnations, more closely in touch with popular everyday environmental thinking than organizations like the Sierra Club or the Wilderness Society. To understand post-1960 environmentalism, one must look more closely at the generation of environmental thinkers and doers who struggled to craft an environmental philosophy that recognized that humans "were as gods, and might as well get good at it."

From Technophobia to Counterculture Environmentalism

Popular representations of counterculture environmentalists often include stereotypical back-to-nature communes complete with bearded wilderness advocates "stalking the wild asparagus" and with naked, dirty-faced children draped in flowers.[10] It was not uncommon for younger environmentalists inspired by a renewed interest in the life and writings of Thoreau, Muir, Scott Nearing, and an emerging group of countercultural environmental prophets like Gary Snyder, to drop out and take to the woods, but never in large numbers. During the 1960s and 1970s, many counterculture environmentalists did in actuality reject the modern world of large-scale technological systems in favor of a simpler, more primitive, and environmentally conscious lifestyle.[11]

At the same time, other counterculture environmentalists moved in an entirely different direction. Influenced by a mix of New Left thinkers and western regional traditions, this faction turned its attention to a critical reevaluation of long-standing assumptions about the relationship between nature, technology, and society. In particular, these environmentalists worked to replace the wilderness focus that dominated 1960s environmentalism with a more encompassing ecological sensibility that embraced all of the earth's environments, including cities and the human landscape. In the late 1960s and 1970s, technologically minded counterculture environmentalists helped reshape the American environmental movement, infusing it with a youthful energy and providing it with a new sense of purpose and direction predicated on the idea that America was entering a new phase in its development.

This new phase was envisioned as a "postscarcity" economy, where advanced industrial societies theoretically possessed the means to provide abundance and freedom and reconcile nature and technology if only they choose to do so.[12] Buckminster Fuller was an early proponent of this idea and worked to compile impressive statistics to support his claims. Later, New Left social theorists like Herbert Marcuse and Murray Bookchin, postscarcity adherents, shared the belief that "the poison is . . . its own antidote."[13] In other words, technology used amorally and unecologically created the social and environmental problems of industrial capitalism; therefore, technology used morally and ecologically could create a revolution toward a utopian future. The New Left critics emphasized that social and environmental problems in America stemmed not from a lack of resources, but from a misguided waste of the "technology of abundance."[14] If, these critics argued, the American people could be convinced to abandon their bourgeois quest for consumer goods, then valuable resources could be redirected toward establishing social equity and ecological harmony instead of consumerism and waste. In the late 1960s, postscarcity assumptions fueled a brief period of technology-based utopian optimism that profoundly influenced a generation of environmentalists.

The move away from antimodernism manifested itself in many ways, from Buckminster Fuller–inspired geodesic domes to Steve Jobs and Steve Wozniak developing *personal* computers to put the power of information in the hands of individuals.[15] Working toward similar goals, other counterculture environmentalists and sympathetic scientists and engineers focused on alternative energy, ecological design, recycling, and creative waste management as the best ways to subvert the large industrial structures they viewed as most damaging to the environment. Whether they were building personal computers in their garage or designing composting toilets, the idea that technology could be directed toward shaping a brighter future became a driving force in environmental advocacy after 1970.

The utopian optimism and revolutionary political program of the New Left failed to become a part of the mainstream environmental movement. Consumed with the reactive fight against the Vietnam War and university bureaucracies, the predominantly campus-based New Left movement fragmented and disintegrated in the early 1970s. Further, the New Left was fundamentally antiscience and antitechnology, with science and math departments and research institutes like the Stanford Research Institute as focal points of protest. This bias helped fuel the protest movement but offered few realistic solutions to problems facing Americans of the time. More importantly, the counterculture environmentalism emerging on the pages of *Whole Earth* and in communes and western cities like San Francisco was rarely really in tune with the Left, new or old, or the Right, new or old, and actually drew its inspiration more from a distinctly western libertarian sensibility—a fusion of social liberalism and western individualism with a special appreciation for technological know-how focused on living with the challenging environments of the region. As historian Jennifer Burns explains, "A libertarian sensibility is different than the libertarian movement, which like the Libertarian Party tends toward orthodoxy, rigidity, and therefore irrelevance. But libertarian sensibility is flexible."[16] It is important to note this libertarian sensibility because it helps explain the tendency of the counterculture to embrace individual and small-group tinkering and design with a social conscience as an alternative to participatory politics or reliance on the federal government to solve the environmental crisis.[17]

The relationship between the counterculture, technology, and the environment is complex. Terms like *counterculture environmental movement* are really too encompassing, and it would be a mistake to assume that all of those who considered themselves counterculturalists and environmentalists thought or acted alike. Even among those who advocated the use of technology to solve environmental problems, there was rarely a clear program of action or thought. Often, it seemed as though countercultural environmentalists occupied separate but parallel universes defined by whether they considered technology to be the problem or the solution. Not until the emergence of the ecological design movement of the 1970s did counterculture environmentalists construct something resembling a unified philosophy that united like-minded individuals and organizations under one banner. Nevertheless, what differentiated counterculture environmentalists from other environmental activists in the 1960s and 1970s was a shared desire to use environmental research, new technologies, ecological thinking, and environmental advocacy to shape a social revolution based on alternative lifestyles and communities—alternatives that would enable future generations to live in harmony with one another and the environment.

Obviously, counterculture environmentalists were not the first Americans to debate technology and the environment. The technology/nature debate dates to the beginnings of the Industrial Revolution of the nineteenth century at least. Whereas some Americans looked at advances in science and technology with a wary eye, others tended to view technology as beneficial and benign. This was particularly true for a generation of middle-class Progressive conservation advocates who believed that rational planning, expert management, and science were the keys to a sound environmental future. From amateur conservation advocacy groups to the utilitarian U.S. Forest Service of Gifford Pinchot and his biggest supporter, Theodore Roosevelt, American conservation advocates looked to science for solutions to waste and wanton destruction of increasingly scarce natural resources. For the better part of the twentieth century, most resource conservation advocacy grew from the notion that through science and the march of progress, humans could tame and control all elements of the natural world, stopping waste and maximizing productivity. This type of thinking inspired massive reclamation and irrigation projects and experiments with chemicals to rid the world of unwanted pests and predators. This steadfast faith in technology and the scientific worldview prevailed well into the 1940s.[18]

In 1947, prominent ornithologist William Vogt took time away from his studies of birds and penned a bleak book on the future of America and the human race. The following year, his *Road to Survival* sent a shock wave through the conservation community. Vogt's compelling narrative told a tale of misplaced faith in technology in the dawning atomic age and warned readers of the profound ecological consequences of the spread of industrial capitalism in the wake of World War II. Particularly concerned with the American obsession with progress, Vogt argued that "the rising living standard, as material progress is called, is universally assumed to be to the advantage of the human race. Yet . . . what is the effect of our allegedly rising living standard on the natural resources that are the basis of our survival?"[19] This cautionary tale reflected the central concerns of the dominant group of American conservationists of the 1940s and 1950s, such as the Audubon Society and the Sierra Club, and tapped into older Malthusian fears of resource depletion and scarcity. Conservationists predicated their environmental politics on the assumption that inefficiency, waste, and abuse of resources had created a potentially crippling scarcity.

Conservation-minded critics of technology often found themselves swimming against a tide of reverence for science and industry. From 1945 through the 1950s, conservationists struggled to explain to the public why, during a period of apparent abundance, it was more important than ever to require limits on both production and consumption. Conservation advocates like Vogt framed the debate in stark terms as a war between industrial technology and the environment with the fate of humanity at stake.

The response of the conservationists was tempered by their inability to articulate a compelling and environmentally acceptable alternative to perpetual economic growth and technological mastery of nature. During the Progressive period and into the 1930s, conservationists had played a large role in shaping federal land-management policies and influencing public opinion on important environmental issues. The movement was successful in raising environmental awareness and reversing some basic assumptions about the environment and the need for conservation and preservation. Progressive conservation significantly altered the landscape of the American West by regulating the use and development of millions of square miles of western lands. Despite this success, by 1945 the authority of the conservationists was on the wane. Renewed resistance to conservation programs by western ranchers and extractive industries put conservationists on the defensive. General public interest in residential and business development, which was linked by extensive highway systems, also posed a problem for environmental advocates, who focused most of their attention on restriction of use and regulation of development in the name of conservation.

The members of the major conservation organizations—the Sierra Club, the Izaak Walton League, and the Wilderness Society—tended to be upper middle-class or upper-class urban-based recreation groups. New issues and new leadership in the late 1940s began to shift the focus of the movement away from the simple ideal of efficient use of resources—for the greatest good over the longest period—toward a more focused effort to protect and preserve wilderness in perpetuity. The Sierra Club, guided by youthful board members such as David Brower, took the lead in establishing a new direction for the American conservation movement and began the subtle shifting of political and social consciousness that led away from limited concerns about resource cultivation and protection and toward a more holistic ecological sensibility. Brower used his position as a board member and editor of the *Sierra Club Bulletin* to push the club to take a stand on controversial issues such as wilderness preservation, overpopulation, and pollution.

This shift from conservation to environmentalism required some significant changes in perspective from a new generation of environmentally aware Americans. From the late 1800s through World War II, conservationists advocated a more efficient use of natural resources; like most Americans, conservationists were optimistic that American technological inventiveness could lead to a future of progress and preservation. One significant exception to this general trend was the Wilderness Society. Decades ahead of its time, the organization, founded in 1935 by an unusual collection of land planners, foresters, and wilderness enthusiasts, raised concerns about the relationship of consumer recreation technology and the health of the environment. Specifically, the Wilderness Society founders deplored the impact of auto recreation on the national parks and, more importantly, on forests that experienced radical increases in visitation with each expansion of auto tourism.[20] Unlike their counterparts in the conservation movement, the Wilderness Society members refused to embrace the technological enthusiasm of the Progressive conservationists.

The wilderness movement found its strongest voice in conservationist Aldo Leopold's influential book, *A Sand County Almanac* (1949).[21] Leopold's beautifully written stories of life on a run-down sand farm and his efforts to restore the land provided a blueprint for a new land ethic and model of ecological thinking that inspired a generation of environmental advocates to broaden the scope of their activism. A sharp eye for subtle change enabled Leopold to give voice to those who recognized that the modern world was changing the environment in ways that few even noticed.

The concerns of wilderness advocates sometimes created tensions between private environmental advocacy groups and federal agencies once closely allied with the conservation movement. At the same time, other federal bureaus, especially the water, soil, and wildlife agencies, led the way in raising the public awareness of the consequences of the postwar economic boom on the environment. Early on, the federal government sought to redress pollution as a blight on the environment and a serious public health threat. In 1955, Congress made a cautious move toward environmental regulation with the passage of the first of a series of Air Pollution Control Acts that provided for research that would lead to the regulations on industrial emissions of the Clean Air Act of 1963. The first of many environmental laws aimed at placing restrictions on production, the initial Air Pollution Control Act was important not for what it actually accomplished, but for the environmental politics that supported it.[22]

Clean air and pollution in general was of growing concern for many Americans in the late 1950s and contributed to the emergence of modern environmentalism by clearly demonstrating how unchecked development could affect public health and quality of life. Environmental advocates worried about changes in the post-war world of production and the long-term implications of those changes for the health of the environment. The chemical revolution, atomic science, reclamation, and increases in the scope of extractive industries created new and frightening environmental problems that galvanized a growing segment of the public behind the environmental movement. At the heart of the critique of production after 1945 was a growing fear of technology in general and the atom bomb in particular.

By the end of the 1940s, the very real environmental consequences of unbridled technological enthusiasm were beginning to become clear to the American public. The defining moment came when the Bureau of Reclamation proposed a system of dams for the Grand Canyon and adjacent areas. The Colorado River Storage Project (CRSP) called for a chain of ten dams on the Colorado River. This project, though bold in scope, was not a departure for the bureau: It had been building dams to control flooding and provide irrigation and hydroelectric power with broad support from Congress and the public for two generations.[23]

Once uncritically hailed as technological wonders, by the 1950s dams had become a focal point for growing concern about the environmental consequences of rapid technological progress. The Bureau of Reclamation was now at odds with groups like the Sierra Club and a rapidly growing segment of the public that began to question the economic logic of the monumental dam projects. Controversy over dam building reached a head in 1950 with the announcement that the CRSP called for a dam to be built inside the boundaries of Dinosaur National Monument at a spot known as Echo Park. Led by the Sierra Club and writers Bernard DeVoto and Arthur Carhart, conservationists built a coalition of organizations and individuals to thwart the Echo Park Dam.[24] In 1954, Congress, responding to intense public opposition, killed the project. This coalition successfully defeated the Echo Park Dam project by convincing enough politically active Americans that in the atomic age there needed to be limits on growth and that destruction of even remote areas like Echo Park could potentially affect their quality of life. After the Echo Park fight, many Americans came to view proposed technological solutions to problems with a more critical eye. Most Americans only needed to look in their own backyards to see examples of technology-fueled growth impacting the environment in ways that threatened their own quality of life.

Environmental awareness shifted from the remote wilderness and scenic wonders that had inspired previous generations of environmental advocates and focused increasingly on quality-of-life issues in cities and suburbs. During the postwar period, millions were drawn to suburbs because they seemingly offered a more rural, more natural, life than could be found in crowded cities. As historian Adam Rome has persuasively argued, suburban growth helped build a new power base for environmentalism, but perhaps more importantly, suburban growth, devouring an area the size of Rhode Island every year, proved that new technologies, when harnessed to domestic economic prosperity, could completely reshape the natural world.[25] Irving, Texas, for example, saw its population rise from 2,621 in 1950 to over 45,000 in less than a decade, rendering the formerly rural countryside on the outskirts of Dallas unrecognizable. Suburbanites watched through the "picture window" as forests, meadows, and even hillsides were cleared to accommodate a flood of other suburbanites. As suburbs encroached on forests, meadows, and swamps previously used for recreation, more Americans began to support limits on growth and protection of at least some scenic lands even while they demanded more of the technology that caused the problems.

Of the new technologies of abundance, none shaped the environment more dramatically than assembly-line suburban housing. Tracts of mass-produced homes transformed enormous areas of rural landscape into seas of houses, shopping malls, and highways. Suburban pioneers like William Levitt, builder of the prototypical modern tract suburb of Levittown in New York, were hailed as heroes in the years after the war when serious housing shortages left some returning war vets living with their parents.[26] Suburbs provided relatively inexpensive housing for young families moving into the middle class, and the G.I. Bill and Federal Housing Administration (FHA) offered low-cost mortgages for veterans. Suburbs filled an urgent housing need and promised a level of comfort and modernity unknown to city dwellers. Millions of people moved to suburbs that lined interstate highways spreading ever further into rural areas, often erasing all evidence of what had existed before. The extent of suburban development was so dramatic and transformative during the 1950s and 1960s that it prompted critics like historian Godfrey Hodgson to write about a "suburban-industrial complex" comparable in influence to the military-industrial complex.[27] As suburbanites watched the lands around them converted into concrete and strip malls, many of them also began to experience firsthand the unintended environmental consequences of other aspects of American postwar prosperity. For a remarkably productive group of counterculture designers, the soulless quality of postwar suburban development inspired a significant alternative shelter and design renaissance aimed at uniting modern environmental insights with older craft traditions to create homes that nurtured the spirit and left a light footprint on the land.

The tiny town of Bolinas, California, became an unlikely epicenter of efforts to craft pragmatic responses to the environmental and cultural problems of suburbia. Bolinas became one of the first towns in America to place limits on growth for quality-of-life and environmental reasons. Like many exurban, agricultural towns across the West, Bolinas experienced a boom during the 1950s and 1960s as a wave of countercultural pioneers came to settle what was already a thriving Bohemian artist's colony over the hills north of San Francisco in affluent Marin County. In the early 1960s, suburban sprawl began to threaten the once-bucolic town perched on a Pacific Ocean peninsula ringed with beautiful beaches and excellent surfing. As late as the 1940s, land was plentiful and cheap. The *San Francisco Chronicle* gave away 20 x 100-foot lots to subscribers as part of a marketing campaign. Few of the people who received the free land ever moved to the area, but in the 1960s lots of hippies did and, in the classic western tradition, they became vocal "neonatives" who wanted to keep this beautiful slice of Marin a secret for as long as possible.[28] Bolinas was home to several central figures in the history of *Whole Earth* and in the 1970s became a pacesetting example of the power of citizen science and pragmatic environmentalism in action.

Suburban life brought environmental problems home to a generation of Americans. Also hitting close to home were a long string of disturbing cases of pharmaceutical and chemical disasters. The most visible and distressing of these was the thalidomide crisis. Internationally, doctors learned that the mothers of many babies who had been born with atrophied or missing arms and legs had during pregnancy taken thalidomide, a drug to combat the effects of depression, insomnia, and morning sickness. The thalidomide crisis presented another startling example of the long-term risks and unintended consequences of chemicals on natural processes.

While thalidomide showed how modern medical technology could go terribly wrong, many environmental advocates were more concerned with the problems associated with medical technology that worked too well. After Vogt's unsettling account of the consequences of the ever-expanding world population, many other writers published popular books and articles on the issue of overpopulation. The statistics about world population growth and future prospects were alarming. In 1948, ecologist Fairfield Osborn worried that with a growth rate of 1 percent per year the world population might top three billion by the year 2000. His popular book, *Our Plundered Planet* (1948), was criticized for being alarmist.[29] A decade later, Osborn's predictions seemed conservative, as the world population increased far more rapidly than even the most pessimistic environmentalists had predicted.

In the United States, population growth was staggering. The baby boom between 1945 and 1964 created a massive demographic surge that strained resources and raised questions about the relationship between population and quality of life. By the mid-1960s, the faith in progress through technological innovation was under fire from an increasingly diverse group of environmental advocates informed by a growing ecological consciousness that far surpassed the cautious environmental protection of the conservation movement during the 1940s.

Changes in attitudes about technology during the 1940s and 1950s were an expression of popular culture during a critical period in American history. While

Exurban fears of sprawl, generational conflict, and innovative environmentalism created a sense of excitement and purpose to local politics in the "Town That Saved Itself." Here townspeople protest the widening of the highway to their town at a meeting of the board of supervisors in the early 1970s. (© Ilka Hartmann 2007)

never constituting a mainstream trend, more Americans began questioning the dominant view of technology and progress. The horrifying devastation caused by use of the atomic bomb in Japan was the catalyst for this reevaluation. Once the patriotic fervor surrounding the end of the war subsided, many conservationists and intellectuals started discussing what it meant that humans had the power to destroy the world. Books like John Hersey's *Hiroshima*, published in 1946, graphically depicted the awesome destructive power of nuclear weapons and their impact on humans and inspired a growing segment of society to recognize the far-reaching implications of such technology. Likewise, after years of turning out pro-war propaganda films, Hollywood, along with a legion of science fiction writers in the 1950s, started producing a steady stream of books and films presenting horrifying visions of technology run amok. A generation of Americans born after World War II grew up watching giant nuclear ants or other such mutants of technology destroying humanity in movies like Gordon Douglas's *Them!* (1954). During the 1950s, a slew of creature-feature movies created a subgenre of sci-fi based on nuclear nightmares, with normal creatures exposed to radiation mutating or growing to enormous size and then terrorizing cities. Godzilla, first appearing in 1954's *Gojira*, is the best-known example of this formula. Most of these films end with ingenuity or a more thoughtfully benign technology killing the creature or exposing its fatal weakness, but not before teaching audiences a lesson in the dangers of technological hubris. By the mid-1960s, a growing segment of American society, particularly young Americans, evidenced an increasing ambivalence about technology. During this time there was a widening sense of genuine terror over the evil potential of science without a social conscience.[30] Many older members of the conservation movement also found themselves increasingly alienated from the world of modern atomic science, massive reclamation projects, and postwar consumer technology.

Within the conservation movement, a growing ambivalence toward technology for many quickly grew into full-fledged technophobia. Fear shaped much of the conservationist alienation from the postwar world—fear that the prominence of the hard sciences, the expansion of the space race, and the explosion of consumer technology de-emphasized contact with the nonhuman world. Fears in particular about the consequences of nuclear technology for American society led conservationists like John Eastlick to wonder whether Americans had been "blinded by the fearful brightness of the atomic bomb" and were now stumbling through life with little awareness of the environmental and social degradation that surrounded them.[31]

Despite these concerns, most conservationists continued to use modernist means to express and act upon their antimodernist revulsion. Even as their alienation from postwar technocracy grew, their Progressive faith in government agencies and protective federal laws continued to be a staple of the movement.[32] For most of its history, the conservation movement embraced organizational principles and actions based on the idea of linear progress through Progressive enlightenment. At the same time, conservation proponents tended to view the history of the twentieth century as a steady decline toward chaos and environmental collapse brought on by rampant population growth and unregulated technological expansion.[33] Although these two ideals seemed to be diametrically opposed and irreconcilable, both shared the same roots as direct responses to concerns about the relationship between nature and technology in postindustrial America. By drawing on both traditions, sometimes consciously and sometimes not, postwar conservationists and critics of technology attempted to reconcile dreams for reform with competing fears that the system was beyond repair. They were both hopeful and afraid.

Other critics of postwar society, including a growing contingent of more radical environmental preservationists and a group of prominent European and American intellectuals, were less inclined to search for compromise and more willing to propose far-reaching structural changes. The most stunning of these critiques came from biologist Rachel Carson, whose explosive *Silent Spring,* published in 1962, explained in frightening detail the ecological consequences of humanity's attempt to control and regulate the environment.[34] Carson became the first of many to warn of an impending environmental crisis. During the 1960s, a series of influential books appeared warning of an apocalyptic future if the present course was not altered. Carson's fellow biologist, Barry Commoner, produced several best sellers, including *The Closing Circle* (1971), warning of the dangers of sacrificing the health of the planet for temporary material gain.[35] Carson's jeremiad was just the type of strong medicine that Americans needed at the time to awaken to the environmental reality around them. Thirty years later, ringing the tocsin had lost some of its power and much of its appeal, whereas Brand's inspirational futurism and celebration of the possibilities of reinvention and reform were more in line with American culture, if not the environmental movement.

Three other writers also provided inspiration for a new generation of Americans who were questioning the role of technology in the creation of social, economic, and environmental injustice. Jacques Ellul's *The Technological Society* (1980) asserted that "all-embracing technological systems had swallowed up the capitalistic and socialistic economies" and were the greatest threat to freedom in the modern world.[36] Ellul argued that there was "something abominable in the modern artifice itself." The system was so corrupted that only a truly revolutionary reorientation could stop social and environmental decay.[37] Like Ellul, Herbert Marcuse, in his popular *One-Dimensional Man* (1964), described a vast and repressive world technological structure that overshadowed national borders and traditional political ideologies.[38] Together, Marcuse and Ellul provided a critical intellectual framework for Americans looking to construct alternatives to the scientific worldview.

Perhaps the most influential of the structural critics of the technological society was Lewis Mumford. Mumford began his career as a strong proponent of science and technology. His 1934 classic, *Technics and Civilization*, a central text in the American technocracy movement, influenced a generation and strengthened the popular belief that technology was moving civilization toward a new golden age.[39] Like most progressive thinkers of the industrial period, Mumford envisioned a modern world where technology helped correct the chaos of nature and brought balance to ecology. In *Technics*, Mumford extolled the virtues of the machine and painted a positive picture of how technology could reshape the world to eliminate drudgery and usher in an unprecedented period in history where machines and nature worked together for human benefit. But this prophet of the machine age began to rethink his position in the 1960s. Like Marcuse and Ellul, Mumford became increasingly alarmed about the power of large technological systems. As he looked around at the world of the 1960s and 1970s, he worried that the ascendance of the "megamachine" boded ill for society.[40] The *machine*, once the symbol of progress toward a more balanced world, began to emerge as a metaphor for describing a seemingly out-of-control capitalist system.[41]

The preoccupation with technology and its consequences became one of the central features of the 1960s' social and environmental movements, and of the counterculture in particular. In 1968, Theodore Roszak released his influential study of the youth movement, *The Making of a Counter Culture*.[42] Roszak maintained that the counterculture was a direct reaction to "technocracy," which he defined as a "society in which those who govern justify themselves by appeal to technical experts, who in turn justify themselves by appeals to scientific forms of knowledge."[43] The counterculture radicals of the 1960s, Roszak argued, were the only group in America capable of divorcing themselves from the stranglehold of 1950s' technology and its insidious centralizing tendencies. Roszak's position on technocracy was similar to those of Ellul and Marcuse. For Roszak, the most appealing characteristic of the counterculture was its rejection of technology and the systems it spawned. Charles Reich, in his controversial best seller *The Greening of America* (1970), also highlighted the youth movement's rejection of technology as a fundamental component of the counterculture ideology.[44] Both Reich and Roszak cited technocracy's bureaucratic organization and complexity as the central evil. From their perspective, and a growing number of the younger generation, the problem with America stemmed from the realization that there was nothing small, nothing simple, nothing remaining on a human scale.

This mind-boggling bigness and bureaucratization likewise concerned British economist E. F. Schumacher, whose popular book *Small Is Beautiful* (1973) became a model for decentralized humanistic economics "as if people mattered."[45] Of all the structural critiques of technological systems, Schumacher's provided the best model for constructive action and was particularly influential in shaping an emerging counterculture environmentalism. Unlike more pessimistic critics of the modern technocracy, Schumacher provided assurance that by striving to regain individual control of economics and environments, "our landscapes [could] become healthy and beautiful again and our people . . . regain the dignity of man, who knows himself as higher than the animal but never forgets that *noblesse oblige*."[46] The key to Schumacher's vision was an enlightened adaptation of technology. In *Small Is Beautiful*, Schumacher highlighted what he called "intermediate technologies"—those technical advances that stand "halfway between traditional and modern technology"—as the solution to the dissonance between nature and technology in the modern world.[47] These technologies could be as simple as using modern materials to construct better windmills or more efficient portable water turbines for developing nations. The key to intermediate technologies was to apply advances in science to specific local communities and ecosystems. Schumacher's ideas were quickly picked up and expanded upon by a wide range of individuals and organizations, often with wildly different agendas, that came together under the banner of a loosely defined ideology that became known as *appropriate technology* (AT).

Appropriate technology emerged as a popular cause at a 1968 conference held in England on technological needs for lesser developed nations.[48] For individuals and organizations concerned with the plight of developing nations, Schumacher's ideas about intermediate technologies seemed to provide a possible solution to the problem of how to promote a more equitable distribution of wealth while avoiding the inherent environmental and social problems of industrialization.[49] Appropriate technology quickly became a catchall for a wide spectrum of activities involving research into older technologies that had been lost after the Industrial Revolution and the development of new high-tech and low-tech small-scale innovations. The most striking thing about the move toward appropriate technology according to historian Samuel Hays was "not so much the mechanical devices themselves as the kinds of knowledge and management they implied." Alternative technology represented a move away from the Progressive faith in expertise and professionalization and toward an environmental philosophy predicated on self-education and individual experience.[50] Appropriate technology also represented a viable alternative to wilderness-based environmental advocacy.

Ideas emerging from the New Left also bolstered the AT movement. Eco-anarchist Murray Bookchin's writings were particularly influential. Bookchin provided a critical political framework for appropriate technology by situating the quest for alternative technologies within the framework of revolutionary New Left politics. In books such as *Our Synthetic Environment* (1962) and *Post-scarcity Anarchism* (1971), he argued that highly industrialized nations possessed the potential to create a utopian "ecological society, with new ecotechnologies, and ecocommunities."[51] From this perspective, the notion of scarcity, a defining fear of the conservation movement, was a ruse perpetuated by "hierarchical society" in an attempt to keep the majority from understanding the revolutionary potentialities of advanced technology.[52] More than most New Left critics, Bookchin also clearly linked revolutionary politics with environmentalism and technology. "Whether now or in the future," he wrote, "human relationships with nature are always mediated by science, technology and knowledge."[53] By explicitly fusing radical politics and ecology, the New Left provided a model for a distinctly counterculture environmentalism. From the perspective of the New Left, pollution and environmental destruction were not simply a matter of avoidable waste, but a symptom of a corrupt economic system that consistently stripped both the environment and the average citizen of rights and resources.[54]

Although the utopian program of Bookchin and the New Left ultimately failed to capture the hearts of most environmentalists, it did help establish a permanent relationship for many between environmental and social politics. This linking of the social, political, and environmental in the 1970s paved the way for new trends of the 1970s and 1980s, such as the environmental justice movement. For inner-city African Americans and others who felt alienated from the predominantly white middle-class environmental groups like the Sierra Club or the Wilderness Society, the New Left vision of environmental politics provided inspiration.[55] By connecting ecological thinking with urban social issues and radical politics, the New Left introduced environmentalism to a new and more diverse group of urban Americans who had felt little connection to the wilderness- and recreation-based advocacy of the conservation and preservation movements.

For counterculture environmental thinkers like Brand, the New Left critique raised important questions about technocracy but only bolstered the growing small-scale technological fascination of many counterculture environmentalists like himself.[56] The AT movement represented a different direction from the radical politics of the late 1960s and into the 1970s. By then, the campus-based New Left movement was primarily a movement against the Vietnam War. New Left politics on the campus focused on striking back at the Pentagon, IBM, AT&T, and other representatives of the technocratic power structure. Escalating violence, renewed scarcity fears, and a host of pressures both inside and outside the campus-based movement caused the New Left to fracture and ultimately collapse. Disillusioned by the failure of the revolution, many who were part of both the New Left and the counterculture began to move away from radical politics. Technological enthusiasts like Brand had never felt a connection to the political side of the New Left and paid serious attention only to the philosophies and ideas promoted by the most innovative and pragmatic of this generation of creative thinkers.

By the early 1970s, the neo-Luddites in the American environmental movement had ceded ground to a growing number of appropriate technologists. This new group of counterculture designers, environmentalists, scientists, and social activists looked to new modes of alternative living that recognized the liberating power of decentralized individualistic technology. The AT movement was varied and diffuse, with much disagreement even among its adherents as to how to define their loosely affiliated sensibility. The term meant different things to different groups, but there was general agreement that an appropriate technology had the following features: "low investment cost per workplace, low capital investment per unit of output, organizational simplicity, high adaptability to a particular social or cultural environment, sparing use of natural resources, low cost of final product or high potential for employment."[57] In other words, an appropriate technology was cheap, simple, and ecologically safe. The proponents of appropriate technology also agreed on the basic idea that alternative technologies could be used to create more self-sufficient lifestyles and new social structures based on democratic control of innovation and communitarian anarchism. For supporters of appropriate technology, the most radical action one could take against the status quo was not throwing bombs or staging sit-ins, but fabricating wind generators to "unplug from the grid."

The AT movement represented something new. More than simply an alternative to a perceived technological crisis, it was a reenvisioning of environmental activism from the ground up. The move toward appropriate technology represented a significant break for the counterculture and the environmental movement. A new breed of young environmentalists built on the ideas of Schumacher, Bookchin, Marcuse, and others to craft a very different agenda from their technophobic predecessors in the conservation movement and peers in the wilderness movement of the 1960s. This new agenda would find its best expression in the pages of the *Whole Earth Catalog* and among the alternative designers that Stewart Brand gathered to his new venture. These designers wanted to fight fire with fire; they wanted to resist technocracy and frightening nuclear and military technology by placing the power of small-scale, easily understood, appropriate technology in the hands of anyone willing to listen. Stewart Brand was not a designer, but his life experiences and education uniquely positioned him to become the most significant promoter of the pragmatic environmentalism that early AT proponents envisioned.

The Futurist

"Stewart Brand, a thin blond guy with a blazing disk on his forehead too, and a whole necktie made of Indian beads. No shirt, however, just an Indian bead necktie on bare skin and a white butcher's coat with medals from the King of Sweden on it." So Tom Wolfe introduced Brand to America in his counterculture classic, *The Electric Kool-Aid Acid Test* (1969).[58] Every generation seems to have a handful of individuals whose lives uncannily intersect with many of the key moments of their times. These prescient few stay two steps ahead of their peers, creating and riding the crest of important trends. They tend to be iconoclasts, amateurs, dropouts, and eccentrics free to pursue their passions outside the mainstream. Stewart Brand was one of these rare people. He was present at, and instrumental in, the creation of the American counterculture, the birth of the personal computer, the rise of rock and roll, the back-to-the-land and commune movement, the environmental movement, and a critical reorientation of western politics. He was an experienced LSD veteran before practically anyone had heard of the drug and advocated a new view of the earth that set a standard for how six billion people view their home world. Born a midwesterner, in his adult life Brand became a westerner and the quintessential Northern Californian. His publishing empire influenced a generation and reflected important

Twenty-eight-year-old Stewart Brand as one of the Merry Pranksters, San Francisco, California, 1966. (Ted Streshinsky/ CORBIS)

changes in American politics and culture. To understand *Whole Earth* and the emerging environmental sensibility it so perfectly captured, one needs to understand at least a little about Stewart Brand. Brand's personal philosophy shaped the particular countercultural blending of pragmatic environmentalism with a traditional western regional libertarian sensibility, social liberalism, and technological enthusiasm that characterized the catalog.

Brand was a big idea man who appreciated other big thinkers. Intense and abiding intellectual curiosity defined his early efforts and shaped the rest of his life. A brilliant original thinker and excellent student when he wanted to be, Brand was not an academic. He appreciated "doers" more than professors and traditional scholars, who struck him often and from an early age as pedantic and far too authoritarian for his likes.[59] Despite his strong traditional educational background, Brand became the leading voice of his generation for self-education and a renaissance of amateur research, craft, and design. During his life, he surrounded himself with intellectuals generally not affiliated with universities. He appreciated thinkers who spent their time working on practical applications for their ideas rather than teaching or publishing through traditional academic paths. More than "anybody in the history of counterculture movement," *Whole Earth Review* editor Peter Warshall recalled, Brand was able to "attract everyone from Herman Kahn to Jane Jacobs to Roshi Baker" to his projects; "he was remarkable that way."[60] Not many people are able to sustain the intensity of their youthful curiosity throughout adult life. Brand was one of the exceptions. What made his intellectual curiosity even more unusual was the depth of his eclectic interests and how regularly he successfully converted his ideas into actions that worked their way into broader American culture. Brand did not like the futurist label that the popular media gave him, because it was too tainted by its association with dreamy utopians and cranks.[61] But his remarkable ability to ride future waves that very few others saw coming placed him among the most clairvoyant of his generation.

At times, he risked his physical and mental health by driving himself to the brink of complete exhaustion and collapse. His saving grace was his ability at key moments in his life to recognize his limits and move on. When he successfully launched an idea, he was relatively quick to delegate responsibility and shift to something new, only to return later to the old with renewed energy and ideas. More importantly, he was willing to try crazy projects that seemed doomed to fail and often did. In a career of remarkable successes, Brand also had some spectacular failures. He often seemed better at bucking up and learning from failure than he was at enjoying success. These tendencies enabled him to work on a wide variety of significant projects during his adult life and take intellectual and entrepreneurial risks that paid off more often than not. He was willing to walk away from success and stay the course after failure, perhaps the perfect recipe for a successful life.

One of the consistent characteristics of the *Whole Earth* publications in all of their incarnations was the breadth and depth of reading that informed the articles and reviews. *Whole Earth* was a catalog that was all about books. The book-centered design of *Whole Earth* came directly from Brand's voracious appetite as a reader and student of how things work. Born in 1938 in Rockford, Illinois, the youngest son of parents educated at MIT and Vassar, Brand learned early the value of a good education. According to family friend Dick Raymond, Brand got his iconoclastic worldview at home.[62] At age sixteen, he left the Midwest for good to attend two years at Phillips Exeter Academy in New Hampshire and then, in 1956, he headed west to California to study biology at Stanford.[63]

At Stanford, Brand worked with renowned conservation biologist Paul R. Ehrlich. In the 1960s, Ehrlich was a respected young ecologist interested in butterfly populations and working on some soon-to-be controversial theories of overpopulation.[64] Eight years after Brand graduated, Ehrlich gained international attention with the publication of sensationally bleak treatises on overpopulation and the prospects for the future. *The Population Bomb* (1968) was hailed by many in the environmental movement as a signal work that highlighted what many at the time assumed would be the crux of a future fight for quality of life on earth.[65] Erhlich painted a dark picture of ecological Armageddon in the near future if steps to reduce population were not implemented. Even so, he argued that it was probably too late: "The battle to feed all of humanity is over. In the 1970s and 1980s hundreds of millions of people will starve to death in spite of any crash programs embarked upon now. At this date nothing can prevent a substantial increase in the world death rate."[66]

Although many reviewers then, and since, found Ehrlich's predictions overblown at best and ridiculous at worst, his view tapped into the fundamentally pessimistic mood of American environmentalism and into genuine concerns about overpopulation and famine in the 1960s. Ehrlich, like many futurists in the second half of the twentieth century, saw nothing but trouble in the looming twenty-first century. No one could fault Ehrlich, then or now, for voicing serious concern about population and environment, but his views contributed to the pessimism of American environmentalism as it approached a critical turning point at the end of the sixties. Worse, neo-Mathusians lent an unfortunate social bias to environmentalism at a moment when reformers around the world were searching for more inclusive solutions to ecological concerns.[67] As a student at Stanford, Brand shared his mentor's worries about population and the environmental health of the planet, and later organized a major population happening called Liferaft Earth, although he gradually moved toward an opposite view later in life. He did not seem to share, however, Ehrlich's profoundly pessimistic view of the future. Even as he was training to become an ecologist, Brand maintained his fundamental optimism and belief in the power of people to fix big problems.[68]

One of the most interesting aspects of Brand's later publishing career, and a key reason his books were so popular, was his thoughtfully positive view of the future. Brand's affirmative view of human nature and human agency fueled his research and formed the foundation of all his later work. At a time when many scientists and environmentalists were losing faith in humanity, Brand highlighted human ingenuity and inventiveness. One of the key insights of Brand's later career does appear to have its roots in this formative period working with Ehrlich: the appreciation of long-term thinking, and the realization of just how limited our understanding of the evolution of large systems over time really was.[69]

Biology students in the 1960s learned a different view of the world than previous generations. The species-by-species approach of biological research and preservation efforts of individual species gave way to the more holistic view espoused by the nascent field of ecology.[70] Convincing the American public to appreciate really long-term thinking and whole-systems integrity was one of the most striking successes of the postwar environmental movement and efforts to preserve American wilderness. Insights gained during his studies at Stanford fostered Brand's lifelong interest in the long view.[71] "I was bit early on by a series of biologists," Brand later wrote, "Ed Ricketts (via John Steinbeck's Monterey books), Aldous Huxley (in print and person), Paul Ehrlich, and last and deepest, Gregory Bateson."[72] Brand's close study of biology and ecology provided him with a solid foundation for his emerging environmental philosophy and could have set him on a path toward an environmental advocacy that many with his training were just starting to follow. His passion for literature bolstered this growing interest and enriched his writing. Steinbeck's *Cannery Row* (1945), the source of the Ed Ricketts character, is very much a science book, aimed at the application of a unified field hypothesis, emphasizing interconnectedness. Even as a student, Brand had a talent for reaching beyond disciplinary boundaries and building connections between scholars, artists, and writers others might have missed. Ecologists trained during the early sixties energized the resource protection movement with rising ideas of ecology and a much greater awareness of the interconnectedness of natural systems. What set Brand apart from the ecologically minded environmentalists of the 1960s, however, was his more optimistic interpretation of the disturbing information that ecologists and others working on environmental issues were daily uncovering. Even during his time as a student of biology, Brand intuitively rejected the antimodernism and technophobia common to environmental thinking of this period. Ehrlich's basic idea in *The Population Bomb* was that "population times affluence times technology equals impact." Brand later advocated the reverse, "population times technology equals reduction of impact."[73]

Ehrlich was one of the best-informed and articulate voices of the environmental movement, and his views were perfectly in keeping with the wilderness-based advocacy of the major environmental organizations of the late 1950s and 1960s. During the 1960s, men and women, conservatives and liberals, environmentalists and antienvironmentalists all used the rhetoric of population control to make various points about the proper relationship between people and the environment. The liberal position most often voiced by environmentalists argued that overpopulation places undue strain on ecosystems, ultimately leading to environmental crisis. Therefore, population control must be a cornerstone of environmental advocacy.[74] Outside certain religious denominations, few disagreed that rampant population growth endangers ecosystems and the environmental health of the planet. Yet, population alone cannot explain the environmental crisis of the late twentieth century. Serious overpopulation tends to be a Third World phenomenon. On the other hand, industrially sophisticated nations, where overpopulation is less of a problem, are disproportionately responsible for resource depletion, pollution, overdevelopment, and overconsumption. The population control debate, therefore, has often centered on the politics between developed and undeveloped nations.[75]

But students like Brand took a different message from Ehrlich's biology training. The scientific method and close examination of species and ecosystems appealed to him, but not the conclusion that people, and particularly technology, were the problem. Brand sucked up his biology training, but without most of the romantic fervor or millennial doom and gloom that captured a generation of biologists and ecologists and drove the mainstream environmental movement throughout the twentieth century. Brand gained an abiding appreciation for the nonhuman world during his years at Stanford, but "never understood the romantic wilderness-based environmentalism."[76] Environmental concerns loom large in Brand's life works, but he rarely defined himself as an environmentalist and, when he did, it was always with a critical caveat about the problematic nature of environmentalism as a label and ideology.

Stanford researcher and future *Whole Earth* mentor Dick Raymond had known Brand most of his life as an old family friend. During a summer researching tarantulas for a biology project, Brand lived in Raymond's basement and the two formed a strong friendship despite the difference in their ages and personalities. Like many people who met Brand in his younger years, Raymond was impressed by the confidence of the obviously brilliant student. Even as a student, Brand had a strong will and the self-assurance that comes from a first-class education and privileged background. Raymond was not put off by Brand's attitude; rather, he was impressed with the enterprising young man, who was obviously a deep thinker loaded with ideas and the ambition to make something happen in his life.[77]

Brand's years at Stanford deeply influenced his environmental thinking and established a scientific sensibility that provided an epistemological foundation for his future studies and projects.[78] Equally influential were the two years (1960–1962) that followed graduation when Brand enlisted in the U.S. Army. Brand had prepared for military leadership in the Stanford Reserve Officers' Training Corps (ROTC) program and eagerly enlisted after graduation. After airborne training, Brand moved to a leadership position as a second lieutenant in charge of basic infantry training. This learning experience left a positive mark and gave Brand skills that would serve him well in his later business ventures. He recalled, "At government expense I was trained in leadership and small-unit management."[79] Many interviews over the years reflected Brand's appreciation for his time in the military. Unlike the disgruntled Vietnam vets who joined the antiwar movement after their return or used the counterculture as an escape from their military past, Brand never regretted his time in the military and had a view of Vietnam

very different from many of his peers. "I did not care about Vietnam," Brand recalled. "I was in the military and trained troops to go over there. I wanted to
check it out, but it was too early."[80]

Brand's satisfaction with his time in the military reflected his fundamental intellectual pragmatism and libertarian-leaning conservatism that grew out of his
views on Communism and the Cold War during his time at Stanford.[81] His military experience also tapped his autodidactic tendencies by offering opportunities
to learn during his training and as an instructor. Brand always seemed just as
eager to learn how to make or take apart things, learn or teach new basic skills, as
he was with serious academic study. It might seem hard to imagine one of the
iconic figures of the counterculture as an eager recruit, but army life fit Brand's
personality and politics like a glove. Politically, at the time, Brand was in favor of
the U.S. military policies and in particular the fight against Communism. As a
young undergraduate at Stanford, he concluded that Communism was a threat to
freedom in general and his personal freedom in particular.[82] Despite these concerns and his enthusiasm for his military career, a key component of his personal
philosophy involved avoidance of organized politics in favor of individual action.
This simple philosophy, which tapped into the individualistic traditions of his
adopted region, was central to the *Whole Earth Catalog*, where politics always
took a back seat to ideas and tools.

After working as a troop trainer, Brand was assigned to a post as a photojour-
nalist out of the Pentagon. He received more training that would serve him well in
his later artistic and publishing pursuits. After finishing his two years in the military, Brand headed west again, to the San Francisco Bay, and eventually landed in
Menlo Park next to Stanford. He started an off-and-on study of art and photography at San Francisco State University before embarking on several crucial years
as a central figure in the San Francisco counterculture. Between 1962 and 1968,
Brand rode the cresting wave of the counterculture like few in his generation. He
looms large in the flowering of the psychedelic heyday of sex, drugs, and rock and
roll. Lots of people claimed to have been in the thick of things during this celebrated time of cultural revolution; Brand really was there, virtually everywhere.
His countercultural résumé is stunning.

Journalist John Markoff has argued that LSD and the counterculture were at the
heart of the American personal computer revolution and that Brand was a key
player in this union of hippies and geeks.[83] There was certainly something very
interesting happening at Stanford and Menlo Park during the mid-1960s, and
drugs were the glue that held together a remarkable community of multidisciplinary innovators. In the collective memory of the 1960s, drugs loom large, as though
everyone under thirty was stoned for a decade; "if you can remember the 60s . . ."
the saying goes. Drug use is often dismissed as a gearshift in American culture

driven by middle-class college kids getting their kicks from a new poison. As historian David Farber, though, has argued, "The way some people used some drugs in the Sixties era facilitated their purposeful exit from the rules and regulations that made up the culture they had been poised to inhabit."[84] LSD use in particular cannot simply be dismissed as pointless escapism, as many thoughtful people directly linked their use of this mind-altering chemical to specific thought processes and ideas that were later brought to fruition in technical innovations, artistic endeavors, and thoughtful business models. It might be tempting to write off the six years that Brand spent traveling the physical and psychic road with the counterculture as a last gasp of late youth, but these experiences were clearly critical in shaping the rest of Brand's life and particularly his environmental philosophy.

Brand's psychedelic journey began in 1962 when he signed up for a legal LSD study at the International Foundation for Advanced Study in Menlo Park.[85] The experience had a profound impact on the twenty-four-year-old and sent him on a remarkable path.[86] Between 1962 and 1966, he traveled the American West on an iconic journey of self-discovery partly chronicled in *The Electric Kool-Aid Acid Test*, exploring many of the themes that came to dominate the *Whole Earth Catalog* and establishing the eclectic worldview, with a hidden logic, that made his later works so compelling. The list of his activities during this creative period is long and interesting, including "sundry multi-media performances" and "public events," most significantly a series of "happenings," including "America Needs Indians," and the "Trips Festival."[87]

In August 1965, Brand organized a "mixed-media happening" called America Needs Indians, which was a "sensorium, a Love Feast, a Celebration." He envisioned and then pulled off an event that was part photography exhibit and part do-it-yourself educational experience that included discussions and interactive exhibits on "making your own Navaho tea, Venison, jerky, berry soup, or moccasins." Attendees could also learn how "peyote loosens association, releases affect, and pressures of the unconscious" and that "a peyote experience is not always pleasant," but "it is always instructive."[88]

Brand's interest in Native cultures dated back to his youth in Michigan. Early journal entries demonstrate an unusually thoughtful study and serious reverence for Indian views on nature and the land even at a young age.[89] Brand renewed his interest in Indian culture after his military service when he got a job photographing Indians for the Department of the Interior under Stewart Udall. For the next six years, he traveled to Indian reservations around the West and developed a deep appreciation for their culture. "One reason I like spending my time with Indians," he wrote during these travels, "is that when I'm with Indians I don't feel that I'm 'spending' my time."[90] In Sheridan, Wyoming, he met his future wife, Lois Jennings, "a Chippewa woman from Washington, D.C."[91] This long-term interest culminated with "America Needs Indians," an innovative multimedia event that used movie projectors, Indian dancers, and multiple sound tracks playing simultaneously to create a sensory experience.[92] Brand was at the forefront of the counterculture rediscovery of the American Indian, and this series of events celebrating Indian culture and MC'ed by Brand set the stage for an alliance between hippies and Native Americans. "By the end of the '60s, Indians had been adopted by the hippies," Brand wrote later, "and to everyone's astonishment, not least mine, it basically worked out. There was a transmission of traditional frames of reference from older Indians to hippies, who were passing it to their young peers on the reservations and a lineage was inadvertently, but I think genuinely, preserved."[93] Brand's study of Indian culture gave him a sense of the broader American West that he might have missed if he had spent any more time than he did in Northern California. The interest in Indians also brought him into contact with *One Flew Over the Cuckoo's Nest* author, Merry Prankster leader, and LSD prophet Ken Kesey.

Brand read Kesey's *Sometimes a Great Notion* in 1964 and, by 1965, was a key member of the "swarming circle" surrounding Kesey when he was engineering his infamous Acid Tests.[94] Kesey's travels and trips are canonical big moments of the counterculture. The Trips Festival was one of those "stroboscopic" Technicolor moments that can make the whole period seem like a cliché that never could have actually happened.[95] By 1966, Brand was in the center of the San Francisco counterculture at its apex. He solidified his place in this history when he helped organize the Trips Festival, one of the most celebrated moments of the counterculture. To make a claim as a true believer and real veteran of the West Coast counterculture, Trips was the credential. This three-day event in late January was the biggest, and best publicized, of the Acid Tests and brought together all of the most creative elements of the Bay Area counterculture under one roof. All three nights were an unqualified success, but the second night, a Saturday, stood out. Brand was in charge of publicity and did a fantastic job. Haight-Ashbury historian Charles Perry captured the memorable scene. "The lines to get in the hall were endless," he recalled. "Two or three whole audiences passed through the place on a single night."[96] Most accounts pegged the number of attendees on that second night at just over six thousand euphorically intoxicated souls who crammed themselves into the cavernous interior of San Francisco's Longshoremen's Hall. Inside they found several stages and a chaotic mix of the sublime and the ridiculous. Brand's Indian show was mostly drowned out by the noise and confusion as first Big Brother and the Holding Company took the stage, followed by the Grateful Dead. The weird mixture of multimedia shows happening alongside the music ensured that the environment of the festival maximized the effect of the acid-spiked punch and packs of LSD passing around the room. To no one's surprise, the event was a countercultural opera where the attendees were the real show, and "the two o'clock close-down left a lot of spectacularly wasted people out on the street."[97] Trips captured all of the familiar countercultural tropes: sex, drugs, music, art, and spectacle. Anyone sober enough to have surveyed the scene might have noticed two other significant, but less obvious, central enthusiasms of the counterculture: money and technology.[98] The Trips Festival was not the first event of the era to unite commerce and technology with the cultural trends of the sixties, but it captured an important convergence of interests better than any previous single happening.

Even in 1966, when the counterculture was at its peak of hedonism and excess, there was no one monolithic countercultural sensibility. Historians have debated the term since Theodore Roszak popularized it with his best seller, *The Making of a Counter Culture* (1969).[99] Roszak came to believe that there were two branches of the movement, the "reversionaries" and the "technophiles." Both were advocates of a postindustrial America based on postscarcity—the notion that we really had all we needed for a sustainable future if we could just open our eyes to human potential. The reversionaries wanted to move toward the postindustrial by returning to a simpler life in the Jeffersonian tradition. The commune and back-to-the-land movements were inspired by this ideal.[100] The technophiles were "industrial utopians" enthused by technological invention and disenchanted only with the decline of craft and individual agency within the technocratic world of mid–twentieth-century industrial capitalism.[101]

By 1966, Steward Brand was straddling the line between the reversionaries and the technophiles. Eagerly enthusiastic about technology and actively working to introduce it in new forms through happenings like Trips, he was also captivated by American Indian culture and sympathetic to the back-to-the-landers, who were his primary audience for the first version of *Whole Earth* two years later. At age twenty-eight, Brand was also developing an environmental sensibility that merged his views on technology, culture, and ecology. In February 1966, "wrapped in a blanket in the chill afternoon winter sun, trembling with cold and inchoate emotion," he had a vision that launched his first significant environmentally oriented national campaign.[102] "It was all because of LSD, see," he recalled. "I took some lysergic acid diethylamide on an otherwise boring afternoon and came to the notion that seeing an image of Earth from space would change a lot of things."[103] Brand's inspiration raised a simple question: "Why haven't we seen a photograph of the whole Earth yet?" Captivated by this question, he researched answers. Finding no satisfactory answer to why the National Aeronautics and Space Administration (NASA) had not yet pursued space images of the whole Earth, he launched an impassioned advocacy campaign that became the stuff of legend. In an effort to get the word out about the importance of space photos, he

Stewart Brand's simple little buttons launched his environmental career in the spring of 1966 and quickly established his reputation as an innovative environmentalist with a remarkable ability to see the potential of technologies to promote a sustainable world. (Courtesy of Stewart Brand)

Why haven't we seen a photograph of the whole Earth yet ?

printed his simple question on buttons: "Why haven't we seen a photograph of the whole Earth yet?" Once he had the buttons, he launched a one-man mission to get the idea out nationwide. "I prepared a Day-Glo sandwich board with a little sales shelf on front," he remembered, "decked myself out in a white jump suit, boots and costume top hat with crystal heart and flower, and went to make my debut at the Sather Gate of the University of California in Berkeley, selling my buttons for twenty-five cents."[104] Brand also mailed the buttons to U.S. senators, the space program, and prominent thinkers he admired like Buckminster Fuller and Marshall McLuhan. Student photographer Glen Deny vividly remembered meeting Brand as Brand wandered around San Francisco State University later that spring while wearing his sandwich board advertising the buttons.[105] Hard to imagine now, but, in 1966, there was no photo of the whole planet Earth from space. Brand felt sure that if people could view their planet from space, it would change "everything." Brand's buttons made their way to the lapels of NASA suits, and "legend has it that this accelerated NASA's making good color photos of the Earth from distant space during the Apollo program and that the ecology movement took shape in 1968–1969 partially as a result of those photos."[106] In 1968 *The Whole Earth Catalog* and *Life* brought the first distant images of the globe to the public.

Historian Neil Maher convincingly argues that Brand's advocacy of a view of the whole planet represented a significant shift in thinking about the relationship between nature and technology.[107] The whole Earth photo finally made widely public after the successful *Apollo 17* mission replaced the iconic Earthrise image captured by the crew of *Apollo 8* in December 1968. The Earthrise image embodied the ideal of space as the "final frontier," a particularly American view of the Earth that reflected the spirit of the Kennedy administration's push into space as a continuation of American history and new battlefront in the Cold War. The whole Earth image, however, "de-centered the United States and replaced it with an image of a more global natural environment . . . debunking the frontier narrative" of the American space program. Brand's recognition of the significance of specific space images and advocacy for the whole Earth image in particular were a harbinger of changing views about the role of technology and the environment.[108]

Brand embraced the space technology that could give us an image of Earth from space, but wanted that technology used for nonpolitical ends. Unlike the image of Earth with the Moon in the foreground, the image of the whole Earth forced all who saw it to think holistically.

The whole Earth buttons were Brand's first experience with environmentalism.[109] It is fitting that his career as a leading voice for the environment began with the advocacy of space exploration, an unpopular position with most environmentalists, then or now, who thought the money was better spent on programs to stop environmental degradation at home. Brand, however, was a different kind of environmentalist, skeptical of the romantic rhetoric of wilderness and consistently undogmatic in his views about environmental issues. Approaching thirty, in 1967, he had spent much of his adult life surrounded by science and technology. His views on sustainability, ecology, and quality of life were shaped by the world of ideas in which he enmeshed himself: Photography, computers, engineering, design, military service, ecology, music, art, and science all contributed to his emerging version of environmental activism. A love of good tools, thoughtful technology, scientific inquiry, and a western libertarian skepticism of the government's ability to take the lead in these areas all shaped his philosophy and uniquely positioned him to break free of the consensus culture of post–World War II America and shrug off social expectations regarding conformity. Brand was a rebel with a cause, and he was not alone. His emerging ecological sensibility in particular was indicative of a trend he was both a part of and helped dramatically expand. This budding counterculture environmentalism was quite different from the stereotypical view of escapist hippie tree-huggers who had little to offer in the way of serious solutions to hard environmental problems. The *Whole Earth Catalog* became Brand's vehicle for uniting the threads of the American counterculture he was exploring and the leading forum for an environmental advocacy quite different from the movement that many Americans were just discovering in the spring of 1968 when Brand sat on the plane flying over the vast expanse of the American West.

CHAPTER TWO

Thing-Makers, Tool Freaks, and Prototypers

> This is a book of tools for saving the world at the only
> scale it can be done, one hand at a time.
> *Stewart Brand*, Whole Earth Ecolog

The day after that fateful plane trip home from his father's funeral in March 1968, when the inspiration for the catalog struck, Brand met with Dick Raymond to discuss the idea for an access service at Raymond's newly formed educational Portola Institute in Menlo Park, California.[1] There could not have been a better time or place than the San Francisco Bay Area for an innovative thinker like Brand to pursue his idea of a new information system. San Francisco in 1968 was in the midst of a well-documented creative renaissance. In addition to the familiar cultural explosion embodied by the rise of artists like Janis Joplin, the Grateful Dead, and the rest of the amazing music scene of the Bay Area, the literary and psychedelic Merry Pranksters, the socially minded Diggers, and the guerrilla theater troops, there were less visible efforts like the Raymond's Portola Institute and many similar small organizations working on blending the cultural renaissance with social and environmental reform. It was, in the words of Joan Didion, a moment in the history of America and the American West in particular when "a great many articulate people seemed to have a sense of high social purpose."[2] The San Francisco Didion uncovered in her classic exposé on the time and place, *Slouching towards Bethlehem* (1961), demonstrated how easily that high sense of social purpose could be squandered on just getting high, but there were plenty of examples of enlightenment turned toward productive and socially conscious means.

The Portola Institute was one of the best examples of how creative communities were coalescing around a loose set of shared social and cultural goals in an effort to create new means for achieving personal and community success. The "institute" was really just Dick Raymond in the beginning, using his own money to explore creative ways to improve education.[3] Raymond founded Portola in 1966 "as a nonprofit corporation to encourage, organize, and conduct innovative educational projects."[4] He had experience in nonprofits from his work with the Dumbarton Research Council in Menlo Park. Dumbarton was a nonprofit "association of independent scholars" that supported the arts and music education.[5] Looking back, Raymond remembered just wanting to build a place for people who were "curious, inventive and creative."[6] Raymond's views on alternative research and philanthropy came in part from frustrations encountered while working for several years at the prestigious but controversial Stanford Research Institute (SRI).[7] Antiwar activists at Stanford dubbed SRI the "Doomsday Works" because of its close ties to the military. An excellent 1970 *Esquire* article on *Whole Earth* referred to SRI as one of "the terriblest Son of a Bitch establishments of them all."[8] Raymond's view was more complicated, but he was happy to move away from the SRI bureaucracy, which he found stifling. After SRI, he wanted to build an alternative organization with "no rules, staff, or goals," an "empty vessel" for people to do what they wanted to and explore their most creative ideas. Further, Raymond wondered, "What comes from emptiness? What happens when an environment is empty?" Raymond hoped that this structureless organization could avoid the problems with rigid hierarchies he had encountered at Stanford. He worried that the law of entropy—social institutions arise, become bureaucratic, and dissolve—could undermine his best intentions, and he wanted to try to fight entropy through intentional disorganization.[9] Raymond and Brand shared an interest in exploring ways to unite research in education and technology with land use and environmentalism. Although he was often discouraged by the hit-and-miss results that his style of organizational management fostered, Raymond remained committed to the idea, much to the benefit of Stewart Brand.[10]

Raymond was one of many in the Bay Area who took the initiative to start organizations, hoping communities would grow around them. He was by all accounts a genuinely remarkable person. Ed McClanahan and Gurney Norman wrote of Raymond in 1970 that he conveyed "an almost tangible aura of Goodness, a beatific quality that is part generosity, part sincerity, part innocence . . . a truly virtuous man."[11] In San Francisco, Raymond became a key figure in an eclectic community of activists who often encountered each other at the Glide Methodist Church in San Francisco. In the mid-1960s, Glide, under the direction of its charismatic minister Cecil Williams, became a sort of counterculture think tank that attracted some of the best and brightest from the Bay Area business and charitable community. Starting in 1963, when Williams arrived, Glide became a focal point for countercultural community activism and engaged spirituality. By the late sixties, the church was renowned for its Glide Ensemble choir and dynamic and colorful Sunday services. Between 1963 and 1968, the church hosted Friday lunches for creative people from the Bay Area. Raymond was a regular attendee, as was Bank of California Vice President Michael Phillips. Phillips remembered the lunches as one of the best examples of the creative possibilities of this remarkable period in San Francisco's history.[12] Glide Methodist Church was the birthplace of the network of talented individuals that became the scaffolding upon which *Whole Earth* was built.[13] Over the years, several Glide alumni,

This page of the *Last Whole Earth Catalog* featured patron saint Dick Raymond and the lunchtime volleyball game remembered fondly by all *Whole Earth* staffers as a bonding experience. Also included were an introspective history of the publication to date and a careful description of production methods. This transparency of method and motive conveys the counterculture desire to move away from standard information-delivery models and institutionalized educational practices. (Courtesy of Department of Special Collections and University Libraries, Stanford University Libraries)

History (continued)

About this time Lois and I started living in the store. Joe and Annie and I, with editorial help from Lloyd Kahn, did the Spring 69 CATALOG production amid the busy din of the store, a cheaper mistake. The CATALOG was twice as big and a dollar cheaper. To clear my head after production I hitchhiked to New Mexico for what turned out to be the Great Bus Race (p 245). Joe and Annie also headed for the desert, pending rendezvous in Albuquerque for the July Supplement production.

You should know that all this time Portola Institute was going through continual interesting changes that someone else is going to have to write about. Dick Raymond did one especially nice thing for us: he protected us from the vicisitudes.

Store and mailorder business was gradually picking up, so we hired Hal Hershey, a friend of the Duckworths who had worked in bookstores. We also hired Diana Shugart, a close buddy of Lois' and mine. At the store we had a chart on the wall that showed our income and expenses for each month. The income was gradually catching up.

While we were having a good July production at Steve and Holly Baer's house in Albuquerque, Hal and Diana were starting to face a heavy current in Menlo Park ("52 subscriptions today!"). Philip Morrison had written kindly of us in the June 69 *Scientific American*. We were being mentioned in a lot of underground papers such as the *East Village Other*. And then Nicholas von Hoffman wrote a full piece on the CATALOG that got syndicated all over the U.S. We were caught. We were famous.

(One interesting note. Of all the press notices we eventually got, from *Time* and *Vogue* to *Hotcha!*— in Germany— to the big article in *Esquire*, nothing had the business impact of one tiny mention in "Uncle Ben Sez" in the *Detroit Free Press*, where some reader asked, "How do we start a farm?" and Uncle Ben printed our address. We got hundreds and hundreds of subscriptions from that.)

Hal and Diana hired more people. Deposits at the bank were more frequent: the bank officers got more polite.

In September Joe and I returned to Ortega garage to work on the September Supplement. Annie had stayed on at Lama, so we hired a Kelly Girl to do the typing. As I was driving up the hill to work one day it suddenly hit me that I didn't want to. Instead of golden opportunity the publication was becoming a grim chore. I considered the alternatives of taking my medicine like a good boy or setting about passing on my job to somebody else. I'm sure !

Richard Raymond, President of Portola Institute

Modern Business Forms

A good source of cheap, functional business stationery, note-o-grams, address labels, business notebooks, etc. Catalog free.

—SB

The Drawing Board, Inc.
256 Regal Row
Dallas, Texas 75221

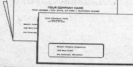

Sound Criticism

What in the Hell is the Big Idea, anyway, threatening to cease publication of the Whole Earth Catalog??? Why, you characters have hardley even started! What if all the telephone companies suddenly

Fidelity Executive

A good source of cheap, functional cardboard files and bins. Also other office equipment.

—SB

Fidelity Products Co.
705 Pennsylvania Ave. So.
Minneapolis, Minnesota 55426

Every noon, volleyball at the store.

440

like dancer-poet Janice Mirikitani, moved into positions on Brand's Point Foundation, founded to support *Whole Earth*, or worked to support various *Whole Earth* and Point programs.[14] Brand later modeled his foundation on Dick Raymond's Portola Institute, which played such a central role in his success.

Raymond was a thoughtful and kind man who must have been blessed with an extra measure of patience. During the mid-to-late sixties, he worked with many young people possessed by the spirit of the age, full of idealism and insights that, more often than not, came from chemicals and needed considerable tempering by the calm wisdom of individuals like Raymond, who had an easy camaraderie with the counterculture swirling around him but remained grounded by his life experience and practical knowledge. In addition to his work with Portola, Raymond was a city planner for Menlo Park, held a master of business administration degree from the Harvard Business School, and knew well how to accomplish large and complicated projects.[15] For the previous six months, Brand had been working on an idea with the manic zeal that he brought to all his efforts. Raymond gave Brand the opportunity to explore his options with no strings attached, launching a productive partnership.

After attending an "Innovation in Education" conference at San Francisco State University, Brand was inspired to work on an "educationists' newsletter" and national series of educational fairs to test his ideas of self-education and alternative activism.[16] Portola's mission was educational, so the idea fit well with the mission of the institute. Raymond encouraged Brand, but funding for an idea as big as Brand's was hard to come by, and the project languished despite Brand's significant energies and commitment. With the "aborted" and "costly failure" of his education fair and "another doomed project," the "E-I-E-I-O Electronic Interconnect Educated Intellect Operation," going nowhere, Brand returned his attention to the idea of an access service.[17]

Six months with little to show for it caused Brand to worry about the reception his idea might receive from Raymond, but Brand decided it was worth a try: "I told him this Access Catalog was what I wanted to do now." He recalled, "Dick listened gravely and asked a few questions I had no answers for (Who do you consider as the audience for this 'catalog'? What kind of expenses do you think you'll have in the first year? What will be in the catalog? How often would you publish it? How many copies?). All I could tell him was that I felt serious enough about the project to put my own money into it, but not for a while yet. I wanted to move into the scheme gradually, using Portola's office, phone, stationary, and finances (which were Dick's personal savings, dwindling fast). He said okay."[18] Raymond said yes, despite Brand's recent failures, partly because of his faith in Brand as an innovative thinker and partly because of his philosophy about business. One of the guiding principles of Portola as it grew and branched into new areas was the idea that big success often came from a "willingness to fail young." Michael Phillips later codified this mantra of the counterculture business community as the first law in his counterculture business classic, *The Seven Laws of Money* (1974).[19]

With Raymond's moral and financial support, Brand set up shop in a cubical at Portola's very modest office space. Brand was not relying on Raymond alone for support. When Brand made the decision to pursue the catalog, he committed his own resources to the project, too. Based on his recent failures with big ideas, he fully realized that he was taking a big risk. "Here I am contemplating," he wrote at the time, "the investment of all my stock . . . all of it. As in, whole life."[20] If there was any hesitation at this point, it did not show. Brand dove into his new project with passion. His work collecting information for his proposed educational fairs gave him a sense of the type of physical layout he might explore in the catalog. It would be book driven, with short reviews, sometimes only one line, and information on where to access materials and sources. The reviews would always be positive. Reviewers would ask simple questions about books or tools: "Does this replace an already existing book or tool? If yes, simply say so. Does it provide the reader with a new tool, including conceptual tools? Is it readable? Try to keep the review to one line to three paragraphs. Use quotes or photos from the book to give the reader a feel for what they would get if they bought it."[21] The goal was for editors to do the culling themselves and waste no time criticizing books or sources they found lacking. This method drove all of the *Whole Earth* publications throughout the following three decades and represented an important distinction between Brand's model of self-education and the adversarial review standards of the academic world, where all too often reviewers worked to build their reputations by tearing down the research of colleagues rather than highlighting emerging insights. The *Whole Earth* method of peer review was the antithesis of academic practice. In the coming years, Brand would encourage readers to become collaborators to a degree that far surpassed the standard letter-to-the-editor sections common then and now. As reader Doug Finley from Berkeley nicely stated, "I feel like I've been carrying on a dialog with Whole Earth for years."[22] This symbiotic collaboration between readers and publishers ensured a rigorous peer-review process for the ideas presented in the catalogs and periodicals, and Brand never balked at receiving and printing constructive criticism on the ideas presented in his publications.

So, Brand started searching Bay Area bookstores for innovative titles. He joined the American Booksellers Association to receive copies of books directly from publishers and started compiling the core bibliography that formed the knowledge base for *Whole Earth*. By the summer of 1968, Brand had a booklist and collection of titles loaded into his 1963 Dodge truck. The first phase of his project could not have been simpler: Physically assemble an alternative library and cart it around to the people who could benefit from such knowledge. With this in mind, he took off with wife Lois on a commune road trip through Colorado and New Mexico. "In about a month the Whole Earth Truck Store did a stunning $200 worth of business," Brand recalled. "No profit, but it didn't cost too much and was good education."[23] The truck store was an abbreviated version of Brand's earlier hope to tour the country with educational fairs. The truck was a store but also a lending library and mobile microeducation fair with Brand's emerging epistemology reified in piles of carefully selected books linked in ways that were becoming intuitive to Brand, though it must have seemed new and intriguing to the commune dwellers lucky enough to encounter the little truck and its enthusiastic and articulate driver.

In August, Brand was back at the Portola office, hiring a staff member and working on the catalog idea. He and Lois were living on land in Ortega Park, California, that Raymond had leased for a Portola teacher's laboratory. By October, with the help of a volunteer and his one staff member, Brand started to compile the first *Whole Earth*. Brand spent the fall of 1968 doing all of the research for the catalog and writing the review copy. The operation was bare bones, and a more rudimentary publishing studio is hard to imagine. Former janitor and artist Joe Bonner built layout tables, and all the work for the catalog was done by hand. Bonner was a great example of the "stone amateurs with energy and enthusiasm" who worked behind the scenes on all the versions of the catalog.[24] Brand had a good eye for talent and, like all good leaders, picked people who had skills he was lacking. When they had the large pages pasted together, Brand carted them down the street to Nowels Publications to be printed. The size of the catalogs was chosen to take advantage of inexpensive newsprint paper that came in large folio size only but was dramatically cheaper than other publication papers. The one thousand copies of the first *Whole Earth* were remarkably polished, considering the homespun character of the effort. Brand trucked the unwieldy catalogs around to likely stores only to meet with much skepticism and little interest in trying to sell them. They were "too big" and, although the eclectic mix of content made perfect sense to Brand, the form and content of the catalog were a mystery to booksellers, who had never seen anything like it.[25]

With the catalog launched, if not to immediate success, Brand rented a storefront just down the street from Portola on Menlo Park's main drag, Santa Cruz Avenue. Brand signed a five-year lease and moved his meager business supplies into the storefront. Handmade tables and displays made the store "a funky pleasant wooden place."[26] Brand hired Annie Helmuth, a young New York transplant who had some publishing experience, to help market the catalog. While Brand worked to research new books and think about the implications and connections of his wide and deep reading, his wife Lois ran the business end of the store and

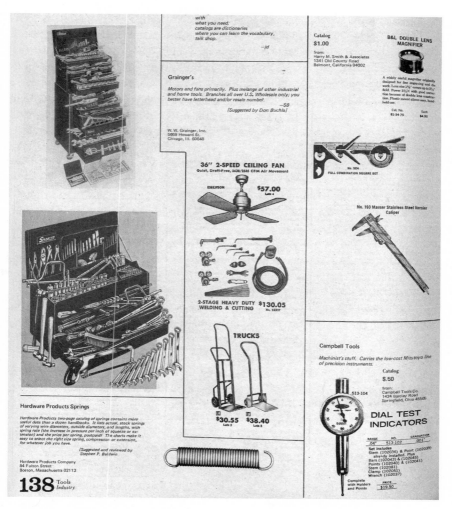

Typical tool page from the *Updated Last Whole Earth* (page 138). Anyone who lived on a commune soon learned that good tools were critical not only to the appropriate technology ideal but to sustain daily life and keep the rain off one's head at night. (Courtesy of Department of Special Collections and University Libraries, Stanford University Libraries)

the catalog. Brand was an idea man with "no business ability" at that point, according to Dick Raymond. Lois was the "hard core of the business," Brand wrote later. "She applied her math background to our bookkeeping, and her sharp tongue to our laziness and forgotten promises. She had the administrative qualities you look for in a good First Sergeant." Lois's contributions in the early years of the publication were clearly invaluable, and it is hard to imagine the catalog living beyond its infancy had she not taken control of the business end of things. Over the years, Brand developed a keen business sense, but in 1968 he was not terribly interested. "At that time, in fact, finances were not particularly on my mind. How to make money was not the design problem," Brand later remembered. "The problem was How to Generate a Low-Maintenance, High-Yield, Self-Sustaining, Critical Information Service."27

Between the fall of 1968 and the spring of 1969, Brand established a management style that he stuck with throughout his career. While he spent his time working on the big ideas, and carefully researching, reading, and writing, he delegated other responsibilities and forgot about them. "Spread responsibility as far as it will go, credit too," became the unofficial motto of the *Whole Earth* shop. Women like Lois played a critical role over the years, a fact Brand readily acknowledged: "In my experience every working organization has one overworked underpaid woman in the middle of things carrying most of the load."28 Lois worked to get the catalog distribution network established in a viable manner. The tipping point came with a distribution deal with San Francisco's Book People and a series of positive national press stories in the summer of 1969.

Later, Brand wrote that a "tiny mention" of *Whole Earth* in the *Detroit Free Press* in response to a reader question about where to get information on starting a farm had the biggest business impact on the catalog by generating "hundreds and hundreds of subscriptions."29 Fame came from an explosion of coverage in publications like *Esquire, Time, Vogue,* and virtually every major newspaper in America and several around the world.30 The *Esquire* story was the most complete, but *Time*'s full-fledged feature story that tried to explain the content of *Whole Earth,* its interesting and mercurial countercultural founder, and its sudden rise to popularity, may have reached more readers. Within six months, Brand and his creation were famous. Brand's reaction to his new celebrity was ambivalent. He had spent plenty of time in the counterculture spotlight, had been featured in one of the most significant books of his generation, and was hardly a stranger to the idea that people found him interesting. On the other hand, he had a certain reticence about his new level of national fame. He later wrote about that pivotal moment, "We were caught. We were famous." *Whole Earth* began issuing quarterly supplements to keep up with demand, and sales reached 60,000 copies. In one short year, Brand's dream of a catalog that would become "an exchange for interesting ideas and heresies" had been realized.31

Success brings responsibility and, for a free spirit like Brand coming off one of the most sustained periods of free spiritedness in modern American history, success can be harsher than failure. The immediate impact of the press coverage was the need for a second edition, quickly followed by a third and then fourth, and so on. "We hired more people. Deposits at the bank were more frequent. The bank officers got more polite."[32] Notwithstanding this newfound success, or more likely because of it, by the fall of 1969, only a year after the catalog design process began, Brand was seriously thinking of moving on. "Everything bad you've heard about fame is quite true," Brand wrote in 1972, looking back on his rise to counterculture glory. "It can throw a personality into positive feedback, where audience demands drive his character past caricature and off the deep end. Its over-rewards can jade a palate permanently."[33] There were, of course, many benefits to fame, and Brand acknowledged how it expanded his access to important and interesting people. *Whole Earth*'s shining moment in the national spotlight ensured that letters from Stewart Brand got answered. This access shaped the next phase of *Whole Earth*'s history as it morphed into the *CoEvolution Quarterly*. But in the fall of 1969, Brand was feeling burned out, and the benefits of fame were not as clear as they would be in hindsight: "In September 1969, as I was driving up the hill to work, it suddenly hit me that I didn't want to. Instead of golden opportunity, the publication was becoming a grim chore."[34] The idea man was ready to move on to the next thing. He told *Time* that he would cease publication not later than 1971, commenting, "If by that time there aren't people and ideas around doing a better job than we have, then we'll have failed."[35]

Whole Earth became so popular because it conveyed a pragmatic and hopeful look at people and tools that gave meaning to the counterculture even as the bloom was fading from the psychedelic moment of the 1960s. By carefully focusing on the practical aspects of the counterculture, *Whole Earth* linked that movement with more accepted cultural traditions and broader trends in American history, redefining the cultural revolution for readers moving away from the excess of the era but still hoping for an alternative future. One of the simplest and most powerful messages woven through the pages of *Whole Earth* was the notion that the personal is political and that individual choice matters. "Countercultural activists," historian James J. Farrell has pointed out, "criticized the depersonalized life of mainstream culture and offered alternatives that promised to create a new world in the shell of the old."[36] While it certainly must have seemed like a radical departure to many who encountered it, *Whole Earth* provided a compelling road map for careful readers who wanted to remake the existing world. Fixing the system rather than abandoning it or building something new from scratch was the critical insight that *Whole Earth* offered environmentally minded readers. The ecological viewpoints collected by Brand supported a humanistic environmentalism predicated on the assumption that cities, technology, and modernity were not the antithesis to an ecologically sound future. At a time when collective action was the norm and the goal of the environmental movement, Brand's ecological individualism geared toward improving modern life was so different that many readers of the catalog would not have recognized it as environmentalism or thought of the catalog as an environmental publication. The environmental ideas in the initial catalogs were mainly implied. The original *Whole Earth* included virtually nothing that was obviously part of the environmental movement. The Sierra Club was the only environmental organization even mentioned and then only as a source of beautiful photographic books and recreational information.

The *Whole Earth Catalog*s and occasional truck store presented real products, not just ideas, and the focus was always on theoretically feasible, if not always reasonable, solutions to real-world problems. For Brand and his cohort, Peter Warshall's *Septic Tank Practices* (1973) was just as revolutionary a book as *Das Kapital*, or more so because of its immediate implications and practicality.[37] Brand's version of practical revolution appealed to the growing numbers of disenchanted New Left radicals who were tired of sitting in crowded meeting rooms and coffeehouses while endlessly debating politics, but who still wanted to somehow subvert the system. *Whole Earth* advanced the compelling notion that, by staying home from the protest demonstration and modifying your toilet or building a geodesic dome or a solar collector, you could make a more immediate and significant contribution to the effort to create an alternative future than through more conventional expressive politics. Moreover, the catalog tapped the energy of the millions of Americans who never cared about the New Left or traditional politics in general—millions who were searching for tangible solutions to immediate problems. Most of these folks were city dwellers hoping for escape, not from the city, but from the mundane; *Whole Earth* offered a tool kit that could aid that escape or at least aid the imagination.

Although good old-fashioned American pragmatism permeated the pulpy pages of *Whole Earth*, Brand was not always a pragmatist. He spent considerable time in the drug-fueled utopian world of intentional communities across the American West. He began with the working assumption that large numbers of Americans would prefer to abandon their current lives and move into self-sustaining, ecologically friendly communities. Millions of Americans were creating a wide variety of intentional communities of every shape and size, and it was not unreasonable to assume that this trend was large enough to last. The first issues of the catalog were aimed at those individuals who were working to use the best of small-scale technology to literally disconnect themselves from the infrastructures of mainstream society and relocate to rural or wilderness areas mainly in the West. At first, *Whole Earth* promoted technologies that encouraged radically detached self-sufficiency as the key to viable revolutionary politics and utopian escape in keeping with the spirit of the late 1960s. The escape to a simple and remote life on the frontier is, of course, one of the most cherished versions of the American dream. In postwar America, the dream was updated by those seeking a more authentic life.

Helen and Scott Nearing blazed a path for the counterculture to follow with their 1954 popular manifesto *Living the Good Life.* The Nearings presented an appealing blueprint for those who wanted to "live frugally and decently" by abandoning 1950s' materialism for a simpler preindustrial life in the countryside.[38] Postwar communards built on a long tradition of American utopianism. The utopian spirit was particularly strong in the American West, where generations of Americans had reinvented themselves in the remoteness of the region. California in particular was a lightning rod for generations of utopians seeking a better life on its golden shores.[39] The 1950s' experiments in simple living inspired a significant back-to-the-land movement in the 1960s and 1970s. Intentional communities, communes, and collective farms sprang up all over the American West in rural communities like Taos, New Mexico. Although they never represented more than a fraction of the counterculture, they loomed large in the popular imagination of the era.

In the early years, *Whole Earth* articulated an appealing vision for those seeking a permanent retreat from the status quo. The first catalog appealed to the dropout school of hippies and back-to-the-landers who took their political cues from the likes of Ken Kesey, who encouraged them to "Just . . . turn your back and say . . . 'Fuck It' and walk away."[40] More importantly, the catalogs articulated an eclectic, but coherent, vision for what you might walk toward after you walk away. This vision integrated ecological awareness, simple living, and spiritual enlightenment into a program for productive and enlightened daily life, even if the retreat was to an alternative urban life.

Although neo-Luddite communards were a distinct minority in a culture that still embraced consumerism and technological progress, there was a general questioning of the Progressive faith in the ability to use large-scale technology to control the environment for perpetual economic growth, and communes were focal points for this questioning. One of the myths of the counterculture commune assumes that a rejection of technology and modernity was an essential part of the equation. Although that could be true, it was not the rule and not the goal of Brand's catalogs. Commune veteran and chronicler Bennett M. Berger argues in his *The Survival of a Counterculture* (1981) that, despite the ubiquitous presence of *Whole Earth* on commune book shelves, counterculture communities generally rejected technology while enthusiastically embracing the tools that were so necessary to their rural life. Tool enthusiasm, however, was precisely what *Whole Earth* was about, and the technophilia expressed by the catalog was perfectly in keeping with the reality of the commune life, if not the Luddite ideal. What was so appealing about *Whole Earth* for intentional community builders was the blending of the primitive and the technological, the time-tested traditions alongside the new wave. Significantly, the catalog also provided a model of enlightened consumption of as great appeal to Brand's generation as it was to their parents. *Whole Earth* was to exurban hippie communards what *Sunset* was to middle-class suburbanites—a guidebook to regionalist living, individualistic recreation, and thoughtful consumption.

The enthusiasm for communes and utopian escape apparent in the first *Whole Earth* faded fairly quickly as the publication matured, and Brand had serious second thoughts about the project of disconnecting from the grid of society and forging outlaw communities. By the early 1970s, Brand realized that *Whole Earth*'s uncritical enthusiasm for self-sufficiency and dropout politics in the first editions may have caused more harm than good. In *Soft-Tech* (1978), he wrote with some regret, "Anyone who has actually tried to live in total self-sufficiency . . . knows the mind-numbing labor and loneliness and frustration and real marginless hazard that goes with the attempt. It is a kind of hysteria."[41]

Commune founder, and eloquent voice for the outlaw community experience, Peter Coyote understood better than most Brand's reflection that the counterculture communes had attracted "the best and worst of men." On his family property, he watched a close-knit community of friends slowly evolve into a collection of "druggies and welfare deadbeats, subsistence farmers, hunters and gatherers, psychedelic pilgrims, dope growers, wanna-be shamans . . . all manner of souls seeking new lives and space in which to invent them."[42] Brand had many positive experiences in communes but mostly as a visitor and traveler. The day-to-day reality of commune life was often very harsh. Looking back, he wrote, "One reason we promote communes is that there's no better place to make all the wishful mistakes, to get your nose rubbed in your fondest fantasies." In *The Last Whole Earth Catalog*, Brand summed up his conclusions about communes with his "Commune Lie":

> We'll let other people take care of us.
> We'll let God take care of us.
> Free lunch. (Robert Heinlein)
>
> The Tragedy of the Commons. (Garret Harden)
> We'll all be honest.
> We'll all be selfless.
> No rules.
> Possessions are bad. Privacy is bad. Money is bad.
> We've got the answer.[43]

The "Commune Lie" reflected Brand's disenchantment with the sloth and he-donism of some who identified themselves as hippies, and the second wave of commune crashers who had little of the thoughtful idealism that built places like the Libre commune in New Mexico. He had certainly taken part in, and organ-ized, plenty of hedonistic escapades, but by the early 1970s he had tired of the one-sidedness of the experience and chafed at the sense of entitlement that many dropouts displayed. Brand was no prodigal—his drug use continued throughout the early 1970s—he just got tired of people who were not working on something. Brand always had a pragmatic and productive side and never abandoned his con-servative faith in individual initiative; hence the reference to Heinlein in the "Lie." Individuals who planned their escape from the mainstream through the pages of later editions of *Whole Earth* discovered a program of action where "choices about the right technology, both useful old gadgets and ingenious new tools, are crucial," but "choices about political matters are not."[44] By 1970, the intentional community movement represented a de facto political ideology that Brand no longer appreciated, although he still valued individuals and groups who elevated personal choices to productive action.[45]

Despite Brand's concerns about an overemphasis in countercultural circles on escapism, most readers of the catalog never took the message literally. The vast majority of the almost two million people who purchased copies of *Whole Earth* in its first three years never left the city, never abandoned society for a lonely exile. The message that most readers got from the *Whole Earth Catalog* was unbridled technological optimism, the idea that innovation and invention with a conscience could overcome even the worst social and environmental problems. It was this message, so profoundly different from the technophobia expressed by a previous generation of environmentalists and critics like Theodore Roszak, that made *Whole Earth* such a significant phenomenon. Brand and a growing community of appropriate technology supporters understood something about "technocracy's children" that Roszak did not: The youth culture of the 1960s and 1970s was, in the words of appropriate tech pioneer and historian Witold Rybczynski, "im-mensely attracted to technology."[46]

In *Whole Earth,* Brand assembled an almost mind-boggling array of informa-tion on tools, science, products, services, and publications ranging from the mun-dane to the downright weird, but all somehow concerned with crafting alternative lifestyles that subverted traditional networks of political, spiritual, and physical energy. For those who encountered the catalog, the experience was often a revela-tion. According to Gareth Branwyn, subsequently a staff writer for *Wired* maga-zine, "I got my first *Whole Earth Catalog* in 1971. It was the same day I scored my first bag of pot. I went over to a friend's house to smoke a joint . . . he pulled out this unwieldy catalog his brother had brought home from college. I was instantly enthralled. I'd never seen anything like it. We lived in a small redneck town in Virginia—people didn't think about such things as 'whole systems' and 'nomadics' and 'Zen Buddhism.' I traded my friend the pot for the catalog."[47] At a time when the New Left movement was dissipating, *Whole Earth* and the alternative technology movement coalescing on its pages provided hope that the quest to construct an alternative environmental and political future was still possible. In contrast to the wilderness movement, the alternative future could be one of cities, technology, and thoughtful consumption, a model with obvious appeal.

Although intentional community dwellers may have been the initial primary intended audience, even the first edition of the catalog had plenty to offer to any intellectually curious urban reader. In keeping with its mission of self-education, the 1968 *Whole Earth* encouraged readers to think globally, as the cover image implied. Appropriately, the first section of the *Whole Earth* was titled "Whole Systems." Whole-systems thinking shaped the form and content of the first catalogs and provided the foundation of all of the *Whole Earth* publications and activities that followed.

Understanding Whole Systems

"Understanding Whole Systems," Stewart Brand wrote, "means looking both larger and smaller than where our daily habits live and seeing clear through our cycles. The result is responsibility but the process is filled with the constant delight of surprise."[48] Of all the interesting sections one finds in the table of contents of the various catalogs, none is more important to understanding the goals of the enterprise and its creator than the always front-and-center "Whole Systems" section. Read this section for the thirty-year life of the *Whole Earth* publications and you will be able to trace the evolution of a cultural sensibility of great significance in the history of American culture in the late twentieth century. The intellectual journey that played out on the pages of the catalogs took readers from Buckminster Fuller to Gregory Bateson and from back-to-the-land communes to space colonies and provides the context for our current ecological dilemma so strikingly illustrated by Al Gore's *Inconvenient Truth* and by the editors of *World Changing* who reimagined *Whole Earth* for the new millennium.[49] Underlying broadly linked ideas that informed *Whole Earth*'s "Whole Systems" section was a solid foundation of extensive research on the part of Brand and his collaborators and contributors. Understanding their interpretations gives contemporary readers a more accurate portrait of the vibrant intellectual community that collectively produced the catalogs. Further, understanding the intellectual roots of the San Francisco intellectual milieu that informed the catalog helps move *Whole Earth* from the realm of counterculture cliché into its rightful place as a storehouse of significant cultural knowledge and icon of an intellectual moment too often reduced to its least common denominator. *Whole Earth* was birthed by the counterculture, but the catalog transcended the confines of that movement very early. It may always be remembered as the counterculture bible or, in the words of *Time* magazine, the "Boy Scout Handbook of the counterculture," but the catalog was much more.[50]

No one better captured the optimistic and eclectic spirit of the early editions of the *Whole Earth Catalog* than the iconoclastic self-taught designer and Harvard dropout Buckminster "Bucky" Fuller. Born in 1895, in Milton, Massachusetts, to a prominent New England family, his great-aunt was Margaret Fuller, the famous transcendentalist and feminist. With a five-generation history of Harvard educations and community leadership, Fuller had all the advantages one could hope for at the dawn of a new century. By the 1970s, Fuller was an old man but still full of radical ideas and an inspiration to a younger generation. Fuller's remarkable personal story is beyond the scope of this book and has been well told elsewhere. A master self-promoter, Fuller generated thousands of media stories throughout his life. He was a lightning rod for reporters, admirers, and critics, and his life has continued to inspire researchers and biographers who paint a picture of a complicated and sometimes tormented genius. A troubled relationship with his successful father grew worse as Fuller managed to get himself kicked out of the family alma matter Harvard, not once, but twice. Fuller was, in the words of one biographer, "an unregenerate anti-academician." Although he failed at school, he was an excellent student, learning engineering and machining techniques on the job as an apprentice millwright and later in the Navy during World War I, where he put his budding technological design genius to the test. While on assignment as a crash-boat commander, Fuller designed a system for hoisting planes out of the water in time to save the trapped pilots.[51]

Fuller has been described by admirers as a poet of technology.[52] For more than four decades, he had been on a personal quest to create a completely new way of viewing design, construction, and the environment. You can't understand the *Whole Earth Catalog* without understanding Buckminster Fuller and his profound influence on several generations of mostly young people who were outlaw thinkers at heart. The first line of the first *Whole Earth* reads, "The insights of Buckminster Fuller are what initiated this catalog."[53] Fuller wanted to reform the "human environment by developing tools that deal more effectively and economically with evolutionary change."[54] Although a prolific designer, Fuller is best known for his revolutionary geodesic domes and his concept of "dymaxion" design. Fuller defined *dymaxion* as "doing the most with the least."[55] For Fuller, the term dymaxion was a catchy way of labeling his philosophy of "comprehensive anticipatory design science."[56] The geodesic dome captured the essence of Fuller's design ideal.

Buckminster Fuller stands in front of his dome built as the U.S. pavilion for the 1967 World's Fair. (Bettmann/Corbis)

Based on complex mathematics and design principles, the geodesic dome was a structure so uncomplicated that almost anyone could build one from materials at hand. The geodesic dome became the preferred domicile for counterculture communes like Colorado's Drop City because the domes were cheap, easy to build, often portable, and environmentally friendly.[57] Fuller's artful designs epitomized the postscarcity ideal of appropriate technologies used to develop alternative communities and alternative societies. At *Whole Earth*, Brand published information on Fuller, Paolo Soleri, Moshe Safdie, and other designers and architects who worked to create alternative realities through design and technical innovation.[58]

Brand and *Whole Earth* played a central role in introducing Fuller to a new generation and promoting him to the status of cult hero. Of the protagonists of this story, design and nomadics editor J. Baldwin was Fuller's best and most careful student, but Brand was his best promoter. Brand's placement of Fuller at the heart of *Whole Earth* ensured that he was a central figure in the counterculture, even though it is hard to imagine anyone less hip than Buckminster Fuller. Theodore

Clard Svenson stands inside a partially completed geodesic dome designed as a "Theater for Psychedelics" at Drop City, Las Animas County, Colorado, August 6, 1967. He holds a painting by members of the commune entitled The Ultimate. When Drop City was organized by a group of artists the previous year, it became one of the earliest rural intentional communities in the country—and the very first to embrace Buckminster Fuller's geodesic dome, using its own "zome" variant as their signature style of architecture. (Courtesy Western History/Genealogy Department, Denver Public Library)

Roszak attributes Fuller's appeal precisely to his uncool grandfatherly appearance in the 1960s and 1970s. He was the crazy-inventor grandfather everyone wished for but no one really had. Although Roszak was no fan of Fuller's, he understood why Fuller became a leading voice for the view that "ingenuity deserved to be celebrated—from the stone axe and American Indian medicine to modern electronics" but was unconvinced that Fuller was an appropriate leader for hippie pragmatists.[59] Roszak argued that Fuller's career epitomized the "blithe eclecticism" of the worldview that *Whole Earth* codified.[60] He sympathized with the optimism of the *Whole Earth* version of counterculture eclecticism, but dismissed Fuller's career as one of "self-advertisement . . . by way of grandiloquent obfuscation, to make much out of little: little ideas, little inventions that could be sensationally clothed in cosmic pretensions," that provided a flimsy foundation for any serious reconsideration of the status quo.[61] It is worth considering Roszak's cutting critiques before examining Fuller's role in the creation of *Whole Earth* because Roszak was not alone in voicing such a strong and critical opinion. His analysis of Fuller reflects the views of many in the counterculture, the New Left, and the environmental movement who did not share Brand's enthusiasm for high or low tech and were profoundly disturbed by the technophilia and pragmatic individualism expressed on the pages of *Whole Earth* and by leading lights like Fuller.

In his wonderful little book *From Satori to Silicon Valley,* Roszak concludes that, much to his distaste, many in the counterculture were "in the market for technological astonishments." Although his arguments are insightful, as is all of his pioneering work on the counterculture, a closer look at the thoughtful way that Brand and others brought Fuller's ideas to bear on their projects does not support Roszak's conclusions.[62] Fuller's design science in particular provided a useful framework for a surging tide of practical research that is still shaping American culture forty years later in the first decade of the twenty-first century.

In explaining Fuller's design science, J. Baldwin points out, "Comprehensive anticipatory design science demands maximum overall efficiency with the least cost to society and ecology."[63] This straightforward, if vague, idea that seems so sensible now was radical for Fuller's day, and his plans and ideas were roundly criticized by the architectural and design establishment for much of his life. Fuller's frenetic enthusiasm, so appealing to younger audiences, alienated him from more skeptical observers. Recalling sharing the stage with Fuller on several occasions, Roszak captured the central criticism that dogged Fuller throughout his career: "I must confess that, though I shared a few platforms with Fuller and did my best to appreciate his books, I never came across anything he said that managed to be, at one and the same time, original, true, significant, and understandable. Worse still, I was never able to distinguish his optimism from plain egomania; I would not have been surprised to hear him announce that he had invented a better tree."[64] Supporters, however, were willing to forgive Fuller's eccentricities as a hallmark of genius and a fair trade for the wealth of ideas he generated.

Fuller was an acute observer of the natural world. Unlike most of his contemporaries, especially in the 1930s, Fuller saw the universe in terms of interconnected triangles and spheres instead of straight lines and boxes and developed a set of complex mathematical theories to turn his designs into reality. The ultimate example of his design ideal was the brilliant and elegantly simple geodesic dome, which consisted of a series of linked triangles forming a sphere that proved to be so strong that it could be built with very lightweight materials and remain structurally sound in virtually any size. The geodesic dome was embraced in the 1950s when Americans, flush with postwar prosperity, had the funds and the desire to modernize the country and develop new technologies for work and home. Fuller's first major commercial success with the geodesic dome came with an important 1952 commission from the Ford Motor Company to enclose the rotunda at its River Rouge plant. The Ford dome legitimized Fuller's design, leading to an explosion of dome building that lasted through the 1970s. As early as 1959, over one thousand domes had enclosed everything from Ford shareholders to bank patrons, soldiers, and airplanes.[65] Fuller's domes became an icon of space-age design. They looked like nothing anyone had seen before and became an architectural expression of "society's dreams of a life liberated from constraints and tutelage" in an age of technological wonders that seemed to herald the dawn of a new age where things ought to look new.[66]

One of the challenges for students of Fuller's ideal of comprehensive design science was that it required flexible holistic thinkers who cast their nets wide in the sea of science, design, and mathematics. Multidisciplinary or, according to Fuller, "omnidisciplinary" thinking was a prerequisite of good designers. Fuller spent over forty years training several generations of students to become omnidisciplinary soldiers in his "design science revolution."[67] For those interested in environmental history, Fuller's design science revolution ideal was arguably his most important contribution to the evolving environmental sensibility that was filtering into American culture during his lifetime. One of the most impressive things about Fuller's career was his ability to foresee trends and developments in technology. He plotted many of these trends on elaborate charts and graphs and was correct in his conclusions more often than not. His ability to successfully anticipate technological developments depended on his comprehensive methodology. By looking at big trends across industries and disciplines, he was able to perceive points of intersection that could lead toward technological advances. Fuller credited his foresight with his bad vision: "I was born cross-eyed. Not until I was four years old was it discovered that this was caused by my being abnormally farsighted . . . until I was four I could see only large

Looking through the distinctive window of a Drop City Zome. (Western History/Genealogy Department, Denver Public Library)

patterns."[68] Fuller's ideal of comprehensive design science, focusing first on the large patterns that unite things, became a central component of an evolving set of ideas coming from a variety of fields that argued "the connections . . . are as important as the objects of our attentions."[69] Fuller's ideal of seeing systems whole reflected his antiacademic personality and dislike of specialization. He influenced *Whole Earth* and its makers in many ways, but none more profoundly than by his example of intellectual curiosity and lifelong embrace of eclecticism and rejection of the obsession with specialization found in formal education.

The "Understanding Whole Systems" section began with several pages of fine print on Fuller, followed by Earth photograph collections and sources on anatomy, geology, and Tantra art. Books by and about John Cage, Christopher Alexander, D'Arcy Wentworth Thompson, and Norbert Wiener were featured in the first example of what became the standard *Whole Earth* format. Books were exploded out onto the large-format pages if they contained interesting illustrations and were always concisely reviewed. The "Whole Systems" section of the first *Whole Earth* provides an overview of individuals and methodologies for understanding complexity and the interconnectedness of "growth or form in any manner."[70] Brand carefully selected the works presented to provide readers with a snapshot of the most interesting tools available for self-directed students who were interested in taking a wider view of the world.[71]

The whole-systems way of thinking as presented on the pages of *Whole Earth* evolved from Brand's thoughtful and broad reading in a variety of areas. When he studied biology at Stanford University, the ecology movement was taking off and Paul Ehrlich was an early pioneer. Systems theory emerged a generation earlier, starting with the work of biologist Ludwig von Bertalanffy in the 1940s and, more significantly, the mathematical feedback theories of Norbert Wiener and others working on cybernetics in the late 1940s and throughout the 1950s.[72] Wiener's 1948 classic, *Cybernetics*, influenced everyone from computer designers to psychologists and was the foundational text of the cybernetics movement.[73] The subtitle of *Cybernetics* was *Or Control and Communications in the Animal and the Machine.* The book presented a future-forward view of self-organizing systems weaving together the technological and biological that was well ahead of its time. At a time when a forty-pound Hewlett-Packard desktop calculator cost $4,500 and was state of the art, Wiener was laying the groundwork for parallel systems of computing that seemed light years away from the bulky chunk of adding machine pictured above him in the 1968 catalog. Wiener's work influenced a generation, and his landmark text with its appealingly universal theories of feedback moved far beyond the field of mathematics. As Brand wrote in that first catalog, "Society, from organism to civilization to universe, is the domain of cybernetics. Norbert Wiener has the story, and to some extent, is the story."[74] The cybernetic thinkers shared a vision of the future that required those committed to change to move the existing system forward, not tear it down.

For the pragmatic wing of the American counterculture, this notion of a nonpolitical revolution of rebuilding toward a postindustrial future based on creative and holistic thinking was hugely appealing. Whole systems was a worldview and epistemology that linked science, technology, ecology, design, and postscarcity and laid the foundation of ecological design and a new type of environmental philosophy more in keeping with American culture than the wilderness-based movement that was at its zenith in the late 1960s. The willingness to boldly envision a bright future achieved through individual agency and intellectual muscle struck just the right tone with a generation of Americans who were searching for alternative motivations to direct their energies. The essence of the counterculture in all its guises was a rejection of traditional motivations: grades, degrees, money, status, progress in the traditional material sense. For some, the rejection was complete and the escape of drugs and hippie abandon the solution. For others who were clearly countercultural, but too creative and productive to truly drop out, the optimism of the whole-systems worldview opened a web of connections and possibilities—possibilities for a revolutionary future based on a reorganization of the economy where new technologies that worked with the environment instead of against it would usher in a postindustrial age of enlightenment and greater equality.

The possibilities of this future became clear to the *Whole Earth* cohort long before it penetrated the specialization-obsessed halls of academia. Brand, Fuller, and the community of ecologically minded outlaws who found a welcome anchor in *Whole Earth*, were not just countercultural pioneers and influential technophiles, they were intellectual pioneers shaping American thought in the second half of the twentieth century. Over the years, many such individuals contributed to *Whole Earth*, but J. Baldwin stood out for his talent as a designer and teacher. His life and work capture the spirit of the appropriate technology and ecological design movements he helped create.

The Prototyper

Of all the talented individuals who epitomized the pragmatic do-it-yourself optimism that helped create the *Whole Earth* publishing network, none better shatters the stereotype of the New Age Hippie than J. Baldwin. He became the master teacher for a generation of counterculture craftsman and chief, "thing-maker, tool freak, and prototyper," for an inventive generation. As editor in charge of tools and nomadics, he wrote several of the most popular articles to appear in *Whole Earth* and *CoEvolution Quarterly*. He was the Zelig of the appropriate technology movement and an important contributor to the ecological design and industrial ecology movements. Born in 1934, James T. Baldwin was the son of a successful AT&T engineer and grew up in the affluent New Jersey suburbs. In many ways, Baldwin's life and work epitomize the rise of an ecological design movement and alternative ecological sensibility within the design community. Ecological design is, quite simply, the application of ecological insights to the design process to enable sustainability. Most commonly associated with architecture, it applies equally to all aspects of material production. Through projects across America and Canada, Baldwin established himself as a leading voice of ecological design before the field had a name. Master craftsman, teacher to generations of design students, and inventor of the pillowdome, Baldwin's career was rich with valuable hands-on experience that he brought to *Whole Earth* and generously shared with hundreds of students between the 1950s and the new millennium. Baldwin was one of the first people Brand approached with his idea for a catalog. Brand recognized Baldwin as one of the most talented of a cadre of outlaw designers and builders who were working through the network of intentional communities springing up all over

Pioneering "Outlaw Designer" and then California State Architect Sim Van Der Ryn (left) and J. Baldwin touring the Kuboto/Smith passive solar house overlooking Lake Tahoe, March 23, 1978. (Photo by Michael Phillips. Courtesy of Department of Special Collections and University Libraries, Stanford University Libraries)

North America during the late 1960s. Whereas Brand was a master of building bridges between ideas, Baldwin was a master builder as well as a deep thinker and crisp writer. As *Whole Earth* evolved, he played a critical role as the voice for sound design of tools, buildings, and publications.

Baldwin learned some hard lessons about design at a young age. When he was eleven years old, his best friend took his father's car for a joyride down their long driveway. As he approached the end of the drive, he swerved and hit a large boulder. The huge and heavy Oldsmobile Rocket 88 was going only ten miles an hour, but the young driver was killed instantly. "The steering column had a big steering wheel with a horn ring, and the horn button was a rocket nose," Baldwin recalled, "made of transparent Lucite aimed right at your heart." This tragedy was a defining event for Baldwin. Why, he wondered, would anyone design a car with a spear aimed at the driver's heart? "It seemed stupid to me," he recalled, "and I decided in tenth grade that I wanted to be a designer."[75] The dramatic death of his best friend was a cruel lesson in how products designed for form over function could be terribly inefficient and dangerous.

In 1955, while he was studying design as a student at the University of Michigan, Baldwin had another pivotal experience when he attended a lecture by Buckminster Fuller. At the time, Baldwin was living in a house full of architecture students. One of the students invited him to attend a lecture by Fuller, and he went on a whim. "I had no idea who he was . . . when he came out on the stage, I thought he looked like one of cartoonist Al Capp's 'Shmoos' because, at the time, Fuller was pretty overweight."[76] The marathon lecture, which lasted into the morning, was a revelation. Fuller's holistic thinking and captivating ideas about design principles and how they could be turned on their head infuriated the architects in the audience but impressed the young Baldwin. He was particularly taken with Fuller's critique of housing designs. "He showed the way a furnace is chosen; you design the building and then say, 'How big a furnace does it need?' instead of from the beginning, saying, 'Let's make the building in a way that doesn't need a furnace, or needs the smallest possible furnace.' In other words, built-in wastrel philosophy was being done as a matter of course, and was indeed being taught at University of Michigan."[77] Not surprisingly, for a teenager, Baldwin was particularly taken with Fuller's credo of living life as a grand experiment. Fuller referred to himself as "Guinea Pig B" and appealingly advised his college audiences not to worry about earning a living. Baldwin took Fuller's advice to heart and decided right then to live his life as an experiment. He became one of Fuller's best students and most capable teacher-translators of the complex theories so tortuously and laboriously posed by the man himself.

Fuller's lectures were demanding affairs. He would lecture nonstop for up to ten hours, never losing his intensity, and all in a language that was literally his own. After a particularly devastating business failure, Fuller refused to speak and was completely silent for the better part of two years. When he emerged from his self-imposed silence, he did so with a new form of speaking about the world that he used for the rest of his life, much to the chagrin of many in his audiences. His obtuse terminology and stream-of-consciousness manner of speaking was difficult for some. As Stewart Brand later wrote about listening to Fuller speak, "Some are put off by his language, which makes demands on your head like suddenly discovering an extra engine in your car—if you don't let it drive you faster, it'll drag you."[78] When Baldwin attended that first Fuller lecture, he was one of the few who stayed to the end, eagerly listening to a language he felt spoke directly to him and his nascent worldview.[79] He was instantly motivated to refine his thinking about design, environment, and even his life path: "He used the term ephemeralization, which, even though I was only eighteen at the time, hit me right between the eyes."[80] It was one of Fuller's most basic concepts, the idea that you could do much more with much less in your life and work: "If you carry more-with-less to the extreme, instead of more with less, you do everything with nothing."[81] Fuller talked about what would later be referred to as zero-energy houses—dwellings that are off the infrastructure grid and produce their own energy and other needs. He drew gasps from the crowd by suggesting that you could build super-light buildings, hang them from zeppelins, and drop them into purposefully created bomb craters. Fuller was loaded with ideas from the common sense to the bizarre, and Baldwin soaked them in. Most of all, he sensed, even then, that Fuller was nondogmatic and not a victim of the type of political "ism" thinking that the young Baldwin already disliked. For the rest of his adult life, Baldwin distrusted dogmatic belief systems and tried to emulate Fuller's rejection of politics in favor of holistic design science. No one who listened to Fuller talk or attempted to read his books would have accused him of oversimplification, but Baldwin found his emphasis on the complex interconnectedness of the world and rejection of simple status quo answers refreshing. Avoiding the trap of oversimplification became a guiding principle for Baldwin as he grew into a thoughtful and careful student of industrial design.

Despite his sometimes grumpy and maniacal manner on stage, Fuller was an optimist of the first order and beloved by those close to him and legions of fans around the globe. One of Fuller's most optimistic ideas was his steadfast disbelief in scarcity. Fuller was an early proponent of postscarcity thinking and exerted part of his prodigious energy researching available world resources and computing resource use to resource need. The fruits of this research are found throughout his vast body of work but are particularly evident in his World Game Institute founded in 1972. World Game was a scenario-simulation experience designed as a countereffort to the war games that figured so prominently in the Cold War.[82] The game was played on giant versions of one of Fuller's great triumphs: the Dymaxion map of the world that hundreds of thousands of people had received as a fold out from *Life* magazine. The goal of the game, in short, was to see what creative solutions for peaceful and productive interaction nations could concoct if given sufficiently good data to make reasoned decisions. World Game came late in Fuller's career, but the central idea—free access to good data leads to good outcomes—was a lifelong goal for Fuller and the idea that shaped the *Whole Earth Catalog* from its inception.

It is hard to imagine a better laboratory for testing the implications of thoughtful design on people and environments than the military. Stewart Brand and J. Baldwin shared intense military experiences that shaped the rest of their lives and set them apart from most of their counterculture friends and fellow travelers. Baldwin points to his three years in the Army's 10th Mountain Division as a critical period in his design education. Stationed in Alaska at Eielson Air Force Base, where he worked in a maintenance yard, Baldwin taught soldiers to drive tanks and snow vehicles, and gained hands-on experience with industrial designs of all sorts. Faced with moving quickly in a frigid environment that could freeze a person's lungs in minutes, Baldwin made an air-to-air heat exchanger "out of old gas masks with big hoses taking the air out of your armpits which kept you from getting too hot and gave you warm, moist air."[83] The inventive Baldwin was an asset to fellow soldiers as he learned about design under harsh conditions. Equally important, he watched time and again as poor designs caused injury or death to the unlucky. The lesson that "bad design can kill" was reinforced during his years in Alaska, where one of his jobs was to clean up human remains after the frequent plane crashes at the air base where he was stationed.[84]

After leaving the army, Baldwin took a job with a major design firm in Ann Arbor, Michigan. He spent a brief, but significant, period learning about the world of industrial design as practiced in corporate America. One of his assignments, to design a fast-food chicken restaurant, led to an epiphany. Baldwin later said that he was not philosophically opposed to fast-food chicken, but he did not like the designs and wanted no part of helping to foist bad design on a landscape already littered with the postwar clutter of fast-food and roadside attractions. Baldwin was teaching design part time at local colleges and thought that "as a 'Coop' designer I wouldn't be able to retain the respect of my students."[85]

The turning point came when Baldwin thought about how he really wanted to live his life: "I had been horrified by the autobiography of Albert Speer, the young designer Hitler asked to be the official architect of the . . . Third Reich." "Had I been in Speer's place," he wondered, "could I have said no?"[86] Baldwin faced a situation that virtually all talented young professionals must deal with at some point: Follow the money and status or follow your heart? "With a newborn child to support, and a chance to rise fast in a world-famous design firm, it was tempting bait. I turned it down."[87] Baldwin never looked back on his early decision to abandon the traditional design path and forge a new one all his own.[88]

Baldwin's next job provided an early opportunity to work in one of the most important areas of a nascent ecological design sensibility—outdoor recreation—and provided the expertise required later for him to become the "Nomadics" section editor for *Whole Earth*. After his abrupt exit from the corporate world, Baldwin was visiting with his old design professors at the University of Michigan when he overheard a student talking about a designer who was building a new kind of tent. Baldwin recalled that he "had seen in a *True Man* magazine or something a dome shaped tent" and was captivated by the simple elegance of the design. After suffering many nights in triangular military-style tents, Baldwin knew that the new design was a major innovation and that he wanted to meet the man responsible for the design. The student directed Baldwin to the shop of Bill

Moss, the designer of the PopTent, in Ann Arbor: "I decided that because of Bucky, I didn't have to look for a job. So I walked to the door and Bill Moss answered the door." Moss hired Baldwin, who spent the next four years immersed in the emerging field of outdoor recreation design.[89]

The late 1950s and early 1960s were a critical period in outdoor recreation design. Baldwin's experience in the army helped fuel his interest in outdoor recreation and industrial design.[90] When Baldwin arrived at Bill Moss's workshop, Moss had perfected his remarkable PopTent but needed a competent prototyper to help work on new projects, and Baldwin's knowledge of the PopTent impressed Moss enough to give him a chance.[91] Like many early outdoor recreation innovators, Moss was motivated to provide better equipment to encourage more Americans to experience the wilderness he loved. His innovative little tent was "among the first nonmilitary tents" and contributed to the broadened appeal of camping and outdoor recreation in the 1950s and 1960s. "In one decade," Baldwin later wrote, "outdoor adventure was transformed from a macho/masochistic exercise to an activity enjoyable by almost anyone."[92] Hunters and anglers were very active during the 1950s in popularizing outdoor recreation. Both hunting and fishing are toolcentric by nature. Outdoor recreation publications like *Field and Stream* and *Sports Afield* dedicated half or more of their space to articles on equipment design and innovation.[93]

What set Moss's work apart was his emphasis on small, light, and elegantly simple designs at a time when other outdoor designers were in chuck-wagon mode. Moss shared an interest in simplicity with a new generation of outdoor enthusiasts who were looking to cutting-edge technology to open new frontiers for leisure and recreation while enabling escape to remote or inaccessible wilderness areas. Moss's PopTent generated interest from Ford, who had a new line of station wagons. Ford wanted to include Moss's nifty little tents along with the station wagons as part of a marketing program aimed at raising awareness of the Ranch Wagon as a proto–sport utility vehicle that would dramatically expand the outdoor recreation market. Baldwin was assigned the job of creating a disposable version of the Moss tent for Ford: "We mocked everything up in Kotex. It was sanitary napkin cloth blown into a fine nylon mesh."[94] The tents were designed to work for a couple of uses and then disintegrate; hardly an environmentally friendly design, but savvy business for both Ford and Moss.

While Baldwin was concerned with the environmental impact of his designs, he was also learning on the fly about how technological innovations could have unintended consequences. Moss was most interested in the innovation part of the equation. The PopTent was the first tent to use fiberglass poles made from a chemical stew. "I assure you that Moss entertained no qualms about polyester resin or trout population dynamics in 1956," Baldwin later wrote. "Now, we would. If after investigation you decide that the resin (and particularly the nasty hardener) is undesirable, what would you use instead?"[95] The Kotex tents led to another assignment to help design camping conversions for Ford's new 1960 Econoline vans, which became a staple of the Ford lineup for the following forty years. The camping conversions were an immediate hit, with 27,000 units sold in the following two years.[96] From the late 1950s through the 1960s, Ford's wagon and van advertising campaigns shifted their focus from depicting the products as utilitarian people and stuff haulers to showing the vehicles loaded with camping gear parked in the midst of some natural setting. Baldwin's recreational work during this time was part of an important design trend that saw seemingly insignificant recreational technologies initially intended for hard-core enthusiasts move into the general marketplace with far-reaching consequences for American culture and economy.

By the 1960s, there was an important convergence of industrial design and outdoor recreation that significantly influenced the rise of environmentalism by providing both greater access while generating questions about the implications of increased activity. New technology facilitated greater access and a radical increase in outdoor recreation. The mass marketing of new outdoor tools also drew in new converts who were as interested in the tools as the activities they were designed for. By the early 1970s, outdoor designers like Moss were helping to create tools with cachet for acquisitive nature lovers who often experienced nature through tools before moving to the real thing. For J. Baldwin, advances in design technology raised important environmental issues. "As always happens," he wrote, "the new designs had an adverse effect on nature: too many people trampling around." On the plus side, "millions of campers, hikers, climbers and boaters did vote for more wilderness, parks, and preserves."[97] During the 1950s, building on the ideas of his mentor Fuller, Baldwin joined a handful of outlaw designers who eschewed traditional career paths and began to develop an ecologically aware, comprehensive design philosophy that formed the foundation of the appropriate technology and ecological design movements of the 1970s and 1980s.[98]

After a successful and productive four years with Moss, Baldwin decided to move away from the Midwest and look for new design challenges. He eventually landed in the San Francisco Bay Area in 1963, where he started work in cybernetics and philosophy with Paul Feyerabend at the University of California–Berkeley.[99] He completed all the required work for the degree, but never graduated because of persistent disagreements with advisors and his involvement in the campus protest movement that exploded during his time at Berkeley. Despite his early conversion to Fuller's philosophy of avoiding all politics, Baldwin could not resist joining in protests that he felt were as much a reaction to institutionalized stupidity as to any particular set of political concerns. He left Berkeley with an expanded mind, a diminished respect for degrees, and a lifelong sense that the stories told in the media rarely reflected the truth of a particular historical moment. This last insight in particular shaped his long career at *Whole Earth*, where he would find a community of people who believed that there was an alternative method for presenting information that required less mediation and more respect for the reader.

A chance encounter, just after leaving Berkeley, led to the beginning of Baldwin's career as a university design instructor. Wandering around the Haight-Ashbury district, where he was living just down the block from the Grateful Dead, Baldwin was unexpectedly hailed from across the street. "This old guy, sleeked-back gray hair . . . probably sixty at the time . . . said with a big German accent, vait, vait, excuse me." Surprised and somewhat worried by the urgency of the stranger, Baldwin started to leave: "I hear puffing behind me and he's come across the street and he's chasing me." The persistent stranger turned out to be Waldemar Johansson, the chair of the design department at San Francisco State University. Johansson had heard Baldwin speak on design at a conference and wanted to offer him a teaching position. It was the type of encounter that would seem ridiculous anywhere but San Francisco in the late sixties with its remarkable concentration of talented people and plentiful prospects for bright young people to encounter opportunity if they walked around with eyes wide open.

San Francisco State University was in the midst of a creative revolution when Baldwin showed up. He reveled in the free-form atmosphere at San Francisco State, where he was given free reign to experiment with his teaching. It was there that he first met Stewart Brand when Brand was just thinking about the idea of a catalog or truck store to provide access to tools for creative people working and living on the fringes and was looking for advice on various topics. Brand found Baldwin in the design department of the university in 1967, where he was rebuilding a giant camera from a U-2 spy plane he had purchased at a military surplus sale for thirty dollars. Brand was in full countercultural mode when Baldwin first met him, with Brand dressed in a toga and top hat. The two men seemed quite different at this point but actually shared a great deal in common. Both were no longer kids: Brand was almost thirty and Baldwin was thirty-six. Both were military veterans who relished their military experience, both were students of design, and both were tool freaks who shared a contagious enthusiasm for teaching others about the acquisition and use of good tools.

Brand wanted Baldwin to collaborate on the design and technology sections of his proposed publication. As an avid catalog reader—Baldwin relished searching for new designs—he was instantly enthusiastic about Brand's idea. As a designer working on conceptual ideas with his students, Baldwin understood that at some point you have to move from paper to reality, and beyond that point you had to create a thing that people needed and were willing to pay to acquire. "Until it's in

SAWS AND SNIPS DEPT. Don't get hung up on tool names. "Tin snips" can cut lots of other things such as leather. The three identical-looking guys are "aircraft snips" and they aren't identical at all. The yellow handled ones cut straight; the green handled ones cut righthanded circles or curves and are just the thing for lefthanded people; the redhandled ones cut left as you might expect.

SIMPLE JIG holds each 2 x 4 inserted for drilling in the same relative grip, and so each will have its hole in the same place. Be sure and blow out chips so part will fit snugly, and always mark jigs so you can detect if they are slowly moving as parts thump into place. CLAMPS are a good thing to have lots of.

J. Baldwin at work with his tools pictured in his popular article "One Highly-Evolved Tool Box," first appearing in the *CoEvolution Quarterly* (Spring 1975). (Courtesy of Department of Special Collections and University Libraries, Stanford University Libraries)

a catalog, it isn't state-of-the-art," he recalled telling Brand that day. "It's state-of-the-conceptual-art, but to be state-of-the-available-art, it's got to be in a catalog." Quoting fellow alternative designer Steve Baer, he said, "You haven't done anything important until you can buy it at Sears."[100] Intrigued by Brand's enthusiasm and obvious intelligence despite his appearance, Baldwin agreed to help by writing some book reviews even if he was too busy to be directly involved. Shortly after that initial meeting, Baldwin found a pink slip in his box telling him he had lost his job at San Francisco State. That same day, he received an unexpected letter from Buckminster Fuller asking if he would be willing to fill in teaching for a year at Southern Illinois University. The timing was perfect, and Baldwin, honored to represent his mentor during his absence, took off for the Midwest.[101]

The year at Southern Illinois was not all that Baldwin had hoped, but it provided an important opportunity to refine his teaching and to work toward turning Fuller's ideas and his own insights about design and American culture into an ecological design curriculum.[102] In both formal and alternative learning environments, Baldwin became an important teacher and mentor for hundreds of students between 1967 and the early 2000s. In the early 1970s, he taught at the infamous countercultural Pacific High School, where dome designer and author of the classic *Dome Books* Lloyd Kahn built a village of experimental geodesic domes with Baldwin and a ragtag bunch of juvenile delinquents who had been sent to the school instead of being incarcerated.[103] Baldwin's sense of humor and careful efforts to explain complex ideas in simple and accessible language made him a natural mentor to the high school kids. Over the years, his ability to articulate Fuller's ideas with direct language and obvious enthusiasm contributed greatly to the growth of ecological design from a shared sensibility among a dispersed group of individuals into an influential movement that transcended the confines of the design community and injected a healthy dose of pragmatism into the American environmental movement.

One of the first formal gatherings of this new breed of pragmatic design-oriented environmentalists, an event called Alloy, was held in a New Mexico ghost town in the spring of 1969. For most of the participants, Alloy began with a serious road trip. Stewart Brand loaded his new Whole Earth Truck Store in Menlo Park with products to sell and set off across the mountains bound for New Mexico. In southern Illinois, J. Baldwin crammed his Citroen with himself and six design students for the two-day marathon drive down to a meeting whose location was still a mystery to him.[104] Brand and Baldwin both converged on Alloy from different directions. For all the attendees, Alloy represented a critical convergence of ideas and the birth of a community and ultimately a movement.

CHAPTER THREE

Baling Wire Hippies

If we are going to muck our way out of the past thirty years of doo-dads and back into some form of less consumptive lifestyles, then we are going to need some kind of technology junk, some group of baling wire hippies who can who can tell us how to convert our broken hair dryers into incubators.
J. D. Smith, 1973[1]

Alloy

Looking back, Stewart Brand wrote, "If I had to point at one thing that contains what the catalog is about, I'd have to say it was Alloy."[2] Alloy the first "programmatic gathering" of "outlaw designers" in America organized by appropriate technologists Steve Baer and Barry Hickman for three days in March 1969.[3] During the late 1960s and early 1970s, New Mexico was the center of alternative technology and countercultural industrial design outside the Bay Area. As Brand wrote, "More of the interesting intentional communities are there," so "more of the interesting outlaw designers are," too. Baer envisioned the event as "a meld of information on Materials, Structure, Energy, Man, Magic, Evolution, and Consciousness."[4] He invited creative individuals representing all of these areas. Speakers at Alloy included J. Baldwin, geodesic dome designer Lloyd Kahn, and Drop City artists. Brand showed up representing *Whole Earth* with his truck store, but more for his own enlightenment and as an enthusiast for any gathering of "outlaws, dope fiends, and fanatics."[5] Brand remembered one hundred and fifty in attendance, a remarkable turnout, considering the remoteness of the site. Attendees came from as far away as New York, Washington, and Canada. "Who were they? (Who Are We?)" Brand preemptively asked readers when he presented the Alloy story in *The Last Whole Earth Catalog.* They were, he answered, "Doers, primarily, with a functional grimy grasp on the world. World-thinkers, drop outs from specialization. Hope freaks."[6] They were J. Baldwin's thing-makers, tool freaks, and prototypers, who intuitively understood whole systems and shared a desire to unite creative design science with ecology.[7]

Baer and Hickman were best known up to that point for their role in the design and construction of the alternative energy structures at the iconic Drop City commune near Trinidad in southeastern Colorado. According to commune historian Timothy Miller, the name for this unusual place came not from dropout or

Scenes from Alloy, in the mountains outside La Luz, New Mexico, March 1969, featured in *The Last Whole Earth*. From a distance the Alloy gathering might have looked like just another of the extended counterculture parties so common in the piñon-covered mountains of New Mexico. Inside the domes, however, gathered a remarkable collection of productive appropriate technology innovators mapping out a tech-friendly environmental ethic decades ahead of its time. (Courtesy of Department of Special Collections and University Libraries, Stanford University Libraries)

from dropping acid, but from what art-colony founders, University of Kansas and Colorado art students Clark Richert, Richard Kallweit, and Gene and Jo Ann Bernofsky, called "Drop Art, which began when they painted rocks and dropped them from a loft window onto the sidewalk . . . watching the reactions of passersby."[8] Starting in 1965, Richert, Kallweit, the Bernofsky's, Bill Voyd, Hickman, Baer, and a host of migrating design enthusiasts used their remote outpost to explore alternative architecture and appropriate technology (AT) applications. Inspired by a Buckminster Fuller lecture in Boulder, Colorado, they constructed an amazing landscape of multicolored domes on seven acres of windswept Colorado farmland. Several of the domes owed more to Steve Baer's designs for a polyhedra dome structure he called a "zome" than to Fuller's classic design. Later the Drop City designs were perfected in the beautifully situated Libre commune in Huefano County, Colorado.[9]

Baer founded a successful design company, Zomeworks, in Albuquerque, New Mexico, to convert his Drop City experiments into viable AT products. Over the coming years, Zomeworks and *Whole Earth* were linked through a long sharing of ideas and funding. Thin and intense, Baer was the embodiment of what *Whole Earth*'s J. D. Smith called "baling wire hippies": tough, practical, resourceful countercultural frontiersmen and frontierswomen who knew tools and how to use them. Like many who attended Alloy, Baer was well educated: He had a degree in mathematics and physics from Amherst College and had also studied at the Swiss Federal Institute of Technology after spending three years in the military stationed in Germany, and was motivated to use his design skills for a good cause.[10] Baer was an early pioneer in alternative energy, with a particular interest in solar energy design, and working with wife Holly and partner, Barry Hickman, wanted to formally bring together other innovators and outlaws like himself.[11] He was one of the first to publish a how-to guide for geodesic dome construction with his 1968 *Dome Cookbook,* which set the standard for the large-format, newsprint counterculture books. Brand got the inspiration for the size and layout of *Whole Earth* from Baer.[12] Baer found an abandoned tile factory outside La Luz, New Mexico, south of Albuquerque, near where an earlier generation of designers from around the world had gathered to test their experimental atomic bomb at the Trinity Site.[13] Baer's site was remote enough to enable the designers to spread out and camp for days, play music, and enjoy a festival-like atmosphere.

J. Baldwin arrived at the Alloy after a two-day marathon drive from his teaching job at Southern Illinois University with six design students crammed into his modestly sized Citroen sedan. The crew took off without any clear directions to the conference other than that it was in New Mexico.[14] On the way, they stopped off at Drop City to see the domes and get directions. Baldwin made the effort to cart himself and his students on this epic trip because, like Baer, he knew that

"Mrs. Oleo Margarine" with daughter "Melissa be" standing in front of a zome made from wood and tar-impregnated wood fiber with distinctive pentagonal windows, at the Drop City commune, Las Animas County, Colorado, December 26, 1966. (Courtesy Western History/Genealogy Department, Denver Public Library)

there were others like himself out there and very much wanted to meet them.[15] "The Alloy conference was amazing," he recalled. "It was the first get-together and recognizing that we were a 'we' of counterculture builders."[16] Baldwin was not enjoying his teaching job and hungered for contact with people who were really getting their hands dirty turning Buckminster Fuller's concepts into hard reality. Alloy was a turning point for him. After his return to Southern Illinois, he resigned his job and took off looking for alternative technology action wherever it might be.

There was plenty of countercultural mayhem at Alloy, and drugs and magic may have distracted some participants from the more productive discussions about technical building plans, the thirty-five dollar house, or how to turn Fuller's *Critical Path* into a practical path. For most of the participants, however, Alloy was a significant effort to articulate an environmental design philosophy. It was one of the birthplaces of ecological design, the movement that became the best expression of the dreams of *Whole Earth*'s human-centered ecology. Brand's observation that these were "drop outs from specialization" and "hope freaks" captured the spirit of the alternative technology as it was just evolving from a shared sensibility and research topic into a full-fledged international movement of profound significance for the twenty-first century.[17]

The AT and early ecological design movements valued broad thinking by creative generalists, amateurs, and technological optimists. The celebratory atmosphere of Alloy and the pure joy expressed by participants over the human potential to take control of our shared environment for community betterment could not be more different than the "apocalyptics and melancholy" and obsession with the impending loss of pristine nature that fueled much of the environmental politics of the late 1960s and early 1970s.[18] Like many of the countercultural happenings of the time, Alloy was a blending of education and entertainment with music, kids working on art projects, poetry readings, a magic show, and the general camaraderie of the campfire atmosphere, which prompted one attendee to observe, "It's amazing that a hundred hippies could come together and even eat at all."[19] Attendee Steve Durkee drew a crowd as he demonstrated techniques for blowing helium bubbles while Bill Pearlman's theater group down from Santa Fe staged performances. Noted photographer turned documentary film maker Robert Frank, best known at that point for his classic photo-essay *The Americans* (1958) and beat-generation documentary *Pull My Daisy* (1959), perched in the background, capturing all the fun and innovation unfolding in the spectacular piñon pine–covered hill country.[20] The resulting documentary and photographs became a touchstone artifact of the ecological design movement.

Alloy represented a remarkable collection of environmental thinkers who were interested in improving urban quality of life and using technological innovation to solve concrete problems like shelter and energy and even taking on some early environmental justice concerns. Alloy was not a clean break from wider environmental politics or thinking at the time. There was a measure of apocalyptic thinking that lent urgency to speeches at the gathering, and issues like overpopulation were of great concern among the ecotechnologists. Still, as the tents were stuffed in their sacks and the Volkswagen camper vans putted away down the dirt roads leading from La Luz at the end of the Alloy conclave, participants like J. Baldwin and Stewart Brand left with a new sense of community and, most importantly, faith in human intelligence to solve the environmental crisis.

Brand left Alloy inspired. The gathering represented everything that *Whole Earth* tried to capture about ecology, technology, design, tools, and community. Alloy provided good evidence of an emerging alternative environmental community dedicated to a different path toward ecological health. Even in the first years of the new millennium, debates about the relationship between nature and culture still resonate and provoke heated discussion among people who share many of the same environmental concerns. At the root of these differences within the environmental community lies a fundamental disagreement about the relationship between human technological development and the ecological health of the planet. In hindsight, Alloy seems well ahead of its time, and it would take decades for the forward-thinking message of these outlaw ecologists to filter into mainstream environmentalism. The most significant early efforts in this direction were aimed at the homes and homesteads of the counterculture generation. Exploring practical applications of AT and ecological design for shelter provided the foundation of the ecological design movement and was a central theme of the *Whole Earth Catalogs* during the 1970s.

Shelter and Land Use

Just behind a restored chicken coop nestled in the rolling hills of Marin County in Northern California rests J. Baldwin's legendary shop truck. No longer mobile, covered in moss, with its paint slowly eroding from walnut acid and moist air, the former bookmobile and shop, once immortalized in a coffee-table book about great workshops, conveys a sense of waiting—waiting for the next phase of a design revolution that went dormant about the same time as the truck but seems more likely now than when Baldwin built his nomadic shop in the late 1960s.[21] Even frozen in place, the nondescript truck seems aimed at a bright future where it will be needed by those ecologically minded doers and inventors who want to literally build a better world one gizmo at a time. For more than three decades, the tool truck, towing its sleek Airstream living quarters/office, was a fixture at many significant moments in the history of ecological design. A boxy plain white exterior hides an amazing interior designed as an AT laboratory by its remarkable owner.[22] Generations of lucky designers, engineers, architects, commune dwellers, and neighbors turned their dreams into reality, or failed and learned some hard lessons about design, inside the remarkably tidy and well-organized mobile shop.

Inside, custom-designed green cabinets hide drawers full of tools organized by what one might use them for. Open a drawer and you may find a hundred or so "pullers," "drivers," or whatever you might need to build a place or a thing. The walls of the truck fold out so that the workspace can expand for bigger projects like solar panels or structural elements of a geodesic dome. "Folks have used this tool set to build hardwood furniture, boats, bicycles, solar collectors, and even whole houses," Baldwin recalled. "It was intentionally designed to be a three-dimensional sketch-pad—a place to make the first physical manifestation of an idea."[23] Baldwin's well-used tool truck embodied the spirit of the *Whole Earth Catalogs* and the tool culture that the catalogs represented. This shop was made to evolve, grow, and accommodate the demands of ecological design. Every tool inside has a story

The neatly organized interior of J. Baldwin's wonderful tool truck. The 2½-ton 1958 Chevy "walk-in" van was a meat truck and then a mobile county library before Baldwin converted it. Originally built to serve as a shop for solar and wind experiments, the workshop evolved into a "three-dimensional sketchpad" that helped birth the ecological design movement. (Photo by the author)

that Baldwin can precisely relate. He can even show you hash marks, sometimes hundreds, indicating how often a particular tool was used over the years. Tools that did not stand up to his always exacting standards were quickly replaced. In addition to being a highly skilled industrial designer, Baldwin was one of the most consistently effective writers and editors at *Whole Earth*. He wrote one of the most popular articles ever published in *Whole Earth*, "One Highly Evolved Toolbox," and his sections of the catalogs and later magazines capture the true sprit of the enterprise. Baldwin's individual contributions and collaborations with Stewart Brand articulated the shared design language that united many of the disparate thinkers that comprised the *Whole Earth* community.

Looking back on the history of *Whole Earth*, Stewart Brand thought that the title of Baldwin's popular article could have been the title of the catalog. Not unlike Baldwin's magical shop-on-wheels, the catalog was one highly evolved toolbox providing a print version of Baldwin's more direct route to an alternative type of design-based environmental advocacy. After two decades of fulfilling *Whole Earth*'s promise to provide "access to tools," Baldwin reflected on the history of his humble truck: "Our portable shop has been evolving for about twenty years now. There's nothing very special about it except that a continuing process of removing obsolete or inadequate tools and replacing them with more suitable ones has resulted in a collection that has become a thing-making system rather than a pile of hardware." Of this quote, Brand commented, "Just so with the *Whole Earth Catalog*."[24]

In 1972, after installing a geodesic dome for Buckminster Fuller on his family's island, Baldwin left the East Coast on an extended trip. On the way west, he stopped in New Mexico, where Steve Baer introduced him to Robert Reines, the son of famed hydrogen bomb pioneer Fred Reines. According to Baldwin, the younger Reines was "trying to do better than his dad" and was "trying to live 100 percent solar" while developing a major new solar project for possible mass-market production.[25] Reines called his operation Integrated Living Systems (ILS). Baldwin stayed at ILS for two years before returning to California to work with *Whole Earth* full time for a while.

It was during his California return that Baldwin acquired his famous tool truck in response to his constant travel and the economic realities facing fledgling AT organizations. Baldwin had noticed that none of the tech pioneers he knew could get adequate funding for their overhead expenses. Grant-giving organizations, like the Pew Trust, provided groups like ILS some overhead but would not pay for the expensive tools that were crucial to everything they did. Baldwin decided that the best service he could provide to the nascent AT movement was to construct a mobile alternative technology laboratory fully stocked with the tools of the trade.[26]

One of the first projects to benefit from Baldwin's wonderful little truck was the Farallones Institute, established by pioneering architect and self-described "outlaw builder" Sim Van Der Ryn in Occidental, California.[27] Van Der Ryn was a professor of architecture at the University of California–Berkeley, author of the best-selling book on waste composting, *The Toilet Papers*, and a pioneering ecological designer.[28] Through his teaching career, he influenced several generations of architects, whom he taught to place ecological concerns at the center of their building designs. During Jerry Brown's administration as California governor (1975–1983), Van Der Ryn was appointed state architect, where he designed passive solar government buildings and helped push tax initiatives that funded AT development in California.[29] In 1969, he founded the Farallones Institute to create a teaching laboratory for ecological design. Baldwin and his tools helped build the sustainable buildings and structures that comprised the modestly sized, but boldly innovative, institute. Farallones was more than just an experiment in green architecture, it was a full-fledged attempt to build a complex of buildings and structures that were as self-sustaining as possible with the technology of the day. At Farallones, the buildings were designed to produce their own energy, recycle waste and reuse it as fuel, and provide for some of the food needs of the institute through organic gardens using the latest techniques in sustainable agriculture.[30] Farallones was just one example of the ways that concerns about architecture were merging with broader environmental concerns.

In the 1970s, it was relatively easy to get people to recognize the ways that ATs offered regionally sensitive and sustainable power production: Giant wind farms in California and solar arrays in New Mexico demonstrated how a large-scale implementation of alternatives might look. Books like Ernest Callenbach's popular *Ecotopia* (1975) presented tantalizing utopian visions of revolutionary societies completely rebuilt with AT as the foundation. Critics like Langdon Winner pointed to the Ecotopian strain of the AT movement as evidence of the potential for AT to become nothing more than an impractical distraction from meaningful political action. But *Whole Earth* generally focused on research and products with practical applications for the average home. The real potential of alternative technology was actually in the home. Home was where complex research could be put into practice, potentially enabling an environmentalism based on simple changes in daily living.

As mentioned previously, suburbanization placed millions of middle-class Americans in close contact with nature and provided inescapable examples of the extent that nature was being daily eaten away by dramatically expanding housing development. This contributed to a greater awareness of environmental issues, leading to successes like the passage of the Wilderness Act in 1964. Counterculture shelter efforts represented a very different response to this trend. Readers of *Whole Earth* in the late 1960s and early 1970s came of age during this period of suburbanization, which partially explains their generation's enthusiasm for environmental issues. For technologically minded readers of *Whole Earth*, however, it was not enough to be aware of the encroachment of human settlement on the land and to respond by placing some areas out of reach. If many suburbanites looked out the picture window and were motivated to protect what nature remained for self-interest or the common good, alternative technologists stood outside looking in and were trying to think how they could make the home a more ecologically sustainable place.

The focus on shelter and domestic technology at *Whole Earth* tapped a deep tradition of placing the reinvention of home at the heart of cultural revolution. During one of his hiatuses from editing the *Whole Earth*, Stewart Brand built a small house for himself. He recalled this time working on his own shelter as by "far the most rewarding effort I'd been involved in since starting the Catalog."[31]

How small was beautiful. Little Hand Made cabin on tiny Bolinas, California, plot captures the ideal of truly simple living and the beauty of the hand-built home. (© Ilka Hartmann 2007)

Brand's experience building his home was hardly unique. Spending some time building your own shelter and working a piece of land was a common right of passage for many in the counterculture and during the 1960s and 1970s, when there was a significant countercultural resettling of the American West. This new group joined a long line of western frontierspeople and, like generations before them, they revisited all the key issues encompassed in a reclaiming of the mythic and very real territory of the American West. Like those who preceded them, these countercultural seekers looked for solutions to old problems with new technology while working to accustom themselves to existing traditions of living with a seemingly rawer nature. Counterculture efforts to bring ecological thinking into home and community design took on a special importance in the American West, where resource scarcity coupled with extremes of heat and cold lent urgency to the search for technological solutions to shelter. Even dedicated urbanites with no previous handcraft experience, like alternative cartoonist and future *Whole Earth* editor Jay Kinney, succumbed to the power of what historian Patricia Limerick has called "fantasies of Western independence and fresh starts."[32] No other aspect of the counterculture, in fact, captured the spirit of the age better than the simple desire to strike out to a new frontier and provide one's own shelter.

From innovations like Fuller's geodesic dome to reevaluations of ancient techniques like adobe building, the counterculture was obsessed with reinventing the ideal of the American home and, in the process, creating autodidactic models for architecture and design in an effort to create practical ecological modes of living. The shelter section was one of the most enduring and popular features of *Whole Earth* throughout the years. At communes and neighborhoods across the world, counterculture designers developed new technologies for ecological living that melded insights from the latest alternative technology research with an older craft tradition. Often these efforts were undertaken by novice builders who learned on the job with mixed results. Their efforts were celebrated not only on the pages of *Whole Earth*, but in an important related literature of alternative architecture and design, and were part of larger national and international trends toward architectural "adhocism" and the "non-plan" movement in the architectural community of the 1960s, in which "the lay 'consumers' of architecture" devised an alternative universe of do-it-yourself construction as part of an effort to "seize control of the forces which manipulate the design and construction of the buildings and cities around them."[33]

It wasn't just hippies who were moving to the West. The massive migration to the western Sun Belt in the decades following World War II provided strong motivation for creative solutions to housing suited to fragile, arid environments. It would take a long time for motivation to turn into action, but *Whole Earth* was on the cutting edge of a cultural ferment and sea change in American architecture while it simultaneously tapped into deep do-it-yourself craft traditions in the American West.[34] The western-focused shelter discussions in *Whole Earth* built on the efforts of New Deal regionalists like Arthur Carhart, who celebrated regionally sensitive design in magazines like *Sunset*, and later decentralist-distributist movement proponents like New Mexico solar architect Peter Van Dresser.[35] In his *The Myth of Santa Fe* (1997), architectural historian Chris Wilson points to Van Dresser as a proto-counterculture environmentalist. Van Dresser started building solar houses in the early 1950s and penned an important statement on alternative design with his *Development on a Human Scale* (1971).[36] Van Dresser's research presaged the countercultural desire to link AT with regional traditions in an effort to ensure that "decentralized economic development and local culture be mutually reinforcing."[37]

In addition to the vibrant New Mexico regionalists, architectural historians point to Drop City, Libre, the Red Rockers, Paolo Soleri's Arizona Arcosanti experiments in hivelike living, the efforts of the New Alchemists to build self-sustaining domestic arks in New England and Canada, Lloyd Kahn and J. Baldwin's Pacific High School experiment, Van Der Ryn's Farallones Institute, Steve Baer's Zomeworks, and, internationally, India's Auroville as significant examples of a countercultural design movement. All became laboratories for countercultural shelter and appropriate domestic technology research that received wide coverage in the popular media.

Efforts to create a countercultural design sensibility also led to popular publications on the expanding universe of self-build.[38] *Whole Earth*'s primary competitor, *Mother Earth News*, founded by John and Jane Shuttleworth in 1971, dedicated a substantial portion of its pages to evolving theories about shelter and became the preferred how-to guide for persistent communards through the 1990s.[39] Of all the self-build publications, none was more significant than Lloyd Kahn's best-selling *Shelter*, a classic of do-it-yourself and hand-built homes from around the world.[40] Kahn's beautifully illustrated *Shelter* was a subtle assault on the replication and conformity of postwar domestic architecture and celebration of precedents that "evolved across centuries or millennia."[41] Kahn founded Shelter Publications in 1970 while working at *Whole Earth* and produced a series of astoundingly successful design classics over a thirty-year period.[42] Prior to *Shelter*, his two *Domebooks* became the instruction manual for Buckminster Fuller's design science revolution and ensured that the geodesic dome or some variant became the standard architectural form for communes, mavericks, and iconoclasts all over the American West. The geodesic dome very quickly became a statement of a sensibility as much as a place to live or work. Forty years later, domes still stand as markers for a geography of difference in the rural West; "Turn left at the dome just off the highway and you'll see the ruins of the commune" is an all too common refrain. Over the last three decades of the twentieth century, architects and designers, inspired by these renegade amateurs, began incorporating counterculture design concepts into complex projects, resulting in the field of ecological design.

Lloyd Kahn's remarkable publishing career reflects the two predominant phases of the counterculture shelter movement, both of which fall within the broader non-plan trends articulated by architectural historians Jonathan Hughes and Simon Sadler. First came the techno-utopian vision of shelter epitomized by Fuller's work with its emphasis on the space-aged geodesic dome built with the latest exotic materials. This trend drove the shelter coverage in the first issues of *Whole Earth* and provided the foundation for a body of literature that dealt directly or indirectly with a countercultural notion of home predicated on new technologies and space-age materials.[43] This counterculture trend represented a continuation of the technologically driven 1950s' vision of home and the latest wave of enthusiasm for the ideal of creating the "home of tomorrow," a term that dated to the late 1920s. In the 1930s, Buckminster Fuller became the leading proponent and designer of the future home, with his Dymaxion House, designed in 1928, becoming the icon of the futuristic home ideal.[44]

The second trend—the move away from high-tech domes and exotic materials and back to traditional craft, biological models, and natural materials—created some tension within the community of ecological designers. Kahn had a complete change of heart about domes and his role in popularizing the idea of the dome as home to a generation.[45] Likewise, Stewart Brand knew that *Whole Earth* had been even more influential in popularizing the work of Fuller and in particular his dome. "As a major propagandist for Fuller domes," Brand wrote with hindsight, "I can report with mixed chagrin and glee that they were a massive, total failure."[46] J. Baldwin had collaborated on a variety of dome projects with Kahn, most significantly the Pacific High School domes, and knew well some of the very significant practical issues with domes: They leak, they are loud, they are hard to furnish. Unlike Brand and Kahn, however, Baldwin both held on to the geodesic dome, with his own pillow dome as a leak-free example of how they could succeed, and provided insightful criticism of the proponents of a more natural design framework who failed to see the ways that their work was predicated on complex technological systems of production.[47]

Baldwin remained enthusiastic about the dome partly because of his interest in the recreational adaptation of geodesics, which played a significant role in shaping the outdoor recreation revolution of the post–World War II period and offered a different model for countercultural nomadic homes and workplaces.[48] Baldwin's background as a recreational designer helped him see the relationship between recreation-tech and environmental concern and how the two might be reconciled through innovations in fixed and portable shelter.

Both Kahn and Baldwin believed that, for ATs to succeed nationally, they had to first succeed in homes broadly defined. They were both consistent and thoughtful critics of utopian AT dreamers and particularly disliked the presentation of AT in *Mother Earth News*, which they thought fostered an unrealistic vision that undermined the long-term success of the movement and provided fodder for critics who painted the whole enterprise as part of a mushy New Age escapism.[49] Both Baldwin and Kahn worked in California and New Mexico to create realistic shelter prototypes for real-world testing. When their ideas failed, they were completely uncompromising in their self-criticism. Kahn was a particularly harsh self-critic who revisited his pioneering dome efforts at Pacific High School and presented his findings of failure with relish and detail for readers of *Shelter*. Appropriate technologists like Kahn and Baldwin had as little sympathy for romantic technologists as they had for romantic wilderness prophets.

It was the first, more technological version of shelter that most appealed to the back-to-the-land and communard wings of the counterculture, with the biological turn in shelter thinking leading to some of the most innovative work toward blending alternative design and changing concepts of home applied to America's urban population. Stewart Brand's Point Foundation later worked directly to support researchers working to apply alternative technology solutions to the urban environment. In the early 1970s, Point board member Huey Johnson funded urban ecological design pioneers, like Helga and Bill Olkowski's research into the "integral urban house" and the New Alchemy Institute's efforts to create the model alternative home with their ark on Prince Edward Island.[50] While the media celebrated the dome-obsessed communards, these researchers were building far more viable models for integrating ecological design into the urban and suburban homes that housed the bulk of the U.S. population.

All of the proponents of ecological home design shared much in common and, notwithstanding the sometimes rancorous debates about domes, worked toward creating a more ecologically sensitive home-design philosophy blending the insights of Fuller's design science with efforts to revive architectural craft and the cultural traditions of indigenous builders. This blending of insights reflected in countercultural shelter research shaped the *Whole Earth Catalog*s. Between 1969 and 1972, Brand shifted the catalogs away from the philosophy of Buckminster Fuller and his techno-utopian design science toward a more biologically based sensibility. The shelter sections featured less on geodesics and more about traditional craft and natural models for architecture.[51]

These counterculture shelter trends are significant because alternative technology linked with older bioregional architectural traditions is perhaps the best example of the think-globally act-locally philosophy that opened doors to individual agency in environmental advocacy that greatly expanded the constituency of the environmental movement. The effort to bring sophisticated design science wedded with the latest insights from ecological research into the home, as architectural historians Christine Macy and Sarah Bonnemaison remind us, provides a counternarrative to the "nineteenth century proposition that nature must be isolated and purified of human activities in order to be protected."[52] Brand's version of this same idea, "We are as gods," fits perfectly with American traditions of property rights and cultural assumptions about individual control over home.[53] As kings or queens of our own castles, we each have at least one small world that we can remake, one opportunity to integrate more ecologically thoughtful technology into our daily lives, and few other efforts can compare in impact. As Brand would later point out in his wonderful critique of modern architecture, *How Buildings Learn* (1994), it is in the home where design, environmental sensibility, and culture are most likely to find a middle ground. Home design is often "classically vernacular," according to Brand, "informal, pragmatic, alive with offhand ingenuity."[54] Starting in the 1950s, the do-it-yourself movement grew into a billion-dollar industry and, by the 1990s, more money was spent by amateur builders than professionals.[55] Thus, the ability to move ATs into the world of the

American home was of critical importance in the 1960s and remains so in the first decade of the twenty-first century.

Countercultural immigrants to the West faced issues with water, Native American cultures, farming, small-town politics, community, and energy. Potter Dennis Park's experience in moving to the remote semi–ghost town of Tuscarora, Nevada, was typical. "I was tense," he wrote, "a raw recruit anticipating small skirmishes with the elements before reaching armistice with nature."[56]

California seemed to attract an abundance of the most talented of this new generation of western immigrants who "came into creative response to the social challenges and possibilities of their era."[57] Individually and in groups, counterculturalists reconfigured the built environment and helped reinvent the ideal of home. One of the constants of American history is the quest for home. During the 1950s, the single-family home took on new significance as a cultural symbol of success, achievement, and a return to normal life in the aftermath of sixteen years of depression and world war. Early in the history of *Whole Earth,* the shelter section captured the spirit of the counterculture and its desire to leave the city and return to the land for a more authentic life. In so doing, the counterculture was tapping into some deep traditions of regionalism. That initial fascination with rural life morphed into a broader concern with the built environment in general and, with the publication of the *CoEvolution Quarterly,* a focus on integrating ecological concerns into the fabric of urban life.

The energy crisis of the 1970s moved the efforts of the ecological designers into the national spotlight and ushered in a brief period of strong federal support for alternative technology research and design. In the 1970s, *Whole Earth* played a significant role in promoting early efforts in ecological design, and several individuals from the *Whole Earth* community played a direct role in California during the Brown administration. By the 1980s, the federal funding had largely dried up, but not before the enthusiasm for alternative shelter trickled into mainstream housing design and development and created a new demand for houses designed with regional environments in mind. Progressive states like California and Colorado fostered these trends, but they were short lived. Solar collectors on Denver roofs became, like leaky geodesic domes, interesting architectural reminders of a green revolution in domestic housing that almost happened but ultimately failed. Bioregional architecture remains an elusive goal in a housing market driven by the rise and fall of interest rates and giant corporate builders who slap up subdivisions as quickly and cheaply as possible. Also, as Simon Sadler reminds us, after the energy crisis "sustainability gained a reputation as something of a hair shirt for architecture."[58] After a remarkably promising period of growth and development of green building and ecological design, America entered a twenty-year period of backsliding that made green developments like the Davis, California, Village Homes an anomaly in a cultural landscape carpeted with McMansions with SUVs (sport utility vehicles) in the driveway.

But it didn't all go away. Architectural historians Christine Macy and Sarah Bonnemaison have argued that the alternative design of the counterculture resulted in a "substantial body of knowledge" that continued to shape architectural design throughout the last quarter of the twentieth century, just not at the briefly high levels experienced during the 1970s.[59] If ecological design was a "hair shirt" for architects, it was not even on the radar screen of fellow travelers in the historic preservation movement. When preservationists did become aware of ecological design, they often were at odds with the goals of appropriate technologists, especially in the urban sphere, even though the historic preservation and environmental movements grew out of the same cultural ferment and shared many assumptions about quality of life.[60]

The convergence of cultural and environmental preservation in America has taken longer than anyone might have imagined, but it should not come as any surprise that some of the best work toward this reconciliation emerged from the counterculture. Countercultural resettlement of ghost towns, urban wastelands, Main Streets, and mountain and coastal towns played a vital role in breathing life back into cultural landscapes that had suffered from the vicissitudes of postwar cultural and economic trends. The obvious meeting ground for these two trains of thought was in the area of New Urbanism. According to historian Michael Wallace, "by the early sixties the people most threatened with urban dislocation and disruption had begun to protest. . . . Amid this ferment, a small band of social scientists, architectural critics, psychologists, and journalists began critiquing the social and psychological consequences of the urban renewal and highway programs."[61] Changing urban patterns can be attributed not only to countercultural impacts, but also to postindustrialization, an unholy partnership, but one which makes sense given the paradoxes of the counterculture.[62]

Whole Earth was an early forum for New Urbanism and Stewart Brand, an insightful observer of the overlapping territory of cultural and natural preservation. In an article titled "Neighborhood Preservation Is an Ecological Issue," he outlined his thoughts on this issue. "What bulldozers do to an existing human community," he wrote, "is no different from what they do to any other climax biological community . . . [A] complex mature meta-life is crippled or killed and replaced by a simplistic early successional non-community . . . [T]he process is always called improvement."[63] This excellent article captures Brand's subtle synthesis of the insights of ecological and cultural preservation that were well established by the 1970s but not clearly linked for most Americans. The article was a response to efforts by the city of Sausalito, California, to dismantle the houseboat community where Brand lived and worked. The Sausalito houseboat war in the late 1960s–1970s highlighted conflicts between environmental and cultural preservationists, and Brand's frustration with this misunderstanding between two like-minded groups is clear in this landmark article.

Brand was influenced by, and was a prominent proponent of, Jane Jacobs's *Death and Life of Great American Cities* (1961), which did for urban preservation what Rachel Carson's *Silent Spring* did for the environmental movement: It served as a keen observer's wake-up call about the unintended consequences of urban homogenization.[64] *Whole Earth*, and later *CoEvolution Quarterly*, disseminated the thinking of the New Urbanists, and Brand was well ahead of the game in providing his readers with clear links between architecture, ecological design, historic preservation, and urbanism. After the initial enthusiasm with rural communes, Brand shifted his focus and the focus of *Whole Earth* toward urban ecology and consistently emphasized the shared message and language of cultural and natural preservationists. Processes that decreased complexity and variety in communities and ecosystems were presented in *Whole Earth* as part of the same problem. Brand's pragmatic environmentalism did not differentiate between human and animal communities, and he was not afraid to point out what he viewed as hypocrisy on the part of environmentalists who would fight to the death to save a seal habitat but remain silent as developers bulldozed more and more "densely textured human" communities.[65] For Brand and many others in the alternative technology movement, ecological design provided a bridge between concerns with cultural preservation and environmental preservation.

Ecological Design

Whole Earth became the early voice for the ecological design and AT movements, providing a forum for researchers who were working outside the mainstream of environmentalism and their respective academic disciplines. It was a small world at first with lots of overlap and cooperation between the various projects. During the early years, virtually all of the significant players in the emerging ecological design movement shared a direct or indirect connection with *Whole Earth*, either as contributors, editors, or readers of the catalog and quarterlies. This diverse group researched new household technologies, such as composting toilets, affordable greenhouses, and organic gardening techniques, along with alternative energy technologies, but were united by a passion for using design as the foundation of environmentalism. While the research of individuals and organizations working in the area of ecological design varied greatly, all of those involved shared the common goal of using technical research to enable simpler, more ecologically sensitive lives and economies of a human scale. They also shared, according to

Baldwin, "insatiable curiosity about virtually everything" and the "willingness to risk all their money and time . . . to see something happen that they think has to happen."[66] Most importantly, they shared a common understanding of the necessity for "synthesis and integration" of design, economics, and environment.[67]

For technecologists like Baldwin, Van Der Ryn, John Todd, and Robert Reines, thoughtful, ecologically informed design was the only viable long-term solution to environmental problems. As Van Der Ryn put it, "The environmental crisis is a design crisis. It is a consequence of how things are made, buildings are constructed, and landscapes are used."[68] Design is inherently human centered and works within the framework of existing economic structures while remaining, mostly, apolitical. Thus, design, by nature, avoids many of the problems of more standard models of environmental advocacy that more often than not were at odds with the daily realities of a sizable percentage of the people that populate the Earth. During the same period that the American environmental movement was awakening to the problems of industrial pollution and working toward dramatically expanded federal regulation of this pressing issue, ecological design proponents and alternative technologists were two steps ahead.[69]

In their manifesto for dialectical design, *Cradle to Cradle* (2002), authors and internationally recognized ecological designers William McDonough and Michael Braungart built on Jane Jacobs's "commerce path" versus "guardian path" model of environmental activism to argue against the conventional environmental wisdom of "reduce, reuse, recycle."[70] Well known for *Death and Life of Great American Cities* (1961) and her work on city planning, Jacobs was an astute observer of environmental politics and provided a compelling model for survivable futures in her book *Systems of Survival* (1992).[71] Jacobs nicely dissects environmental politics and explains why a more pragmatic approach is necessary for the next century. She posits two competing syndromes: the commerce and the guardian. The *guardian syndrome* refers to the familiar story of government environmental regulation and the agencies and organizations that exist to work on behalf of the public interest. This syndrome is "slow and serious" and, according to McDonough and Braungart, "meant to shun commerce."[72] The *commerce syndrome* is the "day-to-day, instant exchange of value." Commerce is, Jacobs argues, "quick, highly creative, inventive, constantly seeking short- and long-term advantage, and inherently honest." Jacobs acknowledges that neither of these syndromes is a perfect moral system and both play a critical role in the quest for a sustainable future. She does, however, argue against any blending of the two, a phenomenon that results in what she calls "monstrous hybrids."[73] Money, she argues, obviously corrupts the guardian. Even worse is excessive regulation, which can cripple commerce and stifle innovation. Although Jacobs wrote mostly about the East Coast, her subtle libertarian views on markets and government meshed perfectly with the counterculture's enthusiasm for commerce and the distinctly western libertarian sensibility evolving on the pages of *Whole Earth*.

McDonough and Braungart specifically highlight the problems with the guardian model of federal "end of pipe" regulation, which punishes the outcome of bad design instead of rewarding good design. They criticize the shortsighted focus of *ecoefficiency* as an "outwardly admirable, even noble, concept" that is nonetheless a flawed strategy for long-term sustainability. Echoing the defining insight of *Whole Earth*, these authors embrace the possibility of a moral economy and argue, persuasively, that the commerce path is central to any honest effort to create a viable twenty-first-century environmentalism. Echoing Brand's "We are as gods," they insist that "being 'less bad,' is no good."[74] Further, the guardian path, encompassing conservation and preservation, is, they argue, all part of a century-long effort to be "less bad." To be less bad "is to accept things as they are, to believe that poorly designed, dishonorable, destructive systems are the best that humans can do." This is the dismal proposition—failure of the imagination—that *Whole Earth* consistently argued against through its presentation of creative alternatives. Before Jacobs coined the term, *Whole Earth* was the national voice for the commerce path. Thoughtful readers could discover that humans were a part of nature but masters of their own destinies, that creative innovators could use their talents for the good of the planet, and that the market economy could be the best tool for solving the environmental crisis of the postindustrial world. The commerce path explained by Jacobs and so elegantly reformulated in *Cradle to Cradle* offers a hopeful model for everyday environmentalism for the next century.

McDonough's 2005 appearance as a keynote speaker at the annual Sierra Club meeting demonstrates how the commerce path has moved to a place of preeminence even in the stronghold of the wilderness movement. The divide between the ecological design and the preservation movements is only recently fading as renewed scarcity fears and global warming have elevated AT and the commerce-path model of sustainability to unprecedented levels of attention and prestige.

But that was in the future. The concentration on AT and ecological design in the *Whole Earth* publications and among the community of contributors and readers in the early 1970s reflected a larger shift in direction in American environmentalism. This shift was bolstered by a long series of events that galvanized public opinion around quality-of-life issues and created a climate ripe for broader cultural acceptance of new environmental thinking based on a reevaluation of design and consumption.

During the late 1960s and through the mid-1970s, bleak and well-publicized environmental disasters linked to flawed design resulted in changing public opinion and growing consensus on the link between environmental health and quality of life. The guardian-path model of advocacy provided the awareness necessary to expand support for the commerce path of AT and ecological design during the later half of the seventies. In January 1969, an oil well off the coast of Santa Barbara, California, caught fire and exploded, releasing 235,000 gallons of oil. Within several weeks, sticky oil tar covered thirty miles of Pacific beach. That same summer, the Cuyahoga River in Cleveland, Ohio, into which tons of chemicals had been released by various industries, caught fire. The river had burned many times before, but this time resonated beyond local embarrassment. These spectacular environmental crises, vividly captured by the media, helped generate strong bipartisan political support for environmental regulation. Building on precedents set by the first Clean Air Acts, Congress passed a series of sweeping environmental laws.

Of these, the National Environmental Policy Act (NEPA) signed by President Richard Nixon in 1969 was the most significant.[75] Through the leadership of Washington Senator Henry Jackson and the Senate Committee on Interior Affairs, NEPA created the first comprehensive legislation for dealing with broad environmental problems. Prior to the passage of NEPA, government environmental policy was criticized for slow and fragmented decision making on environmental issues and a consistent failure to reconcile federal environmental policy with contemporary environmental science. Jackson's committee worked to create a legislative package that would bring order and scientific thinking directly into environmental policy making. A central provision of NEPA was the requirement that all federally funded projects produce an Environmental Impact Statement (EIS), which is a detailed assessment of the ways that a project would affect the environment.[76] The EIS became a very direct means of managing environmental impact according to contemporary scientific standards and forced all government projects to detail environmental consequences. Most significantly, NEPA created the massive Environmental Protection Agency (EPA), which grew quickly in both size and power to become one of the nation's most significant regulatory bodies. The passage of NEPA and the creation of the EPA permanently established the federal government as a caretaker not only of land and resources but also as a potential force for design regulation aimed at improving environmental quality of life.

NEPA also required the EPA to set environmental regulations. During the 1970s, it focused pollution-control efforts on water and air. The Clean Air Act of 1970 set tough new standards for airborne emissions and required factories to use new technologies like air scrubbers to remove the most dangerous pollutants from smokestack emissions. Scrubber mandates were met with strenuous resistance from industrialists such as coal-fueled power-plant owners, who contended that the scrubbers were impractical, expensive, ineffective, and would preclude American industries from competing in world markets. In response, scrubber manufacturers improved the design and reliability of their scrubbers. EPA standards not only forced polluting industries to clean up production processes but also directly supported green technological innovation. The Clean Water Act of 1972 set new technology standards for industries that emitted waste into water and required these industries to research technological alternatives for reducing emissions. The EPA's focus on technology standards helped move the public debate away from the rancorous 1960s' dispute over culpability in the environmental crisis and toward practical solutions. Throughout the 1970s, well-funded public agencies provided economic incentives for innovative technology research whose practical and economical solutions gained public support and helped silence critics of green technologies.

Although NEPA and the EPA seem to be precisely the type of cumbersome guardian-path programs that *Whole Earth* presented alternatives to, this legislation was different. NEPA empowered citizen science and small-scale pragmatic environmentalism like nothing before. Over the coming decades, NEPA, and the EIS system in particular, encouraged decentralized and dispersed amateur research that ultimately shaped everything from city planning to the management of the national parks. For a young naturalist, avid birder, and future *Whole Earth* editor Peter Warshall, NEPA and the EIS opened the door to an unexpected career.

Born in El Paso, Texas, Warshall moved to Brooklyn with his family when he was three. There he learned about nature in the victory gardens at the Brooklyn Botanic Gardens, where he had a small plot that he tended from a very young age. By twelve, he "was the president of the Children's garden" and already a student of the urban nature that he lived in and the rural landscape around Camp Crystal Lake in Roscoe, New York, where his aunt taught nature-study classes for kids exiled from the city during the polio epidemics of the 1950s.[77] Early one morning at the camp, Warshall's aunt and uncle took him out on the lake at dawn. "I had binoculars for the first time and I was watching two pintail ducks come out of the marsh," he recalled. "I tracked them right into the sun and it nearly burned my eyes out." That experience, "with birds right there as the cause," launched Warshall on an accelerated education as a "maniacal naturalist."[78]

He rapidly finished high school, where he excelled, and was admitted to Harvard University at age sixteen, where he worked toward a biology degree. Those first years at Harvard were very difficult for the young Jewish naturalist from Brooklyn. Harvard was a very Protestant culture, and there were few other Jewish naturalists to mentor the prodigy with the unusual background. Also, even at a very young age, Warshall, like many in his generation, was disappointed in the specialization-oriented environment of the American university. "There were so many people in the late sixties who were really alienated from what they could learn in colleges," he recalled.[79] Like Brand, Warshall was deeply interested in the biology he was studying in school but chafed at the culture of specialization

Peter Warshall in the early 1970s. Photographer and fellow activist Ilka Hartmann vividly recalled the long-haired young Warshall silencing a skeptical audience with his "enormous alternative insights" and Harvard credentials, which forced the community to think twice about dismissing the hippies who were their new neighbors. (© Ilka Hartmann 2007)

that he thought stifled the more creative impulses of students who were searching for a holistic interdisciplinary education.

Still, he persevered, and a series of chance encounters led to remarkable opportunities to craft his own educational program of study. Most significantly, "I played a poker game with a bunch of professors at Harvard and one of the professors said, hey, do you want to go to Africa this summer?"[80] The African trip launched Warshall's graduate studies into the group behavior of monkeys, provided a welcome break from Harvard, and opened his eyes to global environmental issues. Reenergized, he returned to work on a Ph.D. in biological anthropology, hoping to apply the insights of that field to his work as a participant observer of animal cultures. A Fulbright Fellowship enabled him to spend a year in France working with eminent anthropologist Claude Levi-Strauss. Levi-Strauss's structuralism, the idea that cultural meaning is produced through multifaceted structural networks, fit perfectly with Warshall's multidisciplinary interests and desire to link biology and anthropology in order to create a more encompassing model for understanding the human/nature relationship. He eventually received his Ph.D. in biological anthropology from Harvard for his thesis "on the group behavior of rhesus monkeys in Puerto Rico."[81]

In 1970, after completing his dissertation research, but with the writing still ahead, Warshall made his way west, stopping for a while to work cataloging "Native American horses that were still running around the Pryor Mountains" of Wyoming.[82] Eventually he landed in the Bay Area and Bolinas, California. Warshall was one of the counterculture migrants who came to Bolinas at the moment when the town was looking for alternative means to fight suburban sprawl. He became a central protagonist in a series of events that put the town back on the map, despite the efforts of its citizens, and contributed to the rise of an amazing Bay Area environmental coalition whose alumni helped shape the environmental policy of California and the West for the next generation. Warshall came to Bolinas in 1970 to visit poet friend Robert Creeley and finish writing his dissertation. "I thought I would sit in Bolinas and do my Ph.D.," he recalled. "And eventually I did."[83] But not before the amazing birding opportunities of the area provided several months of distraction from his monkeys. "During that period of time," he remembered, "I began to switch and become a birder, a maniacal birder, because I no longer had monkeys around me."[84] Bolinas was a birder's paradise with a nearby bird sanctuary and the world-class birding of the Farallones Islands and surrounding Point Reyes National Seashore. Shortly after his arrival, in January 1971, a devastating oil spill left Bolinas beaches coated in Bunker Sea oil and littered with dead and dying seabirds. Warshall joined forces with fellow resident Orville Schell, already an internationally recognized China scholar and significant future contributor to *Whole Earth*, Greg Hewlett, and a remarkable group of environmentally minded young town members to respond to the crisis.[85] They formed the Bolinas Future Studies Institute, which was really just a small community group, to work on pollution issues and devised new methods for cleaning oil-coated birds without killing them.[86]

The experience of trying, mostly unsuccessfully, to save the birds opened the community's eyes to the extent of the environmental problems that faced their little town: "The town looked at itself and saw a couple things. It was being forced by a large engineering firm, plus the State Water Resource Control Board, to sewer-up, and everyone initially agreed because there was a raw sewage outfall pipe going into the lagoon and into the mouths of surfers."[87] The Future Studies

Volunteers cleaning a remote beach in northern California after the San Francisco Bay oil spill, January 1971. (© Ilka Hartmann 2007)

Institute became a countercultural "shadow government" for the town and launched a pathbreaking effort to use emerging insights about watershed management as the basis for a community growth plan. Bolinas was unincorporated, and the community's Public Utilities District "was what really ran the town." In 1971, the representatives of the utilities district were mostly older residents with a deep history in the area. They initially favored the sewage system as a way to improve the quality of life of residents and correct the lagoon-pollution situation. Warshall, Schell, and a core group of the younger generation came to believe, however, that the system would be a mistake on both counts, and several ran for public office to try to stop the sewers. Control of housing growth and unwanted suburban sprawl became the bridge between the generations. "If you put in a large-volume wastewater pipe, that allows development," Warshall remembered arguing. "If you constrict the size of the pipe or the extent of the infrastructure octopus, you can't have development except for septic tanks." Orville Schell and photographer Ilka Hartman chronicled the ensuing political fight in their wonderful portrait of a community in transition, *The Town That Fought to Save Itself* (1976).

The young candidates won control of the utilities board, and an effort was launched to look for alternatives to the sewer system. As the "only biologist in town that was kind of interested," Warshall became a key researcher as the board instigated a remarkable amateur analysis of the sewers, septic tanks, and watersheds in an effort to fight the powerful State Water Board.[88] President Nixon had just signed NEPA, and the Bolinas crew knew enough to focus in on the EIS as a possible tool for their efforts: "No one had written one on sewerage, so we decided we'd write one . . . so we wrote the first environmental impact statement ever written for a sewage treatment facility," Warshall recalled, "and it blew the mind of the EPA." Warshall used the EIS guidelines to systematically undermine the arguments of the state: "I just literally followed the rules of it: indirect impacts, direct impacts, irreversible impacts, long-term and cumulative impacts."[89] The EIS made a damning case against the large-scale sewage system, and Warshall and the Bolinas crew created a model for using citizen science and the toolbox of NEPA legislation to protect a community from unwanted development.

Through the remarkable chemistry of the early seventies in the Bay Area, talented individuals seemed to show up just when their skills were needed. Warshall's training uniquely qualified him to interpret the EIS, and another town resident, Tim Winneberger, a "maniacal supporter of septic tanks," arrived just in time to launch an innovative environmental study of septic systems immortalized in Warshall's AT classic, *Septic Tank Practices* (1973). The survey provided further ammunition to the argument against the building of the sewer system and, more importantly, was the first attempt to use limitations of infrastructure expansion as a practical means of community growth control.[90] For Warshall, the septic tank survey, which required him to closely examine the waste of his neighbors, was a revelation: "I began to see the anthropology of septic tanks . . . I learned about soils, I learned about drainage . . . [C]ollege had not taught me anything compared to that stuff."[91] After the successful survey, Warshall was elected to the Public Utilities District, serving as an elected official for nine years as a member and then director. "Being in public office was my great education," he remembered. "I had to learn what a watershed was, and it's remained a crucial unit of thought for the rest of my life."[92] The story of Bolinas demonstrated that dispersed citizen science empowered by NEPA could foster innovative local responses to complex global and regional environmental problems.

Warshall went on to have a long association with *Whole Earth* as the most consistently environmentally minded contributor and editor.[93] He dramatically increased the coverage of natural history and innovative environmental thinking in the universe of *Whole Earth* publications over the years, starting with the *Whole Earth Epilog* and culminating with his run as the final editor of the *Whole Earth Review*, where environmental issues and natural history dominated the final years of the publication after a period of fascination with the information revolution in the 1990s.[94] Warshall brought the accumulated wisdom from his experiences with NEPA in Bolinas to the community of *Whole Earth* readers most directly in one

Cover art by Arthur Okamura from an early edition of Peter Warshall's appropriate
technology classic. (Courtesy Arthur Okamura)

of the most significant issues of the *CoEvolution Quarterly* in the winter of 1976. This "Watershed Consciousness" issue became *CoEvolution*'s best seller, indicating the extent to which pragmatic environmentalism and commonsense AT solutions appealed to the universe of *Whole Earth* readers who became *CoEvolution* subscribers.[95] In addition to NEPA, several other federal and some state agencies and organizations promoted AT research and expanded the options for citizen scientists who wanted to follow the Bolinas model.

In 1972, Congress created the U.S. Congress Office of Technology Assessment (OTA) to facilitate research on technology and environment. The OTA quietly contributed to significant advances in green technology. Throughout the 1970s, public support for government-supported environmental research reached a high point. Public opinion polls revealed that a high percentage of Americans thought that the federal government should stimulate ATs to improve the environment.[96] In the 1970s, environmental policy and technology were linked in the minds of the public, leading to a brief period of strong federal support for research in areas like solar energy, wind and geothermal power, and a whole universe of alternatives to the technologies of abundance that Americans had so enthusiastically embraced in the years following World War II.

In 1973, public enthusiasm for AT research reached a new level of urgency. That year, after American assistance had helped Israel defeat an Egyptian-Syrian attack, the Arab-dominated Organization of Petroleum Exporting Countries (OPEC) imposed an embargo on oil exports to the West.[97] Oil prices skyrocketed; filling stations ran out of gas, and panicked Americans waited for hours to buy gas. The OPEC embargo revived old resource scarcity fears and reinvigorated the search for alternative sources of renewable energy.[98] President Jimmy Carter responded to the energy crisis by initiating federal energy conservation programs and by providing federal funding for AT research.[99] In 1976, these programs came together under the umbrella of the National Center for Appropriate Technology, which consolidated various programs in an effort to foster cooperation and accelerate research. While the federal government took the lead during the 1970s in responding to growing public desire for research into viable alternatives to the status quo, private and grassroots organizations played an important role, as well. The 1970s saw a dramatic increase in environmental awareness that benefited proponents of all types of environmental research and advocacy.

The celebration of the first Earth Day on April 22, 1970, dramatically demonstrated the growth of environmental awareness during the 1960s.[100] Earth Day celebrations ranged from tree-planting festivals and litter cleanups to large-scale marches and rallies. An estimated twenty million people participated, making it the largest public demonstration in American history. While the conservation organizations of the 1950s and early 1960s were comprised mainly of middle-aged or older professional men, most of the participants in Earth Day were school children and college-age adults. During these years, all of the major American environmental groups benefited from an infusion of new members, many of them young people. The National Wildlife Federation, for example, saw its membership double between 1966 and 1970.[101] Invigorated by young, politically active new members, groups such as the Sierra Club and the Wilderness Society became more directly involved with political causes and lobbying efforts. Rising with the tide of popular concern and support for environmental legislation, environmental groups became a significant force in American politics.

The federal environmental legislation of the 1970s and rise to prominence of environmental organizations represented a major advance for the environmental movement. There was, however, a serious downside to these gains for supporters of environmentalism. The power of the EPA and the uncompromising demands of some environmentalists precipitated a backlash among industrialists, businesspeople, farmers, and workers who felt that government regulation and environmental controls threatened their livelihood. Likewise, some critics worried that NEPA's checks on development and technology constituted a dangerous antimodernism that privileged nature over people. The Endangered Species Act of 1973 became a focal point for advocates of this view. The pro-commerce, pro-technology research programs of the ecological design and AT movements provided an alternative to the perceived or real antimodernism of the mainstream environmental movement.

The energy crisis in particular presented a perfect opportunity for researchers to demonstrate how ecotechnology might offer realistic solutions to pressing environmental issues. Any car-driving adult of the 1970s was aware of how ecological problems of the postwar era were either directly or indirectly linked to the acquisition and distribution of energy. Long lines at gas stations and soaring fuel prices brought home the reality of finite energy resources. This renewed realization that scarcity was once again a real and long-term problem forced counterculture environmentalists to reevaluate the aspects of their technological enthusiasm derived from 1960s New Left notions of a postscarcity world. By the mid-1970s, it was clear that postscarcity was a long way off. The move away from postscarcity politics toward an AT philosophy that recognized scarcity and reformulated utopian radicalism paved the way for alternative technology and ecological design to move into the mainstream. Organizations working in the area of alternative energy, in particular, were poised to provide a new vision of environmental activism to this broadened audience of concerned Americans.

The Soft Path

All of the new and renewed energy technologies featured in the pages of *Whole Earth* became components of what physicist Amory Lovins referred to as the "soft path." Lovins popularized the soft path to energy solutions in a widely read and highly controversial 1976 article in the prestigious journal *Foreign Affairs*.[102] For Lovins and his supporters, the soft path was the moral alternative to an American "federal policy . . . [that] relies on rapid expansion of centralized high technologies to increase supplies of energy."[103] Instead of increasing centralization, soft-path proponents supported decentralized ATs and urged Western nations, specifically the United States, to direct their research toward renewable alternatives and explore the possibility of shrinking the system to provide a more equitable relationship with developing nations. Amory Lovins was one of the most interesting thinkers of the 1970s. A child prodigy who received his first patent at age 17, he went on to join the faculty at Oxford University at 21.[104] At Oxford, he began working with Friends of the Earth consulting on energy policy, quickly becoming a leading world authority. Then Lovins turned his prodigious talents toward alternative technologies. He was an early and particularly effective and influential proponent of soft technologies like passive solar power and the use of traditional building techniques and materials readily available as immediate solutions that could be implemented while researchers like himself searched for long-term answers. He emphasized that the benefits of soft tech were accessible for regular citizens of the Western world and easily transferable to developing nations, as well. Simple passive solar techniques, like painting a south-facing wall black and covering it with glass, could radically decrease the dependence on large energy systems.[105] During the 1970s, Lovins worked worldwide on AT projects, eventually returning to the United States to stay. In 1982, he founded the Rocky Mountain Institute as a center for alternative energy research and as a living laboratory of active and passive solar technology displayed in the institute's green buildings in Snowmass, Colorado.[106]

Not surprisingly, considering his reputation as an innovative scientist and holistic thinker, Lovins crafted an environmental philosophy and plan for action that defy easy categorization. His wide-ranging research on energy, ecological design, "natural capitalism," and transportation reflects the curiosity and inventiveness that characterized the counterculture environmentalists. In his landmark book *Natural Capitalism* (1999), coauthored with wife Hunter Lovins and longtime *Whole Earth* contributor and entrepreneur Paul Hawken, Lovins presents his perspective on the various environmental mind-sets.[107] Borrowing from biophysicist Donella Meadows, the authors argued that four worldviews shape perceptions of the environmental/economic dynamic: "Reds, Blues, Greens, and Whites."[108] To summarize, the *Blues* are "mainstream free-marketers," who have a "positive bias toward the future based on technological optimism and the strength of the economy." The *Reds* are socialists. The opposite of the Blues, they focus on labor and the human condition and rarely address the environment. The

Greens are environmentalists who "see the world primarily in terms of ecosystems, and thus concentrate on depletion, damage, pollution, and population growth." The *Whites* are the "synthesists," who "do not entirely oppose or agree with any of the three other views."[109] This last category provides a useful way to think about the community of counterculture thinkers of the *Whole Earth* network. The synthesists shared an optimistic faith in human ingenuity and distrust of ideology, preferring, instead, "a middle way of integration, reform, respect, and reliance."[110]

The ecosynthesists also shared a faith in local dependency and local action, arguing that "environmental and social solutions can emerge only when local people are empowered and honored." One of the most powerfully presented environmental ideals that permeated the 1970s editions of *Whole Earth* was the notion of local dependency. The focus on local dependency gradually evolved into the concept of bioregionalism, but always raised questions about where one should seek solutions to environmental problems. "Is it preferable," Stewart Brand asked, "to be dependent on institutions we don't know, and which don't know us, or on people, other organisms, and natural forces that we do know?"[111] Unlike traditional Greens or proponents of the guardian path of federal regulation, ecosynthesists like Lovins thought that politics and ideology were far inferior to the think-globally, act-locally ideal that required deep local knowledge combined with broad multidisciplinary synthetic wisdom. This was particularly true in the quest for alternative technology solutions. For the technology to work, it needed to address broad problems while remaining adaptable to very specific local conditions and needs.

Soft-path proponents could point to several significant energy technologies with long and productive histories that perfectly fit with the ideal of easily accessible renewable energy for a modern world readily adaptable to fit the needs of specific communities or situations. In fact, most of the soft-path solutions to modern energy problems were not new; they were retooled versions of preexisting technologies. None of these older technologies better captures the spirit of the soft-path energy movement than the venerable windmill. The use of wind as a source of power dates to antiquity, when humans first harnessed the wind to power ships and soon after as an efficient means for the mechanization of food production and irrigation. For thousands of years, cultures all over the globe relied on wind power to mill their grains, drain their lowlands, draw water from aquifers, and saw their lumber.[112]

In America, the windmill became an emblem for self-sufficiency as farmers and ranchers moved into the arid plains of the West and mastered the technology of the windmill in order to survive far from established services and energy sources. Americans quickly discovered that windmills could be fabricated out of a wide variety of locally available materials and constructed cheaply from mail-order plans. As early as 1885, windmills were used to generate electrical power in

One of twenty-five windmills built at the Auroville appropriate technology community in India. Nothing better illustrates the blending of old and new technologies aimed at sustainability than the windmill. (Courtesy of Department of Special Collections and University Libraries, Stanford University Libraries)

the United States. Early researchers learned that windmills were an excellent source of electrical power on a small scale, and even small windmills could easily provide enough electricity for a home or small business. Preexisting windmills could be retrofitted with electrical generators and provide power to a remote farm or mill while retaining the capacity to pump water or grind wheat.[113] Although many adopted the windmill as a permanent source of power, wind energy never became the standard that many thought possible, and wind power faded from view for most of the twentieth century.

The energy crisis of the 1970s renewed the interest in wind energy. One reason that wind energy never went mainstream was because of an inability to regulate the source. The power from wind generators ebbed and flowed, and the fickle winds never maintained a schedule. This made wind a poor sole substitute for hydro-electric or coal turbines, which could sustain a constant and manageable flow of energy for large systems and power grids. Soft-path supporters, however, were unconcerned about the problems of wind power for large systems. On the contrary, they were looking for sources of power that were better suited to small systems.

Like E. F. Schumacher, Lovins and other soft-tech proponents believed that the ability to construct small-scale, self-sufficient systems provided individuals and communities with a closer connection to the earth and a greater degree of control over their lives. The windmill was the type of technology that could enable one to use the latest research in electric power generators and new materials like fiberglass to build machines that produced no pollutants and provided essentially free and limitless energy. For soft-path proponents, the potential of the windmill was both practical and political. Disconnecting yourself from the power grid was the first step toward a cleaner environment and a move toward reevaluating all of the large systems that dominated the economy and daily life of developed nations. The key to the politics behind soft-path and AT science was the insight that real change came not from protest, but from constructing viable alternatives to the status quo, starting with the basic elements of human life: food, energy, and shelter. Lovins was certainly an outlaw designer, but his remarkable résumé lent credibility to the AT movement and caused both opponents and supporters to articulate their energy positions carefully. Brand approved not only of Lovins's ideas but also his terminology: "'Soft' signifies that something is alive, resilient, adaptive," Brand mused, "maybe even lovable."[114] In 1966, a band called the Wilde Flowers changed its name to Soft Machine and released a well-received album of the same title. The next year, poet Richard Brautigan published his poem "All Watched Over by Machines of Loving Grace," which perfectly expressed the counterculture desire to unite nature and the machine: "I like to think (it has to be!), of a cybernetic ecology, where we are free of our labors, and joined back to nature, returned to our mammal brothers and sisters, and all watched over by machines of loving grace."[115] By the mid-1970s, soft-path energy research into solar power, wind, geothermal heat, biogas conversion, and recycled fuels moved to the forefront of the environmental and alternative technology movements.

At the same time that a growing number of environmentalists were exploring different paths toward decentralization through renewable energy development, others were working within the second area of the outlaw edge: information technology. For Brand, alternative energy was important, but developing trends in information technology tapped a deep and consistent vein of interest that, at least early on, he saw as a point of convergence. As he later expressed it, "Information technology is a self-accelerating fine-grained global industry that sprints ahead of laws and diffuses beyond them."[116] Brand was intrigued by what he called the "subversive possibilities" of technologies as diverse as recording devices, desktop publishing, individual telecommunications, and especially personal computers (PCs). Moreover, his interest in information technologies meshed well with his environmental thinking. He came to equate the two in this way: "Rafting a wild river makes the body sing with the old dangers, gives the body a sure sense of itself and frees it to explore unfamiliar hazards such as immersion in computerized 'virtual reality.'"[117] In addition to his many accomplishments, Brand was a pioneering participant in the PC revolution and was one of the select group who were there at the creation. He was among a group of counterculturalists who had a deep respect for innovators like Steve Jobs and Steve Wozniak, who were designing and then using their computers to push what Brand referred to as "the edges of the possible and permissible."[118] Like Lovins and the soft-path proponents, alternative information technology was viewed, perhaps somewhat naively at first, as a means of personal empowerment. Brand's own experiences with early desktop publishing and immersion into the early world of computer gamers lent credence to this optimism. The mandate at Apple was to "build the coolest machine you could imagine," something so different that people would rethink the role of the machine in modern life.[119] The very naming of the products suggested that these machines were somehow more natural than the computers of old. Although innovations in computing facilitated the creation of alternative information networks like *Whole Earth* and greatly enhanced the impact of dispersed citizen environmental science, PCs became one of the most insidiously polluting technologies of the late twentieth century.[120]

Old computers were identified by acronyms and numbers, whereas new computers were named Apple and were accessed through a "mouse." This was friendly technology, designed to be nonthreatening and easy to use. The specifics of how information and communications technology could become weapons in the war against the status quo were never clearly articulated by AT proponents.

There was a general sense among optimistic counterculturalists, however, that the PC and other new technologies were intrinsically radical and could change the world simply by existing. The details could be worked out later. In the meantime, the contagious enthusiasm and inventive genius of the counterculturalists inspired a technological revolution that ultimately transformed the American economy in unanticipated ways and created some ironic ideological paradoxes for the alternative technology pioneers who helped spawn that revolution.

For many in the counterculture of the early 1960s, computers had represented the epitome of all that was wrong with technology in the service of technocracy. During that era, computers were giant humming machines that were immensely expensive and required a high level of technical expertise to operate. They were, according to New Left critics, the heartless mechanized brains of oppression, used by IBM and the Pentagon to design weapons of destruction and quantify the body counts in Vietnam. Neo-Luddites dismissed the computer as a malevolent machine of centralization and dehumanization. Critics such as Theodore Roszak argued that computers were nothing more than "low-grade mechanical counterfeits" of the human mind, devices propagated by the "most morally questionable" elements of society.[121] Many of the first purchasers of *Whole Earth* would have agreed with these critiques. They certainly would have had a hard time conceiving a role for computers in their utopian back-to-nature communes.

But other counterculturalists, including Brand, were quick to recognize the potential of the new wave of microcomputers and personal information technology to link individuals and organizations working to transform American society. The widespread dissemination of information was essential to the project of constructing alternatives and transforming society. Long before most, Brand realized that computers had the potential to help build a new cybercommunity and transmit knowledge electronically the same way *Whole Earth* did in print. What, these pioneers wondered, could be more alternative than an electronic frontier, an alternative universe where individuals separated by huge distances could share ideas, images, and thoughts with thousands of other like-minded people all over the world? AT enthusiasts were some of the first Americans to go on line, and the *Whole Earth 'Lectronic Link* (WELL), founded in February 1985, became one of the early attempts to create a virtual community.[122] But this was in the future. In 1969, Brand remained very much engrossed in the innovations that he and a growing crew of collaborators were working out on paper in the catalogs and an expanded series of supplements. The supplements were free-for-alls of the drug and sex side of the counterculture, prominently featuring the work of underground comics pioneer R. Crumb and other emerging regional artists and writers. The now sought-after supplements were an important aspect of the celebration of regionalism that characterized *Whole Earth* throughout its long publication run.[123]

Brand's interest in information technologies was a logical outgrowth of his short, but brilliant, career as a counterculture promoter of events and happenings. The Alloy conference in March 1969 inspired Brand to use the platform of *Whole Earth* to expand his audience and capitalize on his newfound celebrity. In October 1969, Brand teamed up with Trips Festival promoter Jerry Mander to organize a major public "Hunger Show" to increase awareness of overpopulation and world hunger. He called his project Liferaft Earth and from the start it promised to be the type of attention-grabbing countercultural happening that Brand was a master at staging.[124] More significantly, Liferaft Earth provided a model for the type of direct environmental activism that would characterize Brand's Point Foundation that operated behind the scenes of *Whole Earth* in the 1970s.

Liferaft Earth

The life raft was an apt metaphor for the type of countercultural environmentalism Brand was promoting. Building from Fuller's notion of Spaceship Earth, he argued that we were all stuck on the planet with some significant constraints outside our control but, most importantly, many things within our control.[125] The message of Liferaft Earth was take control of your environment, change what you don't like, use your head and all the tools and technology you can muster to solve collective problems. Brand was keenly aware of population issues and, in 1969, still convinced that overpopulation was a serious issue: "There's a shit storm coming. Not a nice clean earthquake or satisfying revolution but pain in new dimensions: world pain, sub-continents that starve and sub-continents that eat unable to avoid each other. The consequences will dominate our lives."[126] Brand's hope was that Liferaft Earth would draw attention to the population issue dramatically, but without using the in-your-face tactics of the political protest movement and without the Malthusian overtones that tainted much of the discussion from environmentalists at that time.

Liferaft Earth was a rare example of Brand directly taking on a politically charged issue and instigating action rather than simply presenting debate in print, and it was a harbinger of the activism he would promote with his Point Foundation in the coming years. Following Buckminster Fuller's advice, Brand decided from the beginning that *Whole Earth* would avoid politics. Population issues were certainly political, but Brand wanted to use another tactic that he thought would move his event out of the political realm; he wanted to frame the debate as part of a game. That same year, Buckminster Fuller inaugurated his answer to the environmental crisis: the World Game.[127] In the typical Fuller fashion of no half-measures, the goal of the World Game was to "make the world work for 100% of humanity in the shortest possible time through spontaneous cooperation without ecological offense or the disadvantage of anyone."[128] Fuller's dream called for giant versions of his Dymaxion world map installed where national and world leaders could literally play out game scenarios that required players to understand the distribution of world resources and redistribute them as needed to create global equality. Fuller had collected immense amounts of data and rightfully pointed out that there was plenty to go around. This postscarcity optimism was hugely appealing to advocates of AT.

Giant maps did spring up on college campuses, and Fuller founded the World Game Institute in 1972 to facilitate larger gatherings and advocate for the World Game scenario. Unfortunately, according to J. Baldwin, the high idealism of the game was often "smothered by a pleasant, utterly ineffective touchie-feelie crowd that felt personal enlightenment had to be realized before larger problems could be addressed."[129] Of course, it is easy to criticize the whole exercise as the product of monumental political naïveté from someone who intentionally refused to acknowledge politics as a viable form of communication and management. On the other hand, it is hard to knock a project with such optimistic high morals so passionately and genuinely pursued on behalf of the greater good. Like many of Fuller's ideas, the World Game was a viable idea, but seemingly executed without any serious consideration of how to deal with the political or economic realities of the time.

Brand, too, was thinking hard about "how to be able to change the games that Peoples play." Like Fuller, Brand wondered how game scenarios might be used as an alternative to the political protest he so disliked. "The strategy of game change is: you don't change a game by winning it or losing it or refereeing it or observing it," he wrote in his preface to a large section on Liferaft Earth in the *Last Whole Earth*. "You change it by leaving it and going somewhere else and starting a new game from scratch."[130] Between 1968 and the mid-1970s, Brand continued to work on ideas for how to use games to bring people together, release aggression, and facilitate a more authentic interaction than participatory politics could provide.[131] The idea behind the Liferaft Earth game was to get a large group of players to sign on for a public fast. Participants would starve themselves for a week with the media providing a national "stadium" for a hunger Olympics. "How many of us arrogant world-shapers *knew* hunger?" Brand asked his friends. Filmmaker Robert Frank signed on to create a documentary of the event as he had with Alloy. Brand's friend Wavy Gravy (Hugh Romney) organized entertainment for the volunteers while Brand wrangled with Bay Area municipal leaders unwilling to let him stage his event. Part of the problem was that fire marshals had issues with Brand's proposal for a huge polyethylene inflatable pillow as the center piece of the show. They were worried it might explode in flames, roasting the participants alive. Dick Raymond stepped in to ease the fears of the insurance company and "all the others who needed hourly placation." What they ended up with was a parking lot behind the Hayward, California, office of a poverty program run by Bill Goetz, a man Brand had never met but who nonetheless turned out to be a generous and understanding host.

Undaunted by the decidedly unhip location and the tightness of the time frame, the project was launched with good media coverage. Instead of the giant pillow, the Southwest Electric Campfire "nomad architects" put together a remarkable inflated polyethylene tube that encircled the whole parking lot. A participant said, "Not until it was fully inflated did we realize it looked exactly like the bulging sides of a rubber raft."[132] The week-long sufferfest could be the subject of a group-dynamics study. Weather and other contingencies conspired to wreck the Hayward site and forced the group to move first to the *Whole Earth* store in Menlo Park, where the atmosphere was "harrowing," and then to a house in the mountains, where things got weirder but more comfortable. A core group of fifty-two, including Brand, stuck with the plan and starved for the whole week as others gave up the misery as the show moved. The moves left many of the audience and most of the media behind and caused the event to lose the spotlight of the single location. In the end, the event cost *Whole Earth* $2,357, and Brand concluded that "publicly it was a failure," but for the participants it was a near-religious experience that certainly none ever forgot. This event was unusual in the history of *Whole Earth* because of its narrow focus on a politically charged environmental issue. It was also the last time that *Whole Earth*'s environmental philosophy so closely overlapped with the issues and concerns of the mainstream movement. After 1969, Brand's personal environmental thinking and the environmental philosophy conveyed on the pages of the catalogs moved in another direction toward a more complex set of ecological and cultural concerns. Over the coming years, Brand retained his willingness to act as impresario for various happenings but limited them to decidedly upbeat events.

The starvation and stress of Liferaft Earth along with the growth of the *Whole Earth* business was starting to take a toll on Brand. "I actually thought," he wrote about that time, "I could fit liferaft earth in between the September Supplement and fall catalog. Setting up the event was even harder than production. Then starving for a week was no way to recuperate. Dumb."[133] Sales of the catalog during the period of spring of 1969 through fall of 1970 were impressive. The original catalog print run was only 1,000 units. That number had grown to 60,000 unit printings one year later, with frequent reprintings required to meet demand. By the fall of 1969, *Whole Earth* was a significant cultural phenomenon. Offers were coming in for distribution rights from major publishing houses, and money was pouring into the Portola Institute faster than it was going out to cover costs and events. The venture was moving forward and picking up steam, but Brand continued to announce in the catalog supplements that production would cease after the summer of 1971.[134] Rather than enjoy his success, Brand was suffering and looking forward to moving on as soon as was practical. Readers responded with correspondence that ranged from disappointment to anger.[135] Gregory Groth Jacobs, an articulate reader from London, wrote demanding an explanation for the untimely "suicide" of what he and others had quickly come to adopt as the voice of their generation. "What in the Hell is the Big Idea, anyway?" Jacobs wrote in a long letter Brand published in the *Last Whole Earth*. "As any designer knows—you cut off the Source, the Inspirator—and you cripple the Product, you trip up the Effort . . . So, goddammit, grow up, or at least justify your suicide."[136] For Brand, the last request must have seemed frighteningly close to his personal reality at that point. In the first months of 1970, he "went over some edge" and found himself adrift and despairing: "It was a nervous breakdown, garden variety."[137]

In the weeks following Liferaft Earth, Brand had put in eighty-hour work-weeks at the catalog while trying to oversee the growing business at the truck store. His doubts about the future of the enterprise coupled with a crazed work schedule were simply too much. Brand "jittered through the January 70 production" and then handed the editorship over to Gurney Norman, who inaugurated a long series of guest editorships that became a tradition throughout the life of *Whole Earth* over the coming three decades.[138] As he had for the previous several years, Dick Raymond provided invaluable support to the foundering Brand during this difficult time. Brand was partially out of commission for the better part of 1970 as he struggled with his demons. The catalog continued to forge on under the direction of first Norman and then J. D. Smith and Wendell Berry. By the beginning of 1971, Brand returned full time, more at ease and with a keen awareness of the price of success.

During the fall of 1970 and into the spring of 1971, Brand and crew worked feverishly to put together what would prove to be their masterpiece and their claim to a place in American cultural history: *The Last Whole Earth Catalog*. For those who fondly remember their encounter with *Whole Earth*, it is most likely the *Last Catalog* that they remember. Between its introduction in July 1971 through the spring of 1974, the *Last Catalog* sold a remarkable 1.2 million copies, bringing in more than three million dollars to the newly established Point Foundation and winning the National Book Award in 1972. The solidly packed *Last Catalog* topped out at almost four hundred and fifty pages, with a durable squared binding that ensured that the catalogs could pass through many hands and move off and on many a commune bookshelf for years without disintegrating like the previous stapled copies. Still, the materials were designed to be cheap and ephemeral. In self-publishing technique, content, marketing, finance, and distribution, the *Last Catalog* perfected the model of information delivery and publisher/

The National Book Award–winning *Last Whole Earth*, first published in 1972. This bulky classic had the highest circulation and is the best-remembered version of the catalog. In both content and design it was the concept perfected. (Courtesy Stewart Brand)

reader symbiosis that Brand had envisioned on that plane ride three years earlier. Brand's introspective analysis of his own brief but meteoric rise as the counterculture publisher demonstrates the remarkable wisdom gained over three years. Truly believing that this was the last *Whole Earth*, Brand explained himself and his model and provided a nice road map for those who might want to follow in his footsteps.

The *Last Catalog* followed the familiar format established in the earlier editions but with much more depth of coverage and many more reviews. This largest version of the catalog to date also included several unique features. There were two large, well-illustrated stories by Brand about Liferaft Earth and Alloy and a serialized short novel, *Divine Right's Trip*, by pinch-hitting editor Gurney Norman. Norman was promised a percentage of the *Last Catalog* profits for this contribution. This proved remarkably lucrative to Norman when the sales of the catalog soared. Environmentalist and agricultural critic Wendell Berry contributed a full page, as did several who wrote on spiritualism or intentional communities. Despite some extra space for commune dwellers, what was notable about the *Last Catalog* was how far it had moved from the first editions. A greatly enlarged "Nomadics" section demonstrated how the focus at *Whole Earth* had shifted from permanent escape from civilization toward temporary escape through recreation, both urban and rural. The greatly expanded sections on technology, craft, and tools were more coherently linked by 1971 and presented a more comprehensive vision for living an alternative life that was more firmly grounded within the existing framework of American culture than the more extreme versions of escape celebrated in earlier versions.

On a practical level, the *Whole Earth* method of researching and presenting information was perfected in the *Last Catalog*. Although much larger, the catalog feels more thoughtful, with no wasted space. Book reviews still formed the bulk of the catalog, and Brand's goal was still to print only positive reviews and only after very carefully sorting the good from the bad. Sorting on behalf of the readers consumed huge amounts of time for key reviewers like Brand and Baldwin, who considered it "utterly unglamorous; it is shoveling shit by the mountainload." The criteria for this sorting was printed on the first page of the first catalog: "An item is listed in the catalog if it is deemed: Useful as a tool, Relevant to independent education, High quality or low cost, Easily available by mail." This sorting relegated some books to "the stove, where their publishers belonged," while the lucky few ended up with a review as concise as one sentence up to a full page for those deemed especially useful or elegant like Edward Allen's 1969 *Stone Shelters*.[139] Besides recapping the exploits of Liferaft Earth and the community creation at Alloy, both two years in the past, the *Last Catalog* featured no new events. Brand had realized that the "unglamorous" work of culling ideas out of an ocean of printed material was the best way to help the communities of readers he served and the causes he held dear.

If Brand had regrets, they seemed focused on how he might have missed the kind of hands-on self-education that figured so prominently in the growing alternative technology and ecological design movements. That spirit of Alloy that brought people together to craft concrete solutions to real problems was very evident on the pages of the *Last Catalog,* but as Brand planned for the catalog's formal demise, he seemed to envy his colleagues who were the protagonists of his publications. "At the beginning of the catalog," he reflected, "I imagined us becoming primarily a research organization, with nifty projects everywhere, earnest folk climbing around on new dome designs, solar generators, manure convert-ers."[140] The remarkable success of the *Last Catalog* did more to promote the expansion of AT thinking outside the small circle of friends that composed the movement as assembled at Alloy than any research program Brand could have devised no matter how interesting. By bringing together ideas, individuals, and organizations that were not united anywhere else, *Whole Earth* helped mold a sensibility about ecological living integrated within the framework of a market economy easily adoptable by average folks. Notwithstanding the personal cost of a heavy workload and the self-doubt that came with his success, by 1972 Brand had achieved his goal to create a "Self-Sustaining Critical Information Service." Now that his creation had taken on a life of its own, all he had left to do was plan his escape.[141]

CHAPTER FOUR

On Point

At 6 am, under a grey mantle with a hint of light in the East,
we departed from Sausalito aboard the "Hurricane" and headed out to sea.
Beginning of a Point board meeting, April 12, 1972

Almost since the first success of *Whole Earth*, Stewart Brand had contemplated its end. Starting in 1969, when the catalog was receiving remarkable media attention and Brand was becoming famous, he began announcing the impending end of the venture. The history of the catalog seemed somehow accelerated from the very beginning. The pace of work was always frantic. Success came very quickly, followed by national and international fame. By 1969, Brand's writings about the project already indicated serious doubts about how long it should continue. His work on the big *Last Catalog* left him wishing for opportunities to move away from the ideal of self-education that shaped the first catalogs and begin to work on projects that could more directly influence some of the trends he valued most.[1] He was most wistful about the alternative technology movement that seemed to be the focal point of promising work by the "doers" he initiated the catalog to serve.[2] Looking back, he realized that all of the time he was promoting tools, he wasn't using any himself: "I was too busy being an editor."[3] He was also thinking hard about what to do with the most powerful tool at his disposal: money. By 1971, he was ready to try some new experiments with the "money tool" and see how he might stick "a needle in the gaseous foundation world."[4] So, in the early summer of 1971, Brand and crew started to "clean out the garage and sell the production equipment," while working with the staff of the San Francisco Exploratorium to set up a "demise party" that would formally and very publicly kill off the *Whole Earth*.[5]

The demise party was the premature wake for Brand's wildly successful creation. By late spring of 1971, Brand was a wreck. Three years of remarkable success and the heavy weight of responsibility that comes with fame knocked the wind out of his sails and left the leading optimist of a generation prematurely aged and a shadow of his former energetic self. Success is a scary thing, often more frightening than failure. For a self-styled dropout who found himself in charge of a growing business with employees to care for and books to balance, it was all a little too much. Only three years into the *Whole Earth* experiment, Brand wanted out—more than wanted, he desperately needed to escape for his mental and physical health. In a letter to Lawrence Ferlinghetti, Brand said he felt like he was "riding a crazy horse"; no longer in charge as the brilliant idea man, he was "riding the runaway catalog, recalcitrant unconscious and stagnant."[6] He had plunged into a deep depression and, on the night of the demise party, he hung onto the microphone for support as he faced a large crowd of friends and admirers; haggard, gaunt, and dressed in a flowing black coat, Brand looked like a haunted man.[7]

The next month, Brand's haggard image appeared on the back cover of *Rolling Stone*. Underneath ran the caption "Three years ago Stewart Brand decided to start the Whole Earth Catalog on $20,000. On June 12th, 1500 people couldn't decide on anything."[8] In typical fashion, Brand had decided that *Whole Earth* would go out with a bang. The "cavernous interior" of the Exploratorium at the San Francisco Place of Arts and Sciences was the venue.[9] The demise party he organized was more reminiscent of his Trips Festival than Liferaft Earth or Alloy: It was a full-fledged countercultural carnival. *Rolling Stone* reporters noted the resemblance but observed that "the Demise party belonged to another age—a future age which often harks back to a past one." If Trips ushered in a countercultural explosion of creativity in San Francisco, the demise party punctuated an ending, not just of the catalog, but of a cultural moment. By 1972, Brand observed, "We no longer had any remnant of a Generational Story to sustain us from without. A healthy uncertainty was afoot."[10]

For most who attended, the party was just that—a party, a celebration of good work and remarkable success. Attendee Fred Moore wrote about the scene, "What a party! All those optical gadgets and illusions, the music, dancing, food and drink, and the people—over a thousand of us! Just seeing us, meeting each other, folks from Vermont, D.C., New Mexico, California, from all over—so many beautiful people getting it on! Nitrous oxide balloons, volleyball, grass, good vibes—an out-of-sight party!"[11] A band called "The Golden Toad made every kind of music from bluegrass to bellydance," and, according to Brand, there was lots of "weirding around." In addition to the nonstop volleyball game, a trademark pastime of the *Whole Earth* crew, there was "boffing—jousting with Styrofoam swords, hectic but harmless."[12]

At ten thirty in the evening, with the party going "full steam," Brand sprung an unanticipated "experiment in participatory granting" on the crowd of staff, readers, contributors, and fellow travelers.[13] Event MC Scott Beach took the stage with Brand and, while lifting a sack up for the crowd, announced that he had two hundred one-hundred dollar bills, "yes, $20,000," and that it was "now the property of the party-goers, just as soon as they could decide what to do with them."[14] Partygoers were invited to the stage to take the microphone and offer their suggestions for what to do with the money. The ensuing debate lasted until six the next morning. A parade of individuals took their turn offering suggestions that ranged from flushing the money down the toilet to sending it to an impoverished country for food, to more specific and self-interested suggestions, like using it to improve the commune, or to the Darwinian solution of throwing it to the crowd.[15] According to observers, "it was difficult to tell" what Brand thought about the mayhem that followed his surprise: "Clad in his cassock, Brand stood on the stage and noted each suggestion on a blackboard."[16] As the night wore on, the visibly exhausted Brand dozed on his arm while the dwindling crowd argued.

For some attendees, the event took on sinister undertones. "Was he simply watching everybody squirm, unable to cope with the curve ball he had thrown them?" they wondered.[17] With a mop of dark wavy hair and a distant look in his eyes, Fred Moore emerged as the tortured soul of the spiritual counterculture. A former dishwasher and future important player in the personal computer revolution, Moore was a key voice in the debate, partly because he was willing to hang in to the end and partly because he was more intensely affected by the idea of free money. He, too, wondered about Brand's motives as it became increasingly clear that the money had "produced a bummer." For Moore, the bummer was visceral: He knew that the giveaway was more than just a lark for Brand, that the money issue tapped a deep current of unease with Brand and the counterculture as it entered a new phase in 1972. "In effect," Moore said, "by making us the recipients, or part of that decision-making process, he is sharing the guilt or corruption of what this money can do."[18] Likewise, Peter Warshall "felt a little distant from the great exuberance of let's give away all of our money." "I was a financial conservative," he remembered, and did not find the picture of his friends, who "were all over the place looking for money wanting to make movies or save the world in a thousand ways," appealing or humorous.[19] But money was not the source of Brand's angst in 1972. He had been responsible for money for most of his life and understood better than most that money was part of the counterculture from the beginning regardless of what idealists thought.

In the *Last Whole Earth Catalog*, Brand clearly spelled out his philosophy on money and admonished skittish readers who thought that somehow they had removed themselves from the ugly world of capital. "You may not think capitalism is nice, and I don't know if it's nice. But we should both know that the whole earth catalog is made of it." "Why am I saying this?" he continued. "Because many who applaud the catalog and wholeheartedly use it, have no applause for the uses of money, of ego, of structure (read uptightness), of competition, of business as usual. All the things, plus others, which make the catalog, and make the elective applauders into partial liars, and me one too if I aid the lie."[20] The demise experiment was Brand's way of hindering the "lie" and forcing his counterculture colleagues to try to come to grips with money, possibly because he had seen the positive power of the emerging counterculture business model he helped create and hoped, contrary to what Moore thought, to share the power as much as assuage his guilt.

At seven in the morning, with the cleanup almost complete and Brand there waiting to sign off on whatever was decided, only Moore and a handful of diehards were left. The group had collected the remaining fourteen thousand dollars, and the rest disappeared into the crowd never to return. The diehards decided that the money should go to the bank and be used for good works to be determined later. Fred Moore was left literally holding the bag. If he had been holding a sack of clinking bottles of nitroglycerin, he could not have been more nervous than he was with that money. It was a cruel irony that Fred Moore ended up with the cash. It is hard to imagine that anyone could have been more tortured by this responsibility than Mr. Moore. As the crowd finally faded away in the early morning light, Moore, "wandered around for a while, bewildered and awed, trying to get riders to accompany him back to Palo Alto and wondering aloud whether he should deposit the money in a night bank deposit . . . then realizing he had no bank account."[21] Moore spent days agonizing over the money and wondering why Brand had inflicted this pain.

"My reasons for perpetrating?" Brand wrote later. "Pure curiosity." Some of the themes of the experiment that were apparent to Brand afterward: "The money kept trying to come back . . . Handling of more than a pocketful of power was new to most, upsetting, educational . . . Ideas were mostly busy—unoriginal, guilt-ridden . . . People who focused on the process of deciding had a much better time." Maybe the most important lesson, and the one that would require several more years of experience for Brand to fully understand, was "Free money is crazy."[22] This last insight appeared later as Michael Phillips's "Seventh Law of Money": You can never really receive money as a gift.[23] Brand, Phillips, and a remarkable small group of board members would learn much more about money and philanthropy over the next three years as they built the Point Foundation, the successor to the Portola Foundation as overseer of *Whole Earth*'s fortunes. Unlike the Portola Institute, which had a wide educational agenda, the Point Foundation would serve, in its early years, as an activist arm for Brand and *Whole Earth,* rather than a holding company and foster parent for housing funds. Although the foundation funded activities that represented the eclectic interests of board members and reflected the contents of the catalogs that funded it, certain areas of interest shared by the board emerged early: most significantly, a desire to use a substantial percentage of foundation funds to support activities aimed at the ecological health of the planet.

One way to see how the pragmatic environmental philosophy of *Whole Earth* presented viable options for alternative environmental advocacy and action is to look at the goals and accomplishments of the Point Foundation. Over time, the foundation served to maintain the evolving *Whole Earth* publications network, and few would know it other than through the brief financial reports at the end of quarterlies and then magazines, but during the 1970s, Point was an active experiment fostering the design science revolution. The foundation was innovative in its constitution and mission. Instead of soliciting applications for grants, Point had directors and "Agents" tasked with finding worthy causes and making a case for support. A precursor to the McArthur Foundation "genius" grants, Point's mission was to support creative efforts, people, and organizations working in a variety of social, cultural, and environmental areas, with few strings attached. Environmental concerns dominated the early years of the foundation, with Point funds launching creative efforts in ecological design and supporting a variety of environmental activities, big and small, significantly ahead of their time and mostly outside the mainstream of the environmental movement.

Like the catalog that funded it, the Point Foundation's giving program reflected shifting cultural perceptions of technology, consumerism, and environmentalism in the early 1970s. Board debates about appropriate uses of funds and discussions about the type of environmental activities that the foundation should support provide insights into the concerns that were shaping an alternative counterculture-inspired commerce path for environmental activism. In light of contemporary concerns with alternative paths toward ecological living in the early years of the twenty-first century, the efforts and past experiences in alternative environmentalism reflected in the history of Point take on a new significance. Point provides insights into the antecedents for current efforts in appropriate technology (AT) and ecological design and one possible answer to the question "Whatever happened to the alternative technology movement?" Or was another, more positive, outcome possible? Although limited in time and scope, and always suffering from an overabundance of enthusiasm and an underabundance of oversight and planning, Point provided an activist model for Jane Jacobs's commerce-path ideal at a time when the alternative technology movement was still predominantly grass roots and had yet to become entangled in the addictive web of federal funding.

As the 1970s progressed, the movement became increasingly dependent on the guardian model and thus wedded to the ups and downs of the American political cycle. In 1980, when Ronald Reagan assumed office, federal funding for alternative technology dried up quickly, and organizations and agencies once flush with government cash found themselves without a funding source or a backup plan. But for a brief optimistic period in the early 1970s, Point offered one example of how it might have worked out differently.

The Moral Economy of the Counterculture

Very early in the heady days of *Whole Earth*'s national success and Brand's grow-ing fame, he started thinking about creative ways to use the profits from his brain-child to do good. The roots of Brand's altruism and philanthropic good works owed much to the example set by Dick Raymond's Portola Institute. As early as the summer of 1970, Brand was working with Raymond to brainstorm ideas about philanthropic activities he might sponsor with funds from *Whole Earth*.

Brand was very unclear on what form this activity might take, other than stick-ing with his iconoclastic worldview and creating something as different as possible from institutional foundation structure. In general, though, he and Raymond agreed that they were "very concerned about the doing-in of the planet" and wanted environmental issues to be a primary focal point for their activities. "Our concern," Raymond wrote about these early discussions, "is with education, soci-ety, ecology, technology, or whatever—almost any point of view reveals the world's distress very plainly."[24] Besides laying a philosophical foundation for the future foundation, most of 1970 was spent exploring a somewhat odd idea that nonethe-less reflected Brand's ongoing interest in fostering a community of alternative technologists. "Stewart Brand," Dick Raymond wrote, "has proposed a far-out project (literally) as a sequel to the *Whole Earth Catalog* (which he plans to discon-tinue next year). He is willing to put a large part of the Catalog's cash surplus into an experiment presently being called 'Mountain Fantasy.'"[25] This proposed com-munity might serve as a "proving ground," Brand argued, to be used for "social in-ventions instead of rocket fuel or nerve gas experiments." After exploring large land purchases and the reality of setting up some sort of permanent encampment, both Brand and Raymond realized that this fantasy was just that, and that their ef-fort needed to be more thoughtful. The demise party provided a reality check and caused both men to think hard about using large sums of money for goalless edu-cational experiments. "So, idealists become pragmatists," Brand summarized.[26]

After the dust settled from the demise party, Brand and Raymond solicited help from friends with a variety of fields of expertise to help shape a new founda-tion. None was as helpful as Michael Phillips. Phillips is perhaps one of the least appreciated, but most interesting, protagonists of the San Francisco countercul-ture. A Bank of California vice president with an impeccable financial résumé and a degree in economics from the University of Chicago, Phillips was also deeply committed to social reform and cultural experimentation. Starting in the late 1960s and extending well into the 1980s, Phillips put his financial expertise to good use as the moneyman for some of the most creative projects and organ-izations to emerge out of the countercultural ferment of early-1970s' San Fran-cisco. He excelled at demystifying banking and finance and played a significant role in shaping an alternative business model of great consequence for the last

Robert Gnaizda (left), California Rural Legal Assistance founder and chief deputy secretary for health, welfare and prisons under Governor Jerry Brown, Point board member and San Francisco Zen Center Abbot Richard Baker Roshi, and Michael Phillips (in white at right), Santa Fe, 1983. (Courtesy Michael Phillips)

quarter of the twentieth century. He codified his elegantly simple philosophy of Buddhism-inspired "right livelihood" in the bible of counterculture business, *The Seven Laws of Money*.[27] Dick Raymond later felt that Michael Phillips's efforts to create a counterculture business network represented "the most productive consequences of the *Whole Earth* and Portola experiments."[28] "It was real," he said, and the Briarpatch Network that Phillips spun off from his work at Point had concrete long-term positive effects with international consequences. Phillips eventually took over the leadership of Point from Dick Raymond and stayed involved, according to Brand, as the "financial wizard . . . and head wagoneer for our anti-foundation" through the 1970s.[29]

Phillips was a thoughtful voice advocating money as the ultimate tool for social good. Like Brand, his familiarity with money and financial responsibility gave him respect for the power of money and a realistic view of the role that it played in the counterculture.[30] Phillips's comfort level with finance was in stark contrast to the tortured ambivalence about money expressed by Fred Moore and other counterculturalists who simply could not reconcile money with their worldview. Unlike Moore, Phillips had no problem with the "money system" and consistently advocated knowledge of the market economy over utopian plans to create something new. "Personally I'm quite satisfied with the current money system," he wrote to Brand, "so my interests are in modifications on a small scale."[31] The idea of Point was to modify the system and get money directly into the hands of innovative people so that they could spend their time working on ideas, not finance. Ultimately, Phillips hoped that a boost from Point would enable counterculture innovators to master the market, not tear it down. Raymond agreed: "I personally believe that cadres of lay people seeking the answers to problems relating to their own lives and a microcosmic sense of scale is one of the best reflected in the *Whole Earth Catalog.* Many of us are shifting our ideology about money—the spending and gathering of it."[32]

After leaving his Bank of California job, Phillips became the financial director for the Glide Methodist Church and embarked on a new career as a leader in alternative business. Phillips had met Dick Raymond through their joint association with Glide and was on the board of the Portola Institute. Raymond brought Phillips in to meet with Brand and advise on their new philanthropic venture. Phillips had a number of suggestions for how to best manage the money and set up the foundation to avoid some of the legal and tax problems that plagued foundations of the time. Along those lines, he suggested that Point be set up as a "supporting foundation" and specifically as a Portola alumni foundation. This alumni designation would enable Point to function with fewer restrictions than an independent private foundation that would be heavily taxed.[33] Phillips knew Brand well through his work with Portola and understood that the activities of Point could raise red flags for the IRS unless they were careful.[34] One of the ironies of the Point/Portola relationship was that while Portola was the parent organization it was also "chronically broke," while Point was flush with revenues from the *Last Whole Earth Catalog.*[35]

Phillips's time with Point is particularly interesting because at first glance his granting practices were the most eccentric of anyone on the board, and yet he, along with environmentalist Huey Johnson, also achieved the best practical results over time. On the eccentric level, Phillips concentrated on funding sexual experimentation through the development of "sex furniture" and art projects like "carving a mile-long sign in the desert with a tractor for observation by satellites."[36] On the practical level, Phillips provided excellent financial advice to Point and a fully realized countercultural economics philosophy.[37] He merged his personal interests with his practical bent through support of groups like Margo St. James's coyote (Call Off Your Old Tired Ethics) prostitute's collective, which operated out of Phillips's private office. Phillip's office became the de facto hub of the Bay Area countercultural business network that was formalized with the Briarpatch. His advice and advocacy were instrumental in the creation of a remarkably successful subeconomy in the region that foreshadowed the rise of alternative business and the greening of the American economy in the coming decades. Phillips and the talented bunch of "Briars" he gathered into his network, despite their long hair and countercultural lifestyles, were totally serious about business and creating a viable new model for enlightened economics that could succeed in the marketplace. One of the soon-to-be supersuccessful people who supported Phillips was Paul Hawken, who spent two decades as a board member or advisor to the Point Foundation while he built several successful national businesses and became an influential advocate for "natural capitalism."[38] Hawken cautioned skeptics in the business community not to dismiss Phillips: "If anyone downtown thinks the Briarpatch is a trendy hippie puffball that will blow away, they might take note that it is coordinated by Andy Alpine, who has a B.A. in Economics . . . and a doctorate in Law," and that "it was co-founded by Michael Phillips, one-time director of Marketing and Planning for Bank of America."[39] Phillips's involvement ensured that Point would start with a strong and savvy financial foundation.

Point was an experiment on many levels. It was an effort to create a model for foundation organization and philanthropic activism informed by the social context of the counterculture. Point became a means for exploring what Michael Phillips called "way out money schemes" grounded in careful considerations of right livelihood, simple living, and British economist E. F. Schumacher's "economics as if people mattered."[40] With Phillips's help, the foundation would have solid accounting practices and investment portfolios to ensure that, if money was wasted, it was never because of sloppy bookkeeping or financial ignorance. Looking back, Huey Johnson said the efforts at Point were a "classic and important way" to think about philanthropy.[41] Further, Point was a meaningful attempt to support social causes, mainly in the Bay Area, related to quality of life and social justice. Most importantly, Point represented a directed effort to help convert the constellation of alternative environmentalism and technoecology presented on the pages of *Whole Earth* into something resembling a more coherent movement.

As financial director for the progressive Glide Church, Michael Phillips organized efforts to bring the philosophy of Point into the city. The 6th Street Park he built for the homeless, shown shortly after opening 1979, was one example. (Courtesy Michael Phillips)

If for no other reason, Point is significant because it moved *Whole Earth* briefly away from presenting others' ideas in synthetic form to taking direct action to change the world or at least a small part of it.

By late fall 1971, Raymond and Brand had assembled an interesting board of directors and began holding regular meetings. These, as one might expect, were not your typical board meetings. Over the years, meetings were held "on salmon boats at sea, in tipis, in Glide's sex room" and "at the Black Panther school in Oakland."[42] Board member Huey Johnson remembered these meetings fondly as a right of passage for "a small-town boy," who "learned more in contemporary new directions in society than I ever could have any other way."[43] For Johnson and board members who had not lived in the counterculture world that Brand so comfortably inhabited, the idea of sitting on the floor of a tipi or sailing out to sea at dawn for a foundation board meeting was an eye-opener to say the least. "Environmentalists didn't get in that pond very often," he remembered.[44] The venues, however, were the least iconoclastic aspect of Point. The organizational structure and creative and intellectual demands placed on board members really set the foundation apart from its staid contemporaries in the conservative world of early-1970s' American philanthropy.

Taking on this world presented a juicy new challenge for Brand, who believed "it would be fascinating to revolutionize the foundation business, like the *Whole Earth* did the publishing business."[45]

The history of American philanthropy in the twentieth century raises complex questions about wealth and power in a country that has never resolved the issue of private versus state responsibility for social welfare. As Waldemar Nielsen demonstrated in *The Big Foundations* (1972), his monumental survey of twentieth-century foundations, "In the great jungle of American democracy and capitalism, there is no more strange or improbable creature than the private foundation."[46] In 1971, Point was poised to add a new species of strange creature to this improbable world—a species that better represented the tradition of American foundations to act as benefactors to innovative "seekers of root causes" and finders of creative solutions than was the "gaseous foundation world of the 1960s." Point became an example of how philanthropy could be changed by a new and enlightened generation.[47] "Being a counter-foundation," Brand explained, meant "dealing individual-to-individual rather than institution-to-project, pursuing leanness, rigor, and surprise in good-doing."[48] For board member Huey Johnson, Point provided an opportunity to create an alternative to a granting system he abhorred. As someone who worked extensively in the foundation world, Johnson was eager for a gutsy mold-shattering experience in philanthropy that would emphasize what he called "radical giving," an alternative to "the enduring trend to safe traditional educational, religious, or medical giving" that he found "disheartening" and "costly to society."[49] The work of Point, though often disorganized and frustratingly underplanned, exemplified the root meaning of the word *philanthropy:* love of humanity. It was the love and respect for people, their products, and the places they lived day to day that differentiated Point from most environmentally minded foundations and organizations of the period.

On November 2, 1971, Brand convened the first meeting of the Point board. They were tentatively calling the new foundation Parnassus from an earlier venture, but shifted to Point within the first month. The meeting included members who would remain active with Point for years, and some who lasted only a meeting or two. Dick Raymond left the initial meetings enthused and fondly remembered these hopeful initial forays into a new venture.[50] Significant early board recruits present at the November meeting were Stewart Brand and wife Lois, Stanford Research Institute engineer and computer designer Bill English, environmentalist Huey Johnson, "radical ad man" Jerry Mander, moneyman Michael Phillips, mentor Dick Raymond, and feminist and *Whole Earth* collaborator Diana Shugart.[51] The board also always included a rotating invited member dubbed "Elijah," brought in by a board member to add to the mix. Brand and Raymond carefully explained the ideas behind Point as a supporting foundation for Portola in charge of *Whole Earth*'s "soon sizable income."[52] The figure that emerged at the meeting was "something around a million dollars" by the summer of 1972. A million-dollar endowment represented a very tidy sum in the early seventies and offered great potential for good works. At the meeting, Mike Phillips explained the financial strategy he had developed for Glide Church's portfolio as a model for Point. Brand discussed his "mountain fantasy" idea and why he had given it up, but not without suggesting some sort of "land meet, along the lines of Alloy," for which "he felt there was a great need."[53] The discussion of Alloy and how Point might foster the sense of technical innovation and community building that meeting represented indicated clearly what direction Brand hoped the new foundation might go.

Brand also presented his key idea for the Point organization: *agency*. He hoped that agency, in all its meanings, would form the heart of Point both literally and figuratively. The idea of a group of "agents who would operate as individuals, finding and carrying out their own 'missions,'" derived from several related desires on the part of Brand and his new board. Board members at that first meeting voiced concerns about how to avoid "getting a lot of hassles about how to invest the money." Even before they launched their efforts, board members clearly understood they would be "pursued because people knew that you had money to give away."[54] The agency idea could insulate board members from money seekers by making the awarding of money proactive and not driven by applications. The meeting concluded with another of Brand's educational surprises. This time he asked each board member to write an essay on the subject "Destination Crisis—or what to do with our power—personal, Point's, national, planetary, or cosmic." Each board member would be paid $250 with a $100 bonus going to the most useful essay.[55]

Brand offered the assignment to share his own personal questions and elicit some insights from his interesting and eclectic new board. He knew the responsibility that went with the money and wanted thoughtful responses to his questions about how it could best be used. Point was a way for him to share both his success and power, but also some of his guilt. "'What should I do with $1,000,000?' is not very different from 'What should I do with my life?'" he wrote in his "Destination Crisis" paper: "We are blessed/cursed with the leisure, money, and access to try damn near anything we can conceive. Naturally this brings down enormous guilt."[56] The guilt would come, he worried, from the impossibility of doing something good in the face of so many other good things that one might do. "How do you set up a system to spread the responsibility and do the most good?" his essay asked. He concluded that the "agent concept" would accomplish this goal most effectively. "The agents," he explained, "are hired for their resourcefulness at doing maximum good with minimum expenditure."[57] By the December meeting of the Point board, the agent idea had been adopted, and

Point directors were sent out to find agents and give them money so that the agents could then seek out worthy efforts to reward. The agent structure of Point was certainly its most unique feature.[58]

Why the cool agent idea? Besides Brand's goals, part of the decision to depart from the traditional foundation model derived from a dysfunctional group dynamic. The Point board was designed to be eclectic in interests and backgrounds, but wildly divergent viewpoints led almost immediately to significant tensions within the board. In particular, pragmatic environmentalist Johnson and idealist Mander "fought constantly" and agreed on almost nothing.[59] Johnson was so disturbed by the dysfunction of the group that, when it was his turn to invite the meeting's Elijah, he brought in psychiatrist Pierre Mornell to provide group therapy.[60] According to Johnson, it was Mornell who suggested that each board member be given a sizable hunk of the money to do with as he or she saw fit, because it was clear that the diversity of the group was so great they would never be able to make a collective decision.[61] Brand credited Mander for "dispensing with group decision" and granting each board member $55,000 per year to dole out with little oversight from the collective. Whether it was dysfunctional group dynamics or just Brand's creativity or both, the agent idea was genius and something that only a foundation that sprang from *Whole Earth* would be willing to adopt. It was a risky strategy, and Brand and Raymond knew it. "No doubt point will do some half-assed granting," Brand wrote in his "Destination Crisis" paper, but, undaunted, he thought that the possible good outweighed the inevitable bad, and he outlined a hypothetical agent scenario focused on how Point might spur research and wider acceptance of solar energy.[62]

Brand's presentation to the Point Foundation board on December 12, 1971, could function as a primer on the latest trends in alternative energy. In several pages, he outlined key trends, highlighted problems, and explained how Point might work to strategically enhance the prospects of solar energy and other emerging technologies. Most of what he suggested were "brokering moves, catalyst actions between otherwise inert reagents." All of his suggestions—funding research aimed at using mine shafts for solar heat storage, targeted funding for pioneer researchers like Steve Durkee and Steve Baer, and publishing research on solar energy in urban environments—were proactive and aimed at creating new technologies that could solve real and present environmental problems. Brand did not support funding, at this time, for efforts that were opposed to something, such as environmental efforts to block, stop, or preserve. Instead he adopted the philosophy of Buckminster Fuller, who argued against "the folly of trying to vainly remove something when all you have to do is introduce something which obsolesces it." This philosophy ensured that AT research, always aimed at instigating change through addition rather than removal, would dominate much of

Point's grant-based advocacy. He concluded his discussion about what Point might become and how it might function with a list of key characteristics for a successful "mission." The last and most important item listed was "Otherwise unlikely." Projects taken on by Point needed to be cutting-edge ventures outside the mainstream of the foundation world. Projects that, "if we don't lend a hand . . . probably won't happen (this eliminates many good ideas, which turn out to be happening anyway)."[63]

Because of the human-centered approach to ecology expressed in *Whole Earth* and the founding documents of Point, AT was a logical funding target for the new foundation. Brand's paper asked his board to think of problems and then ask what science or new technology is needed to solve them. How could Point help shape, if only in a small way, environmental science and technology to better serve what people feel they need? How might one foster anticipatory design science as an alternative to reactive activism aimed at resolving an existing threat? Brand stressed that agents needed imagination and maybe even "clairvoyance" to be able to study current trends and to help identify future needs and emerging trends worth supporting.[64] He pointed to the science fiction that he and many of his contemporaries in the counterculture read voraciously as a way to get into the proper mind-set.

Science fiction as a genre is inherently anticipatory and provided a logical vehicle, within the cultural context of this board, for explaining the type of forward-thinking activism Brand envisioned for Point: "Somewhere in my education I was misled to believe that science fiction and science fact must be kept rigorously separate. In practice they are so blurred together they are practically one intellectual activity, although the results are published differently, one kind of journal for careful scientific reporting, another kind for wicked speculation."[65] In 1975, the New American Library estimated that there were 1.5 million active science fiction readers in America.[66] *Whole Earth* counted many of these fans as readers and contributors. *Whole Earth* and especially the *CoEvolution Quarterly* were obviously informed by trends in sci-fi literature and frequently featured contributions by leading authors. In 1971, Brand wanted to see what might happen when one threw big money at talented people who were engaged in some of the type of "wicked speculation" that drove the genre of science fiction.

Of the original board members, Bill English was the most articulate proponent of funding aimed at technologies that at the time would have seemed like pure science fiction to all but a handful of computer researchers and fellow travelers like Brand. English was a pioneering computer engineer with Xerox and working at the cutting edge of computing and communications technology in Palo Alto. His 1971 essay for Point, "A Cottage Industry," reads like the best sci-fi, except it is all based on actual research into how computers would shape the coming decades. English shared with fellow board members his belief that America was on the verge of an "information explosion" that would usher in "drastic changes" in American culture and life.[67] These drastic changes would center, English argued, on communications technologies that would replace "the usual paper-oriented communications" and enable a "remote working" environment that would transform American corporate and work culture. English proposed several ways that Point funds could further this information revolution. Specifically, he wanted Point to fund individuals and small companies willing to experiment in using evolving computer networks to facilitate remote working or, as it would later be known, telecommuting. "The tools are here today," he wrote, and the "costs are not too high . . . and no doubt it would lead to good publicity." English's remarkably prescient discussion of the future of information technology accomplished, better than any could know at the time, Brand's request that the board members be "clairvoyant."[68]

English's proposals were, along with Brand's and Johnson's, the most focused and specific. English departed, however, before Point could take up any of his remarkable suggestions, but not before plugging the alternative energy research of Steve Baer and pointing toward AT as a best use of Point funds.[69] Brand was most impressed with Huey Johnson's thoughtful list of environmental programs and plans, calling it "a classic piece of research" worthy of publication.[70] In these early meetings, Dick Raymond and Huey Johnson emerged as consistent voices of reason that often carefully challenged the most far-out ideas, like Jerry Mander's suggestion that they buy one million dollars worth of grain and drop it on a dock in India, and moved board discussions back toward the sensible. Raymond in particular seemed most interested in directing discussions toward practical questions of organizational structure and mission. In reviewing the "Destination Crisis" papers, he pointed out that "the most missing thing in the papers was the question 'why'?" As an experienced foundation director, Raymond seemed less concerned with the expected creative and iconoclastic ideas and more with the cavalier attitude about planning and lack of specific mechanisms for managing the agent system toward a productive end. He also worried about how the intensely individualist model emerging from these early meetings would affect the grantees. "Are we implying," he asked, "that we want to shape other people's behavior according to our own values?" Johnson, too, further advocated sticking with at least some standard foundation features, specifically requiring proposals to filter out unqualified candidates for funding. Mike Phillips responded that "people who can write proposals can't do shit," which Brand followed up with "if I'd had to write a proposal, I never would have done the Catalog."[71]

Brand was steadfast in his rejection of any semblance of traditional foundation structure and later had some regrets about how his personal grants turned out leaving him, temporarily, cynical about the whole process of trying to use money to do good.[72] At the initiation of the Point project, though, he seemed happy to keep the atmosphere as experimental as possible. Raymond and Johnson quickly resigned themselves to this set of unusual realities, mostly stopped debating structural issues, and went along with the very interesting flow of discourse that the Point meetings afforded. For his part, Johnson kept meticulous records of the agents he appointed and the projects he funded. He was accustomed to keeping good records and managing large budgets even when flying by the seat of his pants. By the time he joined the Point board, Johnson had an impressive résumé as a nationally respected maverick environmentalist and an already legendary environmental land man.

The Land Man

Of all the original Point board members, Huey Johnson had perhaps the best combination of openness to "wicked speculation" and knowledge of ways to use money effectively to spur concrete results. Johnson's career also made him the most likely member of the board to organize an environmental giving program of national significance. Even Jerry Mander had to agree in early 1972: "None of our ideas for projects are particularly fundable . . . except possibly Huey's."[73] Although little known outside the world of international environmentalism, Johnson was one of the most innovative environmental activists of his generation. In the early 1960s, after completing a master's degree in biology at Utah State University, Johnson started a career as an environmental activist. He lobbied for the Wilderness Act and public lands in general, became friends with leading wilderness advocates like David Brower, Edward Hilliard, and Howard Zahniser, and gained a national reputation as a tenacious and creative environmental thinker. Looking back at Johnson's sterling environmental résumé, one might be tempted to assume that he embodies the conventional environmental thinking as expressed by mainstream groups, some of which he worked with and many of which he supported. A closer look, however, reveals that Johnson's creativity and innovation in environmental advocacy were more complex than they first appeared and perfectly captured the type of environmental pragmatism that *Whole Earth* promoted.

In 1963, after several years of advocacy, Johnson was appointed Western Regional Director for the Nature Conservancy. He was the eighth employee hired by the young group and the only Nature Conservancy employee west of the Mississippi.[74] He quickly established a western focus for the Conservancy through a series of gutsy land deals that solidified his reputation as the land man for the environmental movement. Between 1963 and 1972, Johnson engineered over fifty critical western purchases for the Conservancy. Johnson, according to Conservancy historian Bill Birchard, "took people's breath away with his brazen land deals."[75] Several of those early transactions became legendary. In 1969, Johnson

Huey Johnson and his son "providing their own protein."(Courtesy Huey Johnson)

agreed to purchase, on behalf of the Conservancy but without board approval, seven thousand acres of prime land in Hawaii connecting "the crater of Haleakala National Park to the sea." Enraged board members demanded that Johnson raise the money to fund the deal, which he did in short order, to the amazement of conservative board secretary Elting Arnold. In another remarkable case, Johnson worked with a group of teenage "hippie, long-haired" Montana brothers who were heirs to a 35,000-acre ranch in prime cattle country. The brothers found Johnson and told him they wanted to preserve the family ranch for future generations. These "greasy-looking" young men faced serious resistance from the powerful ranching lobby, a wealthy ammunitions manufacturer, and their own uncle, who were working through the courts to take the ranch from the brothers whom the uncle had written off as hippie losers. Johnson agreed to take on the project and, in characteristic "swashbuckling style," arranged, with the help of environmentalist and rifle site maker Ed Hilliard, several backroom meetings with the munitions man who was the linchpin in the effort against the brothers. The deal reached a climax with a courtroom drama that pitted the young hippie environmentalists against the powerful ranching lobby. Eventually the governor of Montana, Forrest H. Anderson, became involved and sided with Johnson, tipping the balance and scoring a victory for the boys and the land-trust preservation movement.[76]

In November 1969, Johnson helped organize the thirteenth conference of the U.S. National Commission for the United Nations Educational, Scientific and Cultural Organization (UNESCO). The theme of the meeting was "Man and His Environment: A View toward Survival." Sixty prominent environmental thinkers were invited to present papers addressing this theme, with an eye toward suggestions for action. The meeting brought together politicians, conservationists, students, and an eclectic mix of prominent thinkers. *Whole Earth* current or future contributors attending the meeting included Huey Johnson, Jerry Mander, Margaret Mead, Stephanie Mills, and Gary Snyder. The gathering provided a bridge between the mainstream environmental movement and the alternative efforts emerging from the *Whole Earth* circle of fellow travelers. Johnson edited the papers for a book, *No Deposit—No Return* (1970), that became an important marker of a changing environmental culture at the beginning of the seventies.[77]

The essays included many traditional discussions of preservation of public lands, resource management, population growth, and wilderness. Notable, however, was the large concentration of essays on public health, technology, Spaceship Earth, alternative design, and issues that formed the heart of the environmental ethic conveyed through *Whole Earth*. Johnson's contribution was a thoughtful manifesto for environmental pragmatism and "implementation" tools and strategies to achieve quantifiable environmental goals while avoiding political entanglements. He argued in favor of the creation of special interest groups organized to use the market and legal systems as primary strategies with political process only as a backup. "Many would argue," he wrote in 1969, "that such a hard-driving implementation is dangerous for democracy. I don't agree." Johnson had seen his market-driven methods work well at the Nature Conservancy without the complications of traditional political advocacy. His experience with the Montana ranch acquisition confirmed that implementation could circumvent political polarization and achieve results before public debate even began. Further, like all of the participants at the UNESCO meeting, he felt a strong sense of urgency that fueled his desire for expediency over ideology and politics.[78] "Instead of being manipulated by the system," he concluded, "we can participate in it by implementing changes."[79] That same year, Johnson won the prestigious American Motors Conservation Award for "his role in the preservation of more than 30,000 acres of natural areas in Hawaii," marking him as one of the emerging leaders of the national movement.[80]

Johnson's hard-nosed environmentalism came from hands-on experience as a conservationist and hunter and life lessons learned as a small-town boy who never lost site of the issues faced by those who lived close to the land. By 1970, Johnson had redefined the methods of land-trust activism and established the paradigm for the "action-and-results–oriented dealmaker" who would enable the Nature Conservancy to grow into the largest international environmental organization by the turn of the millennium. Johnson's now-legendary toughness—Stewart Brand once called him "a thug for good"—made him an appealing candidate for national leadership.[81] Convinced by his string of successes, the Nature Conservancy offered Johnson the presidency, but he turned it down to stay in San Francisco. Like a generation of innovators who migrated to San Francisco because they were "square pegs who were uncomfortable back home," Johnson thrived in the "dynamic, futuristic" atmosphere of the Bay Area. His experiences in the American West convinced him that the region was where he wanted to stay and where he would be able to continue to innovate. Thus, he turned down offers to go back east, deciding instead to "send ideas East" while staying on the cutting edge of environmental pragmatism in San Francisco.[82]

The offer from Stewart Brand to join Point came at a perfect time for Huey Johnson. He was already working with Dick Raymond on the Portola Foundation board and was interested in trying to put into practice some of the ideas he articulated in his UNESCO paper. In his "Destination Crisis" essay for the Point board, he outlined a plan for aggressive environmental implementation backed by an "Eco-Renaissance" model of patronage: "During the Renaissance, private patronage rescued artists from the limits of institutionalized religion . . . [W]hat is needed now is patronage for environmental activists, which will enable them to create an Eco-Renaissance in both thought and action." He reiterated his support for the creation of environmental special interest groups with well-funded agents to combat political stagnation and the considerable lag between environmental crisis and public concern leading to political action: "Special interests seem to be an acceptable pragmatic method of running an otherwise impossibly complex system."[83] The Point money provided an opportunity for Johnson to take his experiences and transfer them to a new generation of activists with funding to back them up. These activists, in turn, could "demonstrate that it is still possible, within our system, to implement ideas," even in periods of crippling political polarization, and show that "our hope for survival lies in the decisive leadership of informed individuals."[84]

Johnson considered his time at Point invaluable and transformative for his illustrious environmental career that followed his time on the board.[85] Looking back, he likened his intense experience as a Point board member to a "college education and a subsequent two-year meander around the world."[86] The learning pace was "frantic," he recalled, and the experience forced him to "try and keep up with the deluge of information" generated by the diverse board. One of the significant insights to come out of his work with Point from 1970 to 1972 was the importance of shifting the focus of the environmental movement from wilderness areas to urban centers, which was consistently one of the core messages of *Whole Earth* even in its early editions. Johnson appreciated this "courageous thinking about the future," and the efforts of *Whole Earth* and Point to push environmentalism toward a transition uniting efforts in preservation and AT. He later recalled, "The reason that I am on the Board is that this foundation is the most relevant beginning point for this transition to the future that I can access easily. And it is another sign for optimism."[87]

In 1972, Johnson founded his own land-trust advocacy group: the Trust for Public Land. Johnson's success with the Nature Conservancy and intense work with Point encouraged him to break out on his own and form the Trust for Public Land to further his model of market-savvy environmental advocacy.[88] A central feature of Johnson's environmental philosophy was his belief that "unless we can make the cities more livable it is going to be very difficult to preserve natural areas."[89] His work with Point, which brought him into closer contact with grassroots urban environmentalism, AT proponents, and alternative design activists, convinced him that "city dwellers are really going to be the ones making the decisions on the use of natural resources" and "that unless some traditional environmentalists moved into the urban areas and helped the cities," larger goals of preserving the "integrity of the total environment" would fail.[90]

Johnson's new focus provided a bridge between alternative technologists like John Todd and the New Alchemists who were working on urban agriculture as part of their research program and foundations and advocates who supported more traditional environmental activities. Through his Point funds, Johnson was able to link this research with specific urban activists and projects trying to put the research into practice.[91] The Trust for Public Land was a pioneering effort in urban environmentalism, and Johnson became an articulate spokesman for the "livable communities" movement that helped shape a civic environmentalism and emerging ecological design philosophy that influenced trends in New Urbanism, urban ecological design, and green consumption in the coming decades. In the early 1980s, he worked to organize the American Green Party to bring his ideas more directly into the political sphere.[92] By the end of the 1990s, the Trust for Public Land had banked 1.3 million acres of open space in American urban centers.

In large part because of Johnson's thoughtful arguments at Point board meetings and carefully planned program of environmental giving he outlined for himself and his agents, Point was very "heavily environmentally directed" by the end of 1972.[93] Brand and Raymond also made environmental activities focal points of their granting, in keeping with shared interest in environmental issues and alternative energy technologies in particular. Brand tended to focus sharply on solar energy and the handful of design outlaws, like Steve Baer, who were engaged in research that involved advanced engineering while concentrating on practical products that could move quickly into the consumer market at a reasonable price point. Jerry Mander, one of the most socially aware board members, also used a substantial percentage of his funding for international efforts to save the whales, as did Brand.[94] Although Huey Johnson and Jerry Mander frequently clashed over structural matters and philosophy, Johnson later credited Mander for forcing the Point board to deal with the New Left and the socialist viewpoint. As a radical ad man, Mander was the voice of the New Left on the board.[95] He was also the most concerned with the melding of technology and environment that came so easily to Brand and other members of the board. Mander's environmentalism tended to reflect his socialist politics and in particular his strong views on technology.

Mander was deeply concerned about the counterculture technological enthusiasm. He faulted environmentalists for the limited nature of their critiques of technology, which, he argued, "contributed to the problem by failing to effectively criticize technical evolution despite its obvious, growing, and inherent bias against nature."[96] When technological discussions at board meetings turned to computers and information technology and the work of Bill English, Mander was even stronger in his criticism. When English and Raymond enthusiastically recommended Point funding for computer access and other programs to aid the information revolution, Mander was incredulous. Computers, he angrily commented in one particularly tense meeting, are "anti-ecological, anti-democratic, pro-autocratic . . . and as dangerous as centralized broadcast television."[97] Mander later penned a monumental critique of the "failure of technology" in his widely praised *In the Absence of the Sacred* (1991) as a counterpoint to the technophilia that he felt had poisoned the legacy of the counterculture. Despite Mander's concerns, by the beginning of 1972 Point was strongly established as an innovative new environmental foundation with a strong interest in AT.

For Brand and Johnson, this was as it should have been, but other board members were concerned about this direction. Feminist social activist Diana Shugart, in particular, worried about the predominance of environmental issues at Point. "I hope you don't take this as personal criticism," she wrote to Brand in a letter voicing her concern, "because it's not—I'd just hate to see Point turn into the Sierra Club."[98] Brand replied, "So far, the 'ecology' agents are working best."[99] The 1972 Point report reflected a great deal of organizational soul-searching about future directions for the foundation. Many of the concerns regarding environment versus social causes rose out of a particularly interesting board meeting in November 1972, where "Elijah" Ernest "Chick" Callenbach presented the ideas he was developing for a science fiction book about a utopian future. Callenbach delivered his ideas to the Point board as an "Ecotopian Militant."[100] As a "visitor from the year 1995," Callenbach spawned a surprisingly introspective conversation about decentralized organizational structures and their efficacy. The marathon day-long meeting ended without any clear plan for changing Point's structure, but a general understanding that alternative environmentalism and ecological programs were likely to continue to be the glue that held the diverse board together and provide Point some consistent direction. Callenbach's counterculture environmental classic, *Ecotopia*, would not appear for another five years, but Point was already exploring his vision of an ecotechnological western utopia.

Some of the discussion about diversification of Point activities must have taken hold because 1973 saw significant changes in the board composition and granting practices, which retained a preponderance of environment-related ventures but added a greatly expanded number of social and cultural grants. The board added Nathan Hare, editor of the *Black Scholar* and social activist who worked on behalf of inmate rights, among other issues. Hare focused his grants on social causes in the Bay Area and educational and arts programs. Richard Baker Roshi, from the San Francisco Zen Center, also joined the board. Roshi's center was a beneficiary of Point funding and an influential meeting place for several board members. Jerry Mander and Michael Phillips continued their support for a wide range of social programs in the Bay Area. Brand, Johnson, and Raymond all continued to focus on environmental issues but added a wider array of social causes to their giving programs during 1973.

By 1973, Point had established a clear record of environmental giving that was having a significant impact on grassroots efforts in AT and ecological design not only in the Bay Area and the West, but also nationally and internationally. Compared with the activities of groups like the Sierra Club and the Wilderness Society, Point was a very small player in the world of nonprofit environmentalism. Still, because of the sharp focus on an emerging alternative model of community-based pragmatic environmental advocacy and AT research, Point's history offers a look at a different path for ecological living than that offered by the larger organizations of the day. For the environmental innovators who received Point funds, the foundation offered a lifeline and enabled some of the most interesting examples of counterculture environmental research:

Select Point Environmental Grants through summer of 1974:

- Bolinas Septic Research
- Stockholm Environmental Conference
- Richard Nilsen Applied Ecology Center
- Bill and Helga Olkowski—urban agriculture and recycling

- University of California–Davis Environmental Law Society
- Institute for Applied Ecology—urban agriculture initiative
- Auroville Society
- Center for Environmental Action
- Friends of the Earth—whaling initiative
- Interpretive Ecologics—environmental education for urban youth
- The New Alchemists
- Zomeworks
- Bill Bryan—Northern Rockies initiative
- Charles Watson—Nevada Outdoor Recreation Association
- Greg Archbald—national alternative land policy research
- Joan McIntyre creation of Project Jonah—antiwhaling
- Robert Chrisman—ecology of a black village
- Integrated Living Systems Labs—alternative tech
- Unify—urban nature institute

Fostering Ecotopia

Point's records for 1971–1974 show an environmental giving program that consistently favored individuals and organizations working outside the mainstream of American environmentalism. Several board members did provide support to traditional environmental organizations such as the Sierra Club Legal Defense Fund and the Nature Conservancy, but these grants were the exception. More typical was support offered to alternative technologists like John Todd and the New Alchemists, and Steve Baer and his New Mexico alternative technology company, Zomeworks; or to William Bryan and Charles Watson's efforts to build politically centrist coalitions and redefine the agenda for western public lands preservation;[101] or to early environmental justice efforts like those of Interpretive Ecologics, an Oakland-based environmental education group for low-income urban kids and minorities with an urban focus.[102]

Despite the variety of interests, alternative technology drove much of the larger funding efforts. Early on, Raymond, Brand, and Johnson wondered whether Point ought to establish a standing alternative technology agent, "a visionary or two, the likes of Steve Baer or John Todd," to direct Point's funding in this particularly important area.[103] Not surprisingly, Point worked with the town of Bolinas to provide funds in support of community efforts to turn the whole place into a laboratory for AT, with money for solar collectors for private homes, schools, and businesses.[104]

Michael Phillips observed that "the Federal Government" would be "releasing about $40 million for solar research within a year or two," indicating that he was not sure this area really needed Point funds when the government was going to take the lead. Brand and Raymond, however, retained a healthy skepticism about the government's wiliness and long-term ability to push the type of outlaw design that they valued. Although there was rarely consensus, which was not a goal anyway, the board did agree on enough to launch some collective funding efforts aimed at broadly accepted environmental concerns.

One such effort was dubbed Life Forum, which was described as an "activist's shelter" primarily concerned with fostering unique environmental programs. Several board members contributed large amounts to Life Forum, and the board as a whole agreed to provide $50,000 in 1973.[105] During the spring of 1972, Point spent significant Life Forum funds on a "misbegotten effort to liven up" the United Nations Conference on the Environment in Stockholm. Brand organized the effort to accompany "sundry poets, Indians, radical scientists, and the Hog Farm traveling commune" to Sweden at great expense. Brand later dismissed the effort as a waste of resources, with the exception of the work of Point grantee Joan McIntyre's attempt to promote a United Nations–sponsored whaling moratorium.[106] McIntyre's Project Jonah received considerable funding from several Point board members over the years, and Point became a significant player along with Greenpeace in making "Save the Whales" an international cause that captured remarkable popular media attention and contributed to the growth of international environmentalism.[107] Other projects under the Life Forum umbrella included funding for California Steam, a steam-driven auto project, and Charles Watson's Nevada Outdoor Recreation Association (NORA). Most of the Life Forum projects also received additional funding from individual board members.

NORA, headed by Charles Watson, was a Nevada Bureau of Land Management (BLM) advocacy group working to change American perceptions of the public domain, that most nebulously defined and sketchily protected segment of the public lands.[108] Huey Johnson described Watson as "one of the most effective activists of our time" and a "catalyst for saving millions of acres of public land."[109] Watson was a vocal proponent of a public lands advocacy that came to be known as "Commons Ecology," the effort to "give Earth's unappropriated government lands a heightened environmental priority."[110] Watson recognized early that the BLM controlled thousands of square miles of remarkable natural areas of great value to the American public rather than simply storehouses of commodities to be liquidated as quickly as possible.

Nevada contained the second largest concentration of BLM lands after Alaska, with an amazing forty-eight million acres of public domain BLM managed. In 1958, Watson, along with twelve supporters, created the Nevada Public Domain Survey (NPDS) to inventory these holdings and identify especially significant areas for possible preservation. The NPDS was well ahead of the curve of American culture in valuing vast desert areas that had long been considered wastelands. Brand was a more likely supporter of this effort than it might at first appear. Like many in the California counterculture, Brand was a frequent visitor to the wild deserts of Nevada, where the isolated hot springs of the Black Rock Desert, outside the town of Gerlach in the northwest corner of the state, provided blank stages for counterculture happenings and individual escapes. Virtually unknown to most of America, Black Rock was a Mecca for countercultural types whose desert gatherings became legendary, eventually leading to the annual Burning Man festival of technologically enhanced hedonism that draws thousands to the stark area each year.[111] Burning Man weirdly captures the blending of counterculture technological enthusiasm and love of the desert that was not quite in line with the other types of outdoor recreation supported by NORA. Still, Watson and his fellow Nevada activists built a strong partnership with the San Francisco counterculture as they prepared to fight for the preservation of areas like Black Rock. In arguing for the preservation of such stark desert landscapes as "wilderness," Watson was questioning the foundation of the preservation movement. The NPDS and subsequent NORA advocacy was a fight against "choice morsel" hierarchies of natural beauty that directed the wilderness movement and the preservation efforts of the National Forest Service and National Park Service throughout the twentieth century.

Watson received tens of thousands of dollars from Point at a very critical moment in his long fight for the Nevada deserts and America's public lands. "Words cannot express," he wrote Brand after receiving an unexpected large grant from Point, "the feeling of gratitude for your generous grant . . . at this vital time."[112] Point's support for NORA was instrumental in its successful efforts on behalf of Black Rock and the fight to save Red Rock Canyon outside Las Vegas, which set the tone for a reconsideration of BLM lands picked up by such notable environmentalists as Lady Bird Johnson and Justice William O. Douglas. In no small part because of the efforts of NORA, the BLM drafted an "Organic Act" as part of the 1976 revised Federal Land Policy and Management Act (FLPMA), which provided for wilderness status and permanent protection for lands in the public domain.

The creation of BLM "wilderness" is an excellent example of the convergence of traditional environmental activism with a counterculture informed alternative pragmatic ecological sensibility. Watson and his allies presented a case against wilderness purists and for a careful reconsideration of what lands were worth preserving. The 1970s' fight for Red Rock, on the edge of Las Vegas and now one of the most visited BLM sites in America, illustrated another key component of Watson's advocacy—a reevaluation of the relationship between natural areas and urban centers. Watson and T. H. Watkins clarify this branch of environmental pragmatism in their 1975 book *The Lands No One Knows*.[113]

A young ecologist named William Bryan was another significant Point grantee who straddled the line between traditional environmental advocacy and Point's

The Red Rock Conservation Area was 16 miles from Las Vegas in 1960 when Charlie Watson was beginning his advocacy on behalf of this now completely encircled gem. Watson spent much of his time in the 1950s and 1960s trying to convince members of the Sierra Club and other wilderness advocates to change their position on BLM and desert lands. (Courtesy Nevada State Museum)

pragmatic alternative. Bryan received his Ph.D. in 1971 from the University of Michigan after completing a master of environmental education degree.[114] The University of Michigan was known in the late 1960s and early 1970s for innovative environmental graduate programs focusing on community activism and education and guerilla-style environmental organizing. Bryan's dissertation focused on grassroots environmental organizing and individuals who used guerilla tactics to gain strength from a starting point of weakness. For this research, Bryan worked with Ralph Nader, David Brower, and Huey Johnson, among others. Bryan assumed he was training for an academic career, but his time interviewing Johnson about emerging tactics in environmental advocacy changed his mind. Johnson was impressed with Bryan and thought that Bryan was just the kind of smart young blood who could be successful as a point man for environmental action in areas desperately in need of leadership. According to Bryan, Johnson told him to ditch the academic career and "get out in the real world and get some battle scars."[115]

Bryan was amazed by Johnson's skilled maneuvering between the established world of institutional environmentalism and the rough-and-tumble grassroots community organization milieu in western states where toughness in the face of significant opposition was a requirement. He especially appreciated Johnson's emphasis on social change and human-centered activism as a key to long-term success in environmental advocacy. Bryan was not a fan of the institutional movement of the early seventies and was never a member of any of the national environmental groups.[116] He also had a hard time understanding fellow graduate students and activists who seemed to love nature more than people. Johnson, working through Point, made Bryan an offer he could not refuse. He proposed

William Bryan speaking during
early 1970s NRAG community
organizing campaign. (Courtesy
William Bryan)

that Bryan become a Point agent with a salary of $18,000 a year to go to Montana
and start a community-based environmental advocacy alliance for the Northern
Rocky Mountain states.

Bryan accepted the offer and arrived in Helena, Montana, in the winter of
1972. His timing could not have been better. Montana had just convened a consti-
tutional convention and passed a new constitution that completely upended the

power structure of a state that had been controlled almost completely by the Anaconda Mining Company. Bryan walked into a political free-for-all the likes of which had not been seen in the West in a generation. He quickly became a key player in the state and established a remarkably successful model for community environmental organization that he named the Northern Rockies Action Group (NRAG). Bryan had a talent for creating diverse coalitions that changed the environmental politics of a region where the movement had been exceptionally anemic. He was particularly successful at bringing ranchers into environmental coalitions.[117] In his early years, he established partnerships with small ranchers and farmers like Bill Cook, who were open to new ideas about sustainable practices. These efforts would lead much later to the creation of another nonprofit network for sustainable agriculture in the American West: the Cook Center.[118]

By virtue of his timing, skill, and academic training, Bryan integrated the Northern Rockies into the broader regional environmental movement in a remarkably short period. Very quickly, he gave up on his academic career plans and took his model of environmental organization all over the West and later, as a consultant, worked to reorganize the Wilderness Society and other national organizations. Within a year of receiving his Point funds, Johnson was hailing his work as "one of the successes of our activist granting program," and Dick Raymond called Bryan one of Point's "gold medal candidates."[119] Bryan's work impressed the Point board so much that they organized several meetings in Montana and one particularly memorable meeting in Reno, Nevada. Like Johnson, Bryan had little experience with the counterculture before becoming involved with Point. He recalled his surprise at the Reno meeting when some board members starting passing around laughing-gas canisters. Like others who encountered the Point board, he was amazed by the combination of unorthodox behavior and enlightened seriousness.

By 1974, Bryan and his Northern Rockies Action Group were so successful that it created some friendly competition between Bryan and Johnson. That year, Johnson told Bryan that he and his staff would no longer share any of the Trust for Public Land mailing lists with NRAG—Bryan and NRAG were on their own. Johnson also tried to talk Bryan into leaving Montana and moving to California, where his skills were needed for the national movement. But Bryan decided to remain in the Rockies, where he took his environmental skills into the world of outdoor recreation in the 1980s as a pioneer in the sustainable travel and ecotourism industries. Although Johnson and Bryan remained friends during the coming decades, Johnson never got over Bryan's decision to move into recreation. This move, however, placed Bryan perfectly in line with the environmental philosophy of *Whole Earth*, where nomadics was always an important part of the ecological sensibility of the catalog.[120]

E. F. Schumacher and Montana Governor Tom Judge speaking at
NRAG conference on alternative environmentalism in the late 1970s.
Organizations like NRAG and NORA brought funding and insights
from the alternative environmentalism of the Bay Area into the heart
of the American West. (Courtesy William Bryan)

Techno-Arcadias

The best examples of Point-supported ventures that united more traditional concerns with land preservation and restoration and alternative technology research
were two remarkable utopian communities: Auroville, near the town of Pondicherry in the Tamil Nadu State of India, on the Bay of Bengal, and the New Alchemy Institute of Cape Cod and Prince Edward Island.[121] Both of these ventures
received considerable funding from several board members between 1971 and
1974, and both capture the spirit of environmental pragmatism fostered by Point
in spite of the utopian dreams that fueled both of these efforts. *Whole Earth* and
CoEvolution Quarterly were early and thoughtful proponents of environmental
restoration as the yin to traditional preservation's yang. *Whole Earth* showcased
Aldo Leopold's Sand County restoration project immortalized in his prizewin-
ning *A Sand County Almanac* (1949) and helped to popularize the writings of agricultural renaissance advocate Wendell Berry. Leopold is best remembered as a
pioneering advocate of the wilderness ideal in the National Forests, yet it was his
restoration work on a tortured piece of land heavily modified by humans that provided the material for his most inspired writing. Nature writer and environmentalist Barry Lopez argued in *Helping Nature Heal* (1991), his foreword to *Whole
Earth*'s primer for environmental restoration, that environmental restoration of
damaged lands and crippled places was the most American form of environmental activism. "It is no accident," he wrote, "that restoration work, with its themes
of scientific research, worthy physical labor, and spiritual renewal, suits a Western

temperament so well."[122] Restoration emphasized human ingenuity and directly linked technological prowess with environmental health. The Point board consistently favored support for individuals or organizations working on places where people lived and worked. Of all the innovators supported by Point, none better represented the fusion of ecology and technology than John Todd and his New Alchemy Institute.

The New Alchemy Institute was one of the most celebrated efforts in AT and ecological design of the 1970s. In 1969, biologist John Todd and his wife Nancy Jack Todd founded the New Alchemy Institute to explore the question "Was it, in fact, possible to support Earth's population over time while protecting the natural world?"[123] John Todd was frustrated with the "doomwatch biology" prevalent in university biology departments and influencing the often reactive politics of the environmental movement after the passage of the Wilderness Act of 1964.[124] After much soul-searching, he decided to break out and, with his partner, form his own institute where ecological design pioneers could explore positive and proactive responses to environmental problems.[125] The Todds shared with other ecological design proponents "an extreme sense of urgency that the system that we have created has the potential to undo everything that we would like to see happen."[126] The name New Alchemy came out of thin air but perfectly captured not only what the Todds began working toward at their Cape Cod institute but also the core idea behind the alternative technology movement as a whole and the environmental program of Point in particular. John Todd could have been speaking for Point when he told a *New York Times* reporter, "I got tired of ringing the alarm bell all the time. I want constructive alternatives."[127] *Whole Earth*'s J. Baldwin worked extensively with the Todds on Cape Cod, as their soft-technology expert and later as part of the construction team for their ark on Prince Edward Island, Canada.[128] At New Alchemy, the Todds recalled that Baldwin was an influential voice of pragmatism who grounded their sometimes utopian dreams of social justice: "He used to remind people at the Institute that, at the end of every wrench, not to mention far more imposing forms of technology, were the steel mills of Gary, Indiana: implying that social and technological issues are as old as our use of tools and are probably forever fatefully entwined."[129] Baldwin's association with New Alchemy made the group a logical funding target for the Point Foundation.

Stewart Brand and J. Baldwin often agreed on the quality of a project and the skill of those in charge. Brand became the leading supporter of New Alchemy on the Point board. In the spring of 1973, he visited the New Alchemists and left inspired by their efforts. "I stayed with the New Alchemists who are doing great work but are short on funds so they are now gearing to deal with it," he reported to the board at its May meeting. "I'll be making a $16,000 grant to them."[130] By 1973, Brand was using a substantial percentage of his Point funds to support alternative energy research. He was inspired enough by the work being done at places like New Alchemy that he briefly worked on a proposal for "Windmills at Golden Gate Park." His idea was to "make them a showcase of carpentry as well as a museum of technology."[131] The Golden Gate project, sadly, went nowhere but demonstrates the level of interest at Point for fostering public understanding of ATs.

The New Alchemists were definitely on the cutting edge of alternative energy research, but their primary focus was on what the Todds called "living machines," biologically based waste-recycling/energy-producing systems that cleaned the water, purified the air, and provided food for the dwellings where they were installed. When Brand notified John Todd of his large and unexpected award, Todd was excited but a little perplexed. The letter specifically said that the money was for work on "energy use," and Todd wrote back to Brand, asking for clarification: "If I was to take a piece of sun and a batch of wind, mix it with a little water and shit . . . and out the other end came all kinds of plant and animal foods, and climate control for a greenhouse-cum . . . living structure, would you consider that energy use?"[132]

At New Alchemy, huge green tanks of carefully researched algae housed hundreds of tilapia that ate waste and produced more productive waste that fed the algae. These fish were then harvested as food. A visiting *New York Times* food critic hailed the Todds' fish as "a triumph for the sensualist and ecologist."[133] New Alchemy's greatest triumph came with the opening of its most sophisticated research station on Prince Edward Island (PEI). On September 22, 1976, Canadian Prime Minister Pierre Trudeau landed in a helicopter next to the barely finished PEI Ark. The ark was a monument to the Todds' research and a model for the very latest alternative technologies.

Partially buried in a windswept hillside, the PEI Ark was a wood-sided appropriate technological marvel. An entire exterior wall of the ark was situated so as to take full advantage of the seasonal Sun's energy, which was then stored in a 20,000-gallon containment tank buried beneath the residential area of the building. "To maximize the collection surface," Nancy Todd explained, "the topmost solar collector rose straight upward, billboard fashion, along the entire south façade."[134] Inside, exposed ductwork brought warmed air down into living and working areas and to another buried storage area filled with rock that could retain heat during the day and release it slowly upward during the night. The ark was the realization of the dream of integrated ecological design engineering that fueled the early efforts at Alloy and Steve Baer's work at Drop City a decade earlier.

"Living Machine" designed and built by Living Technologies of Burlington, Vermont, installed at Ethel M candy factory in Las Vegas. The Todds' company, Ocean Arks, worked in partnership with Living Technologies to create marketable industrial versions of the residential living machines they perfected at New Alchemy. The Ethel M machine saves up to 20,000 gallons of water per day while providing compost for the desert landscaping surrounding the plant. The Living Machine is part of a popular tour with over 700,000 visitors annually. (Photo by author)

Brand attended the ceremony and later wrote, "I'm not sure this needs to be said, but the opening of the Ark was in fact a moving—even triumphant—occasion."[135] The inspiring opening of the ark was the high point for New Alchemy and what it is best remembered for, but the institute continued to operate and innovate through the 1980s, until their expenses grew with their success to a point that was no longer sustainable by private foundations or the shrinking pie of federal dollars that withered during the early 1980s.[136] Like many of the promising efforts in pragmatic alternative environmentalism of the 1970s, New Alchemy never achieved sustainable financing for its sustainable ecological programs. Unlike many smaller AT groups, the Todds were able to regroup and continue their good work under a new name, Ocean Arks International, and with a new mission to concentrate on producing designs with the market in mind. By the 1990s, Todd-designed or inspired living machines were processing wastewater at facilities around the country.

The Auroville community in India resembled New Alchemy but was much bigger in scope. At Auroville, spokesman Alan Lithman and 350 souls relocated from all over the world to form a remarkable intentional community that became "a laboratory for the Scientist who believes in mysteries."[137] In 1969, when Lithman first arrived at the Auroville site, he "saw nothing but a vacant landscape that slid into the Bay of Bengal." The spot could not have been more inhospitable, "a once-living earth dying back into a moon."[138] In the coming years, Lithman and his colleagues planted over two million trees, converted monsoon-eroded wastelands into fertile agricultural and grazing lands, and made an unusually productive home for over seven hundred inhabitants. From its inception during the winter of 1968, funded by a grant from the United Nations Educational, Scientific and Cultural Organization (UNESCO), Auroville's founders wanted to build communities from the ground up that incorporated the latest research in environmental design and alternative technology. Working with such damaged lands, the Aurovillians were very sensitive to the future impact of their activities and hoped to use alternative technologies for their energy, sewage, transportation, and all other infrastructure, where possible.

By the time Johnson heard about their efforts, Auroville was well launched but still in great need of funding to expand reforestation and alternative energy efforts. Dick Raymond and Huey Johnson gave Auroville initial grants of $5,000, with an extra $1,000 for Lithman so that he could travel to raise funds for the project. Johnson remembered being very impressed with Auroville but losing track of the project after making his initial grant. When Lithman reported back a year later, Johnson was amazed by the difference the Point funds had made.[139] Much to Johnson's liking, Lithman provided a detailed report of every penny spent and careful descriptions of the effect that the funding had on Auroville

"Aurovillians reaching the site," February, 1968. With the help of Point funds, hard work, and technical innovation the Aurovillians were able to convert this barren landscape into a model of land restoration, sustainable economics, and voluntary simplicity. (Courtesy Auroville Outreachmedia)

projects. When converted to rupees, the Point dollars went very far and the results of even small grants were impressive. Johnson recalled the Auroville project as one of the most remarkable successes of his time at Point. "Twenty-three acres of casuarinas trees have been planted at five sites," he proudly reported at a board meeting in the summer of 1973, and "every penny has been accounted for . . . [T]his demonstrates our usefulness is making seed grants."[140]

Surveying Auroville twenty years after its founding, environmental restoration expert and *CoEvolution* editor Stephanie Mills observed a wide array of alternative technologies and sustainable building techniques in the community: "Throughout Auroville are windmills, biogas plants, and solar panels, along with a bunch of household-level AT experiments, like Tency Baeten's lagoon pond system for treating his home's black and gray water."[141] The efforts to restore the soil through creative planting and careful management of erosion from the powerful Indian monsoons had converted the region from moonscape to fertile landscape by the mid-1980s. What differentiated Auroville from other intentional back-to-the-land communities that rose out of the idealism of the counterculture was its focus on integrating agriculture, alternative technology, and urban ecological design. Auroville was built in a rural area but with the goal of growing the clusters of small villages that constituted the colony together to form one large urban center built as a model for ecologically sustainable urban living.

The Aurovillians were, like their *Whole Earth* supporters, a future-oriented group that geared all of their activities toward designing systems aimed toward future sustainability, with little romanticism about returning to a more pure primitive past. Their model was unique and not easily translatable to the rest of the world,

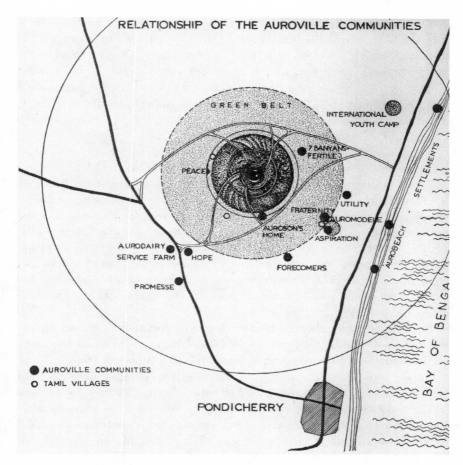

An early map of the ambitious Auroville plan to create a system of linked villages with an urban core all united by appropriate technologies. By 2007 much of the plan had been achieved, and Auroville remains one of the best examples of the Ecotopian dream brought to fruition. (Courtesy Auroville Outreachmedia)

but they outlived most AT groups and surprised critics who dismissed them as New Age utopians who would quickly fade away. Like other pragmatic alternative technologists with roots in the counterculture, the Aurovillians were sustained by their idealism but succeeded because of their faith in human ingenuity and hard work. By the 1990s, Auroville was not a perfect realization of the Ecotopian dream, but it was a good example of what committed amateurs with adequate funding could achieve in one small spot that had once seemed an irretrievable wasteland.

For Point board members Auroville was a bright spot that seemed to confirm the original promise of the *Whole Earth* millions to help push real change in the world and foster a pragmatic alternative path toward the ecological health of the planet. Auroville was hardly a universal model, given its unique spiritual base and location in India, but its longevity provided hope that determined people could create viable models for soft technology–based communities. By the early 2000s, Auroville was still a working vital community providing an international model for viable AT development.[142]

In still more striking ways, the work of Steve Baer and his maverick New Mexico alternative energy company, Zomeworks, captures *Whole Earth*'s tool culture and Point's environmental pragmatism. The singular support to Steve Baer's efforts highlights the evolving environmental philosophy of Stewart Brand as he was about to launch the second life of *Whole Earth* with the *CoEvolution Quarterly*. Point's support for Zomeworks went beyond the normal grant or agent relationship. In addition to grants, Point became a major investor in the company. Early in 1972, Brand made a presentation to the board in support of a substantial initial investment of $30,000 in Zomeworks.[143] From his days with the traveling Whole Earth Truck Store, Brand had encountered Baer's work at Drop City, Colorado, and at the Alloy meeting in 1969.

As discussed earlier, Baer is best known for his contributions to the architecture of Colorado's Drop City with the distinctive polyhedra zomes constructed with car doors that Baer modeled after a children's toy built by his wife Holly.[144] Much to his later annoyance, the zomes became "a cult in the counter-culture" for their "groovy" esthetics and connection to Drop City.[145] Even in his early days of zome building, Baer's true passion was the Sun. Since high school he had been fascinated with solar convection heating and conducted experiments in streams to determine how solar heat changed temperatures, depending on angle. Lewis Mumford's writings on the individual's relationship to technology, Farrington Daniels's *Direct Use of the Sun* (1964), and the work of New Mexico solar designer Peter Van Dresser were all early inspirations for Baer.[146] Even though Baer's early alternative energy designs, like his car windshield solar chimney at Drop City, were amazing examples of soft technology as sculpture, they did not capture the imagination of the counterculture or its chroniclers the way the zomes did.[147]

Baer shared much in common with Stewart Brand. Both were articulate autodidacts who became spokesmen for alternative technology and the science that shaped it.[148] Both were slightly older ex-military men when they started traveling with the counterculture and never lost the drive for practical action they learned in the service. Most importantly, both men were vocal proponents of a commerce-path environmental pragmatism that stressed access to tools and marketable products. "If you want change you have to make it," Baer told an interviewer in 1973, "and one way is to develop the tools of a new way of life and start a business to sell such equipment . . . [Y]ou have to make the change yourself or you have to shut up about it."[149]

Among Point-supported ventures, Zomeworks is notable for its long track record of successful research and consistent efforts to create viable alternative technology products that did not require "subsidies or elaborate research projects."[150] Zomeworks captured Brand's ideal of effective pragmatic environmental work undertaken by talented iconoclasts producing "brutally simple" designs with marketing in mind. Steve Baer, Brand argued to the board, "has an unusual combination of genius, overpowering energy, and long-haul capability."[151] Brand also liked the fact that Zomeworks' research was "not academic, but . . . dedicated to developing and selling marketable systems."[152] Baer and Brand shared a sense of urgency for solar energy research and a belief that the government would act too slowly and ultimately impede the type of outlaw design by "crusty old gents who design . . . in machine shops or invent all alone and pissed off."[153] Both were also dismayed by how ineffective the mainstream environmental movement had been at advocating alternative technology research and influencing policy at the government agencies. "Despite the outcry about pollution and the energy crisis," Baer wrote to Brand, "there is almost no work at all being done on the direct utilization of the sun's energy."[154]

There were, of course, many people working on solar energy around the American West. Baer's frustration is understandable, given the slow start for AT during a time when land-preservation issues dominated environmental politics, but he was only the latest of a long generation of western and southern solar innovators. A wonderful article published a few years later in the *CoEvolution Quarterly* outlined the history of the 1890s' boom in California solar residential water heating that resulted in the creation of most of the technology that would be refined by Baer and others in the 1970s.[155] Brand and Baer were on the front end of a major push that would gain national attention and significant federal support only after the impending energy crisis became an actual crisis.

During the 1970s, Baer produced and marketed a series of remarkable solar products. The list is long, but highlights of Baer's designs include the ingenious Beadwall system (U.S. Patent 3,903,665) that was invented by Baer's University of New Mexico student Dave Harrison. The bead wall consisted of large double-pane windows that were completely clear to catch the winter sun and provide heat. The windows were fitted with a vacuum system that could then fill the space between the glass completely full of Styrofoam insulating beads during the night to hold heat in the building.[156] Baer and Harrison installed a bead-wall unit on the Monte Vista School in Albuquerque, where passing motorists could marvel at the snowstorm of Styrofoam pumping through the windows. Another notable Baer invention that received wide praise and extensive coverage in *Whole Earth* publications was his ingenious Double Bubble Wheel Engine.[157] This simple engine converted thermal energy into mechanical energy through the action of bubbles of gas moving through alternating chambers of hot and cold liquid. Zomeworks continued to innovate through the present and remains one of the best examples of a steadfastly innovative outlaw design firm that managed to produce real marketable products and work successfully through the lean AT 1980s to emerge in a position to capitalize on the latest wave of scarcity-fueled enthusiasm for alternative energy.

Between 1971 and 1974, Point succeeded in its founding goal to create a counterfoundation with few rules and creativity at the heart of all its efforts. Many of these efforts ultimately failed. Agents and grantees sometimes flaked out and took funds but never followed through on projects, and it seemed like the more money they got, the more likely they were to produce disappointing results. By 1974, Brand was feeling disillusioned with the whole project and jaded by the experience of turning friends into adversaries over money. He concluded that the Point experiment was for him something of a failure, but one full of important life lessons that would shape his future ventures. A landmark year for Point was 1974. Most of the original board retired, according to plan, including Brand and Raymond, Johnson, and Mander. "I don't like the idea of free money," Brand reflected after his retirement. "It's no surprise then that I did a poor job of dispensing it."[158] Michael Phillips became the director and continued to be a significant factor in moving Point away from radical giving and toward a more sustainable program of support for the *Whole Earth* publishing empire as it moved toward a new phase and a new era.

Brand was a harsh self-critic and seemed deeply saddened by the experience of Point, despite some very savvy grants and investments. The key lesson that he learned, and several other early members echoed, was that "it's not enough to give money to someone who's good. They must also be in the grip of an overwhelming idea which is very good; otherwise a hideous paralysis will take them over and, in addition, freeze your friendship."[159] Failures that particularly rankled were the spectacularly expensive trip to Stockholm, a last gasp of the notion that simply by staging a happening, opinions and the tide of history could change. Several other ultimately misguided attempts to spawn creative action from some talented people who took Point funds but simply did not follow through were even more disheartening.[160] These disappointments only seemed to confirm Brand's belief that by the mid-1970s there was no "remnant of a generational story to sustain us."[161] Good ideas and goodwill simply were not enough to change the world. Still, by the spring of 1975, he was able to see some good that had come out of his personal contributions to Point, including support for the

The *Whole Earth Epilog* prominently featured the innovative solar energy research of Steve Baer's Point-supported Zomeworks. This page included his remarkable Beadwall, Skylids, and Drumwalls. (Courtesy of Department of Special Collections and University Libraries, Stanford University Libraries)

New Alchemists and other AT projects, and Joan McIntyre's work to save the whales. In a letter of advice to newly appointed board member Jan Mirikitani, Brand cautioned her to be reasonable in her expectations of the power of money to do good work on its own. "Nothing worthwhile happens quickly in this business," he wrote. "The happy surprises arise from large slow movements." The "business" of foundation work had proved just that, a business, that like all others required that creativity and high ideals be tempered by meticulous management and careful planning. Brand's foundation was a microcosm of the transformation of the counterculture during the early 1970s. No longer a generational response to global trends, the counterculture became the foundation for the rise of a new creative class of entrepreneurs and innovators who drew inspiration from the high ideals of the 1960s even as they moved away from the dreams of utopia or escape that rose out of events like the Trips Festival.

The story of the Point Foundation demonstrates that even a steadfast proponent of individual agency and self-education like Brand was willing to intervene directly when he thought the cause and time were right. Point was designed to provide loosely structured facilitation of a more direct advocacy than the catalog allowed, while avoiding the politics inherent in the foundation world. Point was one of the most significant efforts to fund and provide a more solid foundation for the grassroots AT movement along with a broader set of social and cultural concerns. Point's record, however, was somewhat disheartening. It did not succeed in changing the world or the American foundation system, but was significant nonetheless because of the philosophy of commerce and ecological design that the foundation board collectively formulated. Point provided seed money and major grants for a host of alternative environmental programs, and the board's philosophy of finance and simple living captured by the Briarpatch Network was internationally emulated in the coming decades.

Brand's intense three years establishing Point helped refine his environmental philosophy and shaped his new ventures in the publishing world. The in-depth debates that characterized the Point meetings inspired much of the content of the *CoEvolution Quarterly*, and Point members and agents formed a new core group of *Whole Earth* collaborators. The Point Foundation lived on into the early 2000s as the holding company for the various *Whole Earth* publishing ventures but never again participated in philanthropy the way it had during those first years.[162] Point may have failed to revolutionize American foundations or contribute to environmentalism on a par with the Sierra Club or other much larger advocacy groups, but the AT and ecological design movements were just gaining momentum during Point's early years, and the work of this most unusual foundation provided an example for the types of support systems and advocacy networks that AT required to ease alternative businesses into the general marketplace. The next step was to build on the ideas of "right livelihood" and green commerce fostered by Point and the Briarpatch Network. Brand's next publishing venture, the *CoEvolution Quarterly,* codified the ideals of the Point Foundation and provided a captivating model for ecological living, hybrid politics, and green consumption.

CHAPTER FIVE

The Final Frontier

A lot of people are into the Western myth, you know . . . they
want to invent and discover things for themselves.
Stewart Brand, 1970[1]

In 1975, Ernest Callenbach released his popular science fiction novel *Ecotopia*.[2]
His work as an "Elijah" with the Point Foundation during the early years of his
research contributed to his vision of an eco-technological Western utopia fostered
by coalitions of the type of people who read *Whole Earth* in the real world. Callenbach envisioned an environmental utopia based on countercultural political and
social values and, most significantly, new trends in appropriate technology (AT)
set in Northern California. Callenbach's novel painted a picture of an ecolibertar-
ian revolution resulting in the secession of Washington, Oregon, and Northern
California and the formation of an ecological paradise. *Ecotopia* was an old-
fashioned western dressed in counterculture clothes and full of tough, gun-toting
hippie environmentalists finding freedom on a new frontier.[3] Also, at heart, like
most westerns, it's a love story. Despite an emphasis on collective ownership of
the means of production, the book incorporates much of the western libertarian
sensibility and desire for social freedom and free markets unencumbered by a distant centralized state authority. The characters in *Ecotopia* were not drugged-up
flower children interested in peace and love. They were tough, savvy feminists
and countercultural Ben Franklins fighting, tinkering, and innovating while they
debated politics and culture.

Callenbach's novel reveals much about the environmentalism of the early
1970s and the hoped-for potential of new green technologies. Although environmental utopia was the focus, there was much more at work in this fantastic
tale than a fictionalization of environmentalist hopes. As with other American

utopians, Callenbach spent a great deal of his book talking about markets, consumption, and the politics that he felt best facilitated each.[4] He worked hard in his story to explain the connections between politics, environmentalism, consumption, and authenticity. Callenbach's western utopia was not about a return to wilderness as much as a vision for a future based on an urban/exurban landscape powered by ATs and earth-friendly economies and governments. The Ecotopian vision was aimed at city dwellers seeking a new type of enlightened consumerism that accommodated their environmental concerns, individual creativity, and social politics. The novel is an excellent example of the meeting of Left social values and Right distrust of big government that has played such a central role in the politics of the American West. *Ecotopia* melded the counterculture lifestyle and social values with a strange brew of libertarian thinking, collectivism, states rights, and technologically enthusiastic environmentalism in the same counterculture sci-fi tradition as Robert Heinlein's *The Moon Is a Harsh Mistress* (1966). Heinlein paints a futuristic western set on a lunar colony populated by innovative misfits ready to break from the tyranny of distant centralized authority and realize the potential of thoughtful anarchy. Their battle cry of TANSTAAFL! (There Ain't No Such Thing As A Free Lunch) captured the imagination of a generation of counterculture entrepreneurs who valued hard work and innovation and individuals who empowered themselves through actions rather than words. Heinlein's lunar libertarians shunned traditional politics, celebrated human ingenuity in the face of environmental challenges, and favored programs of action based on individual agency over the tyranny of the crowd—all familiar western and counterculture tropes.

In 1975, when *Ecotopia* first appeared, Callenbach's vision tapped into a growing counterculture libertarian or "hip Right" ethos that fueled a host of counterculture publications and other products that shaped the 1970s and infused popular culture with a new version of the western myth. This revival of western themes came with an interesting politics with deep roots in regional history that complicate our understanding of what constituted *countercultural*. As historian Michael Allen's work on the counterculture music scene and the "Cosmic Cowboys" it spawned demonstrates, the politics of the 1970s counterculture defy easy categorization.[5] Callenbach and Heinlein provided compelling fictional models for the nexus of ideas about environment, politics, consumption, and recreation that filled the pages of *Whole Earth* and its successor, the *CoEvolution Quarterly* (*CQ*).

While Callenbach's and Heinlein's ideas were a product of a particular moment in American history, their use of the myth and symbol of the West as a political tool was in keeping with a long tradition of western politics.[6] What made these books interesting then and now is the way both authors effectively used the traditional mythology of western history, but reconfigured the story to include counterculture insights on environmentalism and technology. At the heart of this new western utopian narrative was what now would be called green consumption. In Callenbach's vision, green consumption was just a part of a revolutionary change. In the real world of the mid-1970s, green business and green consumption were emerging as a very pragmatic backdoor to political influence and a commerce model for environmental activism.[7] Environmentalism, as historian Samuel P. Hays reminds us, shifted after World War II from a primary concern with production to an emphasis on consumption. By the 1970s, Brand was at the forefront of an important new phase in that reorientation of environmentalism toward a more ecological consumption fueled by advances in ATs and innovative business models coming from many directions but disproportionately driven by developments in outdoor recreation, long a meeting ground of nature and culture.[8]

Western utopian visions, often depicted as hopelessly romantic and unrealistic, usually reflect deep truths about the role of myth and perception in shaping political culture. Callenbach's Ecotopians and Heinlein's lunar pioneers joined a long line of real and fictional western visionaries who chose the West as the stage for their utopian plays not because it was a tabula rasa, but because it is a region replete with malleable and powerful myths and imagery that lend themselves well to redefinition of self and reinvention of politics. Western utopians, like Callenbach, have often been dismissed as frivolous escapists, malcontents, or misfits whose seeming obsession with obscure technologies and mundane futuristic consumption detracts from their political message. While this may have been the case for many who occupied the expansive western utopian fringe from both sides of the political spectrum, western utopians, both conservative and liberal, often put ideas into play that later moved into the mainstream, shaping national culture and politics. For utopian thinkers like Ernest Callenbach, technology and consumer concepts proved to be the most compelling aspects of their political agenda.[9] More importantly, ideological movements that appeared utopian in the sense that they represented a hope for a distant, but unlikely, future often masked very pragmatic research and work that were put into practice quickly and efficiently. One of the ways that the counterculture fringe moved toward the center by the late 1980s was by entering the marketplace. By moving their message to the market and situating their innovations within the existing framework of western mythology, they changed the way people think about what it means to be authentic.[10]

It is important to recognize that *Whole Earth* and especially its successor, *CQ*, were western in more than the location of their headquarters. Brand may have been born a midwesterner, but by the 1970s he, like many of his peers who migrated to the Bay Area in the 1960s, had claimed the region as his home. Like so many who came to the region throughout American history, counterculture migrants to the West displayed a fierce loyalty that belied their relatively brief connection to the region.[11] Readers certainly noticed the westernness of *CQ*, with some commending and others condemning this bias. Brand stridently defended the western bias of *CQ*. "We're an idea magazine," he wrote in response to a letter criticizing the western orientation of the magazine, "and ideas are loose on the West Coast . . . No one complains that the *New Yorker* is overly regional in outlook."[12]

The reference to the *New Yorker* is telling. Unlike his San Francisco publishing contemporary, *Rolling Stone* founder Jann Wenner, Brand refused to move his operations east to New York and the publishing establishment. Wenner moved his publication to New York and radically expanded *Rolling Stone*'s readership and marketability. Perhaps because of his intense experience with the *Last Whole Earth*, Brand seemed content to remain in the Bay Area, keep his readership low, and retain the freedom to explore cutting-edge ideas and their western context. Brand was able to succeed in the West because of his remarkable ability to attract prominent thinkers to *CQ* despite its relatively small circulation of twenty-five to thirty-five thousand copies per issue. By the early 1970s, Brand was so well known and respected for his promotion of innovators that he could get writers to work for *CQ* for little or no money when established East Coast publications paid very well. Brand rarely needed help attracting authors but received strong support from New York pop publicist John Brockman. Brockman made his reputation by creating a stable of successful western writers who he introduced to the New York scene and the rest of the nation. Peter Warshall remembered Brockman's remarkable ability to secure publishing deals with only the slightest details on the book. Brockman helped Warshall secure an excellent contract for *Septic Tank Practices* (1979) from Doubleday.[13] In the areas of science, technology, and environment, *CQ* demonstrated that a West Coast publication could be as good as anything coming out of New York even if the circulation might never compare.

Whereas Brand had steadfastly worked to ban politics from the pages of the *Whole Earth Catalogs*, *CQ* was different, much to the distaste of some of Brand's longtime supporters like Gurney Norman, who worried that *CQ* threatened to move *Whole Earth* away from the ideal of a publication that was "about tools, not political abstractions," and toward the murky territory of "narrow political advocacy."[14] Despite these debates, *Whole Earth* had always been infused with politics and owed much of its success to its near perfect reenvisioning of the politics of the American West.

The West played a well-recognized role in the transformation of American politics during the second half of the twentieth century. The rise of western conservatism in particular has been closely studied. Most agree that, after World War II, the massive demographic and economic transformation of the region instigated by the war economy created a new and powerful block of white-collar, middle-of-the-road conservative western voters who played a large, if not defining, role in national politics. The Republican Party shifted its focus to these new westerners and used them as the cornerstone of a very successful political strategy. Republicans were so successful in fact that it appeared that a complete transformation had occurred within the West as powerful progressive and liberal western constituencies like blue-collar miners and populist farmers were replaced by white-color workers and conservative entrepreneurial capitalists. These new groups had little sympathy for unionism or the old liberal causes that had energized the region for so long.[15] Even the environmental movement, which drew much of its strength from western states, was eclipsed for a time, at the end of the 1970s and beginning of the 1980s, by the conservative response of the Sagebrush Rebellion, which was a replay of the western land grabs of the 1940s, with ranchers, business leaders, and conservative politicians allying themselves to attempt to move public lands into private hands.[16]

The Republican Party did an excellent job of recognizing changing demographic and cultural trends in the 1950s and 1960s and capitalized on the growing power of the West. Just as important, conservatives recognized the power of western myths and symbols in political marketing. As historian Robert Goldberg has demonstrated, Ronald Reagan's presidential career epitomized the conservative use of the rugged individualistic iconography of the West as a political marketing tool.[17] Reagan often staged press events where he dressed as a cowboy and demonstrated his riding skills. In so doing, he was participating in a long-standing American political tradition dating back to the log-cabin campaigns of the 1800s. Traditional conservatives, however, were not the only politicians in the post–World War II era to use western iconography for political marketing. Democrat Lyndon Baines Johnson was Reagan's equal at what historian Anne Butler has called "putting on the hat," the long-standing tradition of donning western wear to send a political message.[18] The AT and green consumption that grew out of the counterculture in the West provided a powerful political legacy and new twist to the traditional conservative use of western mythology in marketing and politics. Central to this twist was the blending of green consumption into the intellectual tradition of a western libertarian sensibility. Consumption is political and so are the business philosophies and practices that drive and facilitate it. Most significantly, for this study, the connection between politics and consumption is not always driven by what we think of as traditional conservative or liberal politics. Starting in the 1960s, a generation of significant western entrepreneurs created a new economy and with it a new and influential version of the regional libertarian sensibility that shaped both the Left and the Right. *CQ* became a leading voice for this distinctly western hybrid politics. Although it would never enjoy the popularity of *Whole Earth, CQ* presented readers with a much more thoughtful, politically engaged, and refined vision for sustainable ecological living than the *Whole Earth* format could achieve.

Coevolution

In the fall of 1973, after a summer spent at his hand-built house in Canada, where "nothing happened, not even nothing," Brand decided to resurrect the *Whole Earth Catalog*.[19] He explained all this in the concluding sections of *Whole Earth Epilog* that resulted from this significant change of heart, offering readers several good reasons why, after all the formality of the killing of the catalog at the "demise party," he was launching it again. The obvious reason to restart was that the *Last Whole Earth* remained a best seller replenishing the Point Foundation's money faster, for a while at least, than the board could dole it out. But Brand was rightfully concerned that the information contained in the catalog was, by 1973, hopelessly outdated and becoming less useful by the day. The catalogs could achieve their mission of "access to tools" only if they were up to date. The *Last Catalog* had become such a cultural icon that readers were drawn to it for reasons that transcended the actual mission of the book. Whether the address of publishers, locations of providers, or discussions of issues were out of date or not, the catalog continued to convey a sense of the early-1970s' counterculture better than any other publication. Nonetheless, Brand felt compelled to keep the mission of the catalog practical and tool centered, especially in light of the increasing economic uncertainty of the time. "The North American economy," he declared in the *Epilog*, "began to lose its mind, putting more people in need of tools for independence and the economy as a whole in need of greater local resilience." The most honest reason Brand gave for his resurrection of the catalog was that "after burning our bridges we reported before the Throne to announce, 'We're here for our next terrific idea.' The Throne said, 'That Was It.'"[20] As usual, Brand was really selling himself short with that humble statement. While he may have felt that the previous two years of his life had been less fruitful than he might have hoped, by almost any other standard he had been exceptionally productive. The results of that productivity would appear in the flurry of new publications and significant events, like the New Games Tournament he produced between 1973 and 1976.

Once he made his decision to continue the catalog, Brand launched into the project with renewed energy and enthusiasm. Michael Phillips was running the Point Foundation and officially hired Brand to edit the new version of the catalog.[21] The fact that Brand was hired by his own foundation demonstrates the degree to which he had disassociated himself from Point during his hiatus. Brand quickly set up shop in another in a long chain of buildings he would later term

Whole Earth Epilog

Credits

Editor Stewart Brand (SB)
Managing Editor Andrew Fluegelman (AF)
Office Diana Barich
Copy Editor Pamela Cokeley
Research traffic Andrea Sharp

Art Director Alwyn T. Perrin
Typesetting Evelyn Eldridge, Joe Bacon
Camera Andrew Main
Paste-Up Susan King Roth, Stephen Leaper,
        Katherine Borsody
Logo & cover paste-up David Wills

Stewart Brand, Diana Barich, Pam Cokeley, Andrea Sharp, Al Perrin, Peter Warshall, Richard Nilsen, Rosie Menninger, J. Baldwin, Kathleen Whitacre, Diana Sloat, Doris Herrick, Evelyn Eldridge, Don Burns.

Money

WHOLE EARTH EPILOG

Here's what happens to your $4, approximately:

$1.60	Bookseller
.24	Jobber
.41	Penguin
.74	Printing
.15	Brand
.86	POINT (covers production, The
$4.00	CoEvolution Quarterly, future
	Catalogs, and some grants)

(For the hardcover edition— $9.25— percentages are equivalent. The print cost per book is $2.55.)

Epilog Production Costs

Expenses, September 1, 1973 through Aug. 31, 1974

	Epilog	The CQ Spring & Summer Summer '74
Wages ($2.75-$7.00/hr)	$36,940	$ 9,100
Office Supplies	6,775	1,720
Rent & Utilities	760	–
Phone	1,999	–
Travel & Auto	225	–
Public Relations	2,219	–
Legal Fees	3,720	–
Independent Contractors (& Reviewers)	15,319	2,700
Production Supplies	4,787	1,400
Printing	1,584	10,425
Research	3,245	400
Miscellaneous	1,218	
Tongue Fu	15,000	
	$93,791	$25,745 (Income potentially, $20,000)
TOTAL	$104,536	

By comparison, production costs on THE LAST WHOLE EARTH CATALOG were approximately $30,000.

THE LAST WHOLE EARTH CATALOG

Copies sold July 1971 - March 1974

1971	July-Sept	328,990
	Oct-Dec	344,978
1972	Jan-Mar	128,879
	Apr-June	72,896
	July-Sept	115,595
	Oct-Dec	62,633
1973	Jan-Mar	45,959
	Apr-June	19,386
	July-Sept	34,125
	Oct-Dec	370,500
1974	Jan-Mar	25,403
		1,196,513 copies

Gross receipts $3,163,066. Where it went:

Printing	$920,389
Random House	816,546
Gurney Norman	103,449
Bookworks (Don Gerrard)	118,967
POINT	1,203,745

Between June 1971 and September 1973, POINT made grants totaling $663,541. Details in Summer '74 CoEvolution Quarterly. Grants during the winter of '73-'74 comprise roughly another $400,000. To cover print costs for the EPILOG, POINT had to borrow money. POINT's address: Box 99554, San Francisco, CA 94109.

It was standard practice at *Whole Earth* for the credits pages to include detailed information on money and how it was used in the creation and distribution of the catalog. This page from the *Whole Earth Epilog* also featured a fantastic photo of some of the key contributors to the *Epilog*, including (left to right) Stewart Brand, Diana Barich, Pam Cokeley, Andrea Sharp, Al Perrin, Peter Warshall, Richard Nilsen, Rosie Menniger, J. Baldwin, Kathleen Whitacre, Diana Sloat, Doris Herrick, Evelyn Eldridge, and Don Burns. (Courtesy of Department of Special Collections and University Libraries, Stanford University Libraries) "low road" architecture. Brand always favored utilitarian structures that birth in-genious ventures with no pretense of the formality or structure that characterizes the corporate office environment. In this case, a "moldering crab shack full of junk renting at \$40/month" on the pier in Sausalito, California, became the new *Whole Earth* workshop.22 Brand quickly gathered a new group of talented and enthusias-tic supporters to help assemble the catalog. A contract was secured from Penguin Books, a decision that Brand and the Point Foundation regretted later as there were considerable problems with this relationship almost from the start.23 These problems included disagreements regarding responsibility for production costs and delivery that delayed the project. The mounting costs became a major concern for Michael Phillips and the Point Foundation and proved a harbinger of leaner times for the *Whole Earth* publishing ventures. Even while the *Last Whole Earth* continued to sell very well, it was becoming clear that the salad days were near an end and that future ventures would require much more concentrated marketing efforts and very careful financial management to succeed over the long term.

The *Epilog* turned out well despite the publishing issues. In spring 2005, while giving the commencement address at Stanford University, Apple Computer co-founder Steve Jobs fondly recalled his encounter with the *Epilog:* "It was sort of like Google in paperback form . . . [I]t was idealistic, overflowing with neat tools and great notions." Jobs was particularly struck with the back-cover message, "Stay hungry. Stay foolish," which remained with him over the years and was his advice for the graduating class of 2005.24 Despite the lasting influence of this ver-sion of the catalog on significant figures like Jobs, though, the *Epilog* could not re-capture the massive success of the *Last Catalog.* Brand understood well that a mo-ment had passed and, while the idea of providing access to tools was as vital in the mid–1970s as it ever had been, and maybe even more so, the catalog could never recapture the iconic status it had from 1968 to 1972.25

During the two years leading up to the publication of the *Epilog* and the simulta-neous launching of *CQ*, Brand published two significant independent writing proj-ects in *Harper's* and *Rolling Stone* and a book, *II Cybernetic Frontiers* (1974).26 In *Rolling Stone,* Brand published a pathbreaking article on computer gamers and hackers. The article included photographs by Annie Leibovitz and featured the first use of the term *personal computer* in print. "Ready or Not, computers are coming to the people," Brand began his article. "That's good news, maybe the best since psy-chedelics."27 He went on to explain the appeal of computers for the counterculture and to outline carefully the intellectual bridges between computer technology, cy-bernetics, and the universe of ideas that founded *Whole Earth.* For those with the mistaken notion that *Whole Earth* was a Luddite bible for back-to-the-landers, the *Rolling Stone* article might have come as a shock. Close readers of the catalogs, how-ever, would not have been surprised at Brand's interest in computers as tools.28

Brand's *Harper's* article, "Both Sides of the Necessary Paradox," about British anthropologist and cyberneticist Gregory Bateson, was more suggestive of Brand's evolving environmental philosophy. Brand first met Bateson in 1960 at a Veterans Administration hospital in Palo Alto, California. They did not encounter each other again until 1972, when Brand discovered Bateson's book *Steps toward an Ecology of Mind* and consequently found a captivating new muse at a critical moment in his personal life.[29] Like most of the intellectuals who informed Brand and his publications, Bateson was an iconoclastic deep thinker who defied easy classification. Brand sometimes found Bateson a hard sell. "A good many people I know consider Bateson maddeningly obscure," he wrote in an introduction to an issue of *CQ* with Bateson on the cover.[30] An anthropologist by trade, the imposing 6-foot 5-inch Bateson was best known for his relationship with Margaret Mead and for "conceiving the Double Bind theory of schizophrenia."[31]

Brand found Bateson during a very tough period of soul-searching chronicled with remarkable honesty in the *Epilog*. He hailed Bateson's book as "strong medicine" for those like himself who where trying to link "intellectual clarity and moral clarity" and "evoke a shareable self-enhancing ethic of what is sacred, what is right for life."[32] Bateson's particular take on cybernetics and whole systems that provided a method for linking "mysticism, mood, ignorance, and paradox" struck Brand with such force that he decided to seek the man out. The resulting conversations formed the basis for the *Harper's* article. More importantly, several meetings in 1972 launched a transgenerational friendship that profoundly influenced all of Brand's work for the coming decade: "It was talking with Gregory Bateson . . . that gave me a thread to string my beads on."[33] Bateson, with his deeply philosophical view of life and science, replaced Buckminster Fuller as the inspiration for Brand's publishing ventures, and he featured Bateson's work prominently in both the *Whole Earth Epilog* and *CQ*.[34]

Bateson's influence on Brand is fully visible in the *Epilog,* which begins with a frontispiece quoting the book of Job, a subject of discussion between the two, and continues with a page dedicated to Bateson's work as the new lead to the "Whole Systems" section. "Where the insights of Buckminster Fuller initiated the *Whole Earth Catalog,*" Brand wrote, "Gregory Bateson's insights lurk behind most of what's going on in this *Epilog.*"[35] Bateson loomed even larger in *CQ*, which became Brand's primary focus between 1974 and 1985, when it morphed into the *Whole Earth Review.*

Brand launched *CQ* in the spring of 1974. The first issue was small, only ninety-six pages, and featured longer articles in a much more traditional format than the catalogs.[36] Buckminster Fuller's design science philosophy inspired the founding metaphor for *Whole Earth:* Spaceship Earth. By choosing a biological metaphor, coevolution, as the title of his new publication, Brand was sending readers a clear message about the changes in his thinking. "The paradigm shift," he commented to an audience at a World Game workshop in Philadelphia, "is from an engineering metaphor to a biological metaphor." In trying to explain the change in his thinking, Brand made it clear that design and environment were still his core passions, but that he had returned to his roots as a biologist to rethink the basis for design. He became interested in providing readers with information about the mutually beneficial relationships between nature and culture that the study of coevolution was uncovering.[37] Brand credited Paul Ehrlich and Peter Raven with the actual term that they invented to explain "something terribly obvious but not before formally recognized about living organisms. They spend most of their adaptive effort getting along with other life . . . Life evolves with life."[38]

"The ongoing plot of coevolution," he told his readers in 1976, "is continually shifting mutual optimization." When pressed for some examples of how this thinking was shaping the environmental philosophy of his new publication, Brand pointed to the work of John Todd and the New Alchemists as a model of the design sensibility and tool culture of *Whole Earth* evolving toward the next step with the incorporation of living systems directly into alternative technology solutions. Bateson's insights about science, human nature, and the prospects for the environmental health of the planet meshed well with the concerns of appropriate technologists and ecological designers who were featured in the quarterly and lent a healthy dose of skepticism to the coverage of these issues in the *Epilog* and *CQ*.[39]

CQ, "a California-based peculiar magazine," continued the tradition of presenting tool-centered solutions and pragmatic ideas for resolving environmental problems that seemed more pressing than ever by the mid-1970s.[40] Like its forerunner, *CQ* was eclectic and no one theme dominated the decade-long publication run. But, like *Whole Earth*, an abiding concern with environmental issues dominated the publication. *CQ* editor and chronicler Art Kleiner pointed to the new "Apocalypse Juggernaut, Hello" section of *CQ* as evidence of a primary concern "with forecasting environmental/energy/economic" issues. These topics shaped the remarkable first edition of *CQ* and drove the content throughout the life of the publication.[41] *CQ* built on a continued interest in fostering an alternative environmentalism, with a special emphasis on what contributor Jaron Lanier would later call the "techno-humanist interface" or the union of technology/nature/human nature toward a sustainable future.[42] Significantly, *CQ* linked green consumption and voluntary simplicity, most notably with articles by Paul Hawken and Michael Phillips, to the larger project of reconciling technology and nature toward an ecologically healthy future.[43]

Likewise, *CQ* dramatically increased the coverage of alternative technologists, in the new "Soft-Tech" section edited by J. Baldwin, who became the "source of *CoEvolution*'s ongoing authoritativeness on soft technology."[44] Baldwin brought his own wit and insight to a long series of outstanding articles and was joined by long-term contributors like Steve Baer and significant new soft-tech writers like Richard Nilsen and Day Chahroudi. *CQ* published an important book on AT, *Soft-Tech* (1978), edited by Baldwin and Brand, and gave significant space in virtually all issues to articles exploring the areas of alternative technology and ecological design that were converging in the 1970s toward a philosophy of *sustainability*.[45] Even a quarter-century later, *Soft-Tech* remains one of the best sources on AT and the eclectic thinkers and designers working at the time to build a design-based environmental philosophy. An important part of the sustainability picture presented over the years in *CQ* was the linking of urban health and cultural preservation with environmental health and natural preservation. *CQ* offered a unique forum for uniting the usually disparate and often adversarial pursuits of historic preservation, architectural design, and ecological design.[46]

The second issue of *CQ* (Summer 1974) demonstrated clearly how the quarterly was an extension of the work initiated by the Point Foundation. Foundation board member Huey Johnson contributed an excellent article on the "Land Banking" techniques that he was instrumental in perfecting. The issue also included significant articles by J. Baldwin and Steve Baer on alternative technology and chronicled the remarkable New Games Tournament that Brand had organized the previous year. Finally, this issue formally introduced two contributors who would play key roles in the history of the *Whole Earth* network for the coming three decades: Peter Warshall and Stephanie Mills.

Stephanie Mills had written for *Whole Earth* since 1972, but her role greatly expanded with *CQ*. In the second issue, she chronicled her experience as the recipient of two Point grants to host community-building events on the Parisian model of the salon in "Salons and Their Keepers." Brand wrote of Mills's efforts, "The most powerful instrument of intellectual community organization is the salon . . . a hostess with wit, culinary skill, and access to people who ought to know each other is all it takes."[47] When she wasn't hosting dinner parties, Mills was an environmentalist and used this expertise consistently well when she became a *CQ* editor and later as the author of a notable book on land restoration.

Peter Warshall began his long association with the *Whole Earth* community with the *Epilog* but was formally introduced to readers by Brand in the second issue of *CQ* in an article appropriately titled "Natural History Comes to *Whole Earth*."[48] Warshall met Brand through his Bolinas, California, neighbor Lloyd Kahn,[49] and in *CQ* introduced himself to readers as a "nouveau-conservationist," in homage to mentor and biological province pioneer Raymond Dasmann.[50] Brand gave readers a more complete introduction to Warshall, highlighting his rise from Brooklyn to Harvard.[51] In the first two years of *CQ*'s publication, Warshall worked with Brand to produce pathbreaking articles on bioregionalism, resulting in, among other things, a wonderful "Watershed Quiz" that speaks as much to the early 2000s as it did to the mid-1970s, and a world biogeographical provinces map that *CQ* marketed for several years.

Warshall's insights greatly expanded the think-globally, act-locally ethic that had always been present but was articulated far more clearly after he became an editor. *CQ* continued to feature environmental technologies and tools in long technical articles, some of which fell within the familiar section headings of *Whole Earth* (for instance, "Whole Systems" and "Land Use"), whereas others expanded the reach of the old format into areas like politics. Most importantly, *CQ*, because of the expanded format, was better able to weave together the constellation of ideas that informed the catalogs. It also provided a much-loved sanctuary for those thinking hard about what *CQ* reader and *Washington Post* columnist Henry Allen called "The Problem," which he defined as "the survival and co-evolution of our species and biosphere."[52]

Optimism was one of the things that made *Whole Earth* such a success. In *CQ*, Brand continued to present potential solutions to "The Problem" and introduce readers to thinkers who were optimistic about the ability of people to solve the issues plaguing the world in the 1970s. By the mid-1970s, Brand was thinking hard about the legacy of the sixties and the failure of revolutionary and utopian thinking. "People who organize their behavior around the apocalypse," he argued, "are going to have a tough time knowing who they are when the apocalypse fails to show." Brand felt more than ever that in the post–cultural revolutionary seventies people needed to develop good tools and ideas to create realistic programs of change toward a future of ecological sustainability. "It's the good stuff you need when your idea fails," he advised, and *CQ* would be his new venue for bringing the good stuff to the public.[53]

Brand's concerns about the dreamy revolutionary thinking of the sixties and the prospects for a postapocalyptic future owed much to the controversial Herman Kahn.[54] The fact that both Kahn and Warshall helped shape *CQ* is testament to Brand's ability to bring together unlikely bedfellows. Kahn was a Rand Corporation researcher and controversial proponent of a futurist version of systems theory based on methods of analysis like "multi-fold trends, surprise-free projections, scenarios."[55] In the 1960s, the larger-than-life Kahn, who was both heavy and tall, was a popular and forceful conservative speaker with a quick mind and taste for argument. Louis Menand recently called him a "spielmeister," a "jocular, gregarious giant who chattered on about fallout shelters, megaton bombs, and the incineration of millions" with flashy, graphically intense presentations like an extra scary version of Al Gore's documentary *An Inconvenient Truth* (2006).[56]

Kahn became a surprise celebrity known for his bold and frighteningly optimistic scenarios for nuclear conflict during the Cold War. His fatalistic/optimistic projections (it is going to happen, but the world will survive so you better be ready) inspired, according to *Whole Earth* editor Art Kleiner, the title character for Stanley Kubrick's satirical classic film, *Dr. Strangelove* (1964), whose protagonist was, like Kahn, weirdly gleeful about a nuclear future.[57] In the 1970s, Kahn moved away from his roots in nuclear science and became a spokesman for a boldly optimistic notion of a coming belle époque created by radical technological advances and the "rise of multinational corporations." His futurism was easily critiqued, and often was, but had a strong appeal for those who rejected the notion that technology was evil and capitalism was on the wane. Kahn argued that a new phase of multinational capitalism would help erase political boundaries, spur creative technologies, increase affluence, and solve the looming environmental and social problems that caused liberal-minded people to worry about the future of the Earth.[58] His predictions could be vague, and he was often wrong, but he was right often enough to keep him in demand by audiences who appreciated his buoyancy.

Kahn's views about economy and human agency were particularly appealing to Brand during the period Brand was reevaluating his personal philosophy and seeking out interesting researchers to contribute to *CQ*. Brand seemed receptive to Kahn's cutting critiques of liberal countercultural types who were so disappointed by the failure of the cultural revolution and so caught up in nostalgia for their passing moment that they were losing sight of the positive prospects for the future. Kleiner situates Kahn within a larger "millenarian meme." Kahn's was a secular millennial vision that represented a "postindustrial, countercultural, ecologically aware . . . attitude shift affecting every aspect of society."[59]

Brand took a full dose of Kahn's medicine and encouraged others to try to reconnect with the optimism of the early days of the AT movement captured by events like Alloy. The optimism Brand conveyed through the pages of *CQ* was tempered by a clearer understanding, gained from several years of harsh introspection, that while the future might be bright, reaching it would be more complicated than the sixties' imagination envisioned. "Jay Baldwin and I were talking the other day," he told an audience, "about the nostalgia-for-the-mud that we are all immersed in. 'Oh goody, windmills,' you know. I mean talking about going backward."[60] Brand was not arguing against windmills or the AT they represented, just the apprehension that without revolution all the hard work was just delusion. One of the implicit goals of *CQ* was to dig deeper into the nexus of science, technology, and nature and unpack the notions that had appeared in the catalogs but with more space for analysis.

The new quarterly's expanded format placed added control in the hands of the editors, resulting in a product with a more clearly defined viewpoint and enhanced ability to tease out the tantalizing ideas that peppered the pages of the original catalogs. It also meant that the autodidactic model of the catalogs would not carry over to *CQ* or the *Whole Earth Review* that followed. The 30,000 or so avid readers of the magazine were active participants, sending in letters by the thousands and sometimes becoming direct contributors.[61] Still, and in keeping with Brand's founding goal to "see what would happen if an editor were totally unleashed," the content of *CQ* was closely controlled by Brand and a series of guest editors and more clearly reflected the interests and biases of the editors in charge. The magazine frequently explored the utopian fringe in ways that even loyal readers found disturbing and demonstrated that the confluence of cultural trends that gave *Whole Earth* such broad appeal were too fragmented by the mid-1970s to create another equally cohesive community of readers and writers. *CQ* never captured the huge audience of *Whole Earth,* but playing to a smaller, more informed audience had its own advantages particularly in the coverage of the ecological issues. *CQ* launched significant debates that complicated discussions of environmentalism for a segment of influential, ecologically minded Americans, including first publications of the Gaia hypothesis and more in-depth coverage of bioregionalism, space colonization, green consumption, and simple living.

The 1975 publication of Lynn Margulis and James Lovelock's Gaia hypothesis was one of the notable accomplishments of *CQ.*[62] The Gaia hypothesis (*Gaia* is the Greek name for the "Earth") proposed that the Earth was a living organism or, as Lovelock explained it, the "biosphere is a self-regulating entity with the capacity to keep our planet healthy by controlling the chemical and physical environment."[63] Margulis, a distinguished microbiologist, and Lovelock, a prolific British independent researcher, epitomized the thoughtful environmental heretics that appealed to Brand. Margulis is best known for her theories of cell symbiosis—the idea that organisms cooperate and coevolve—and for her work with Lovelock on the Gaia theory.[64] She became an important contributor to *CQ* and advisor to Brand on the latest insights emerging from biological research. The more iconoclastic Lovelock epitomized the melding of biological insights and research with toolcentric pragmatic design science. In addition to his Gaia collaboration with Margulis, Lovelock was a designer of precision instruments, most notably a series of electron-capture detectors he developed in the 1950s that greatly enhanced the ability to record the distribution of chemicals in the atmosphere. This detection enabled researchers like Rachel Carson to understand the effects of pesticides on ecosystems.[65] Brand appreciated the way these two "vaulted disciplinary barriers and dogmas," and *CQ* presented their work at a time when their ideas were greeted with skepticism at best by the academic community.[66] The Gaia thesis sparked a protracted and engaging debate on the pages of *CQ* that filtered out into college classrooms and the media. Although the Gaia theory remained controversial through the last quarter of the twentieth century, environmentally minded *CQ* readers accepted it with much less controversy than greeted another significant contributor to the magazine.

Space Colonies

The publication of Gerard O'Neill's proposals for space colonies in several issues of *CQ* and then as a special *CQ* book, *Space Colonies* (1977), surpassed Gaia in generating fruitful debate and controversy and offered Brand the best opportunity to push the boundaries of environmentalism he had been exploring since college. Nothing better illustrates Brand's willingness to explore emerging areas of technological research at the cost of infuriating loyal environmentally oriented friends and readers than his promotion of space colonization. Space colonies were the ultimate environmental heresy of their day, but they fit perfectly within Brand's expanding conception of environmentalism and his particularly western viewpoint. Like Lynn Margulis, O'Neill was a traditionally trained scientist with excellent credentials: He was a physics professor at Princeton University and researcher at the Stanford Linear Accelerator Center, who, nonetheless, found himself at odds with the scientific establishment after he became the leading proponent of space colonies. O'Neill envisioned massive, slowly rotating, self-supporting structures anchored in space as the future home of millions of people and launching pads for the exploration of the final frontier.[67] Brand introduced him to *CQ* readers as a "high-energy physicist best known in his field for originating the colliding-beam storage ring, which has been used in nuclear accelerators throughout the world."[68] Brand also pointed out O'Neill's military record and accomplishments as a pilot, which were backgrounds and interests they shared.

Like Margulis, O'Neill found a welcome home at *CQ* after facing skepticism or outright hostility from more traditional publications. His first contact with Brand and *CQ* came via the Point Foundation. In February 1974, O'Neill and three colleagues "brave enough to gather in one place and talk about space colonization" decided to have a small conference on the topic.[69] O'Neill had no luck obtaining funding from the established foundations and started looking for alternative sources. "I was led to a very small and special organization," he wrote, "the Point Foundation of San Francisco. Its office was a tiny two-room shack on the roof of the Glide Methodist Church, reached over duckboards laid across the roof."[70] At this unlikely place, O'Neill discovered an enthusiastic audience: "Point, as I was to learn . . . was designed expressly to encourage innovation and prevent lockstep thinking." After the thinly disguised contempt of the large foundations for his work, Point was a breath of fresh air for O'Neill. "Inside the shack," he remembered, "with the rain pouring down outside, I found a comfortable pair of small offices, lined with books." He was greeted by unassuming Point Foundation Secretary Richard Austin, who appeared to O'Neill refreshingly lacking in the "airs of a 'foundation executive.'"[71] Michael Phillips later joined them for an open and optimistic discussion of space colonization and agreed to award O'Neill the six hundred dollars he needed for the conference. While on the fringes of American

Cutaway view of the design for a giant rotating Toroidal colony. (Courtesy NASA, Space Colony Images AC76–0628)

science, O'Neill was at the center of American culture in the 1970s and a leader in a significant reenvisioning of the mythology of the American West.

In 1957, scientists from around the world were working on projects for the International Geophysical Year (IGY), an international celebration of technology. The United States was planning to launch a three-pound satellite into the Earth's orbit as part of the event. The IGY represented an opportunity for international cooperation in technology, especially in the new realm of space. Scientists from around the world were hopeful that the exploration of space could open new avenues for the sharing of information and begin a move away from the culture of secrecy and competition surrounding weapons technology. This optimistic atmosphere was shattered on October 4, 1957, with the successful launch and orbit of the Soviet satellite *Sputnik,* which came as a shock to the free world because the satellite was a graphic illustration of Soviet technological advancement and the power of authoritarian technocracy to accomplish high science. The little beeping space ball caused an international uproar and instigated an unprecedented peacetime technological race.[72] The scientific, political, and diplomatic consequences of the space race are well known. Most environmental organizations, along with many socially minded individuals and groups, condemned the massive expenditure of resources and personnel on a program that seemed to promise so little for those left on Earth. Brand joined O'Neill and two prominent ecologist brothers, Eugene P. and Howard T. Odum, as one of a select group of environmentalists to celebrate space exploration and argue for the benefits of space technology for Earth environments.

The Odums were cybernetic thinkers who worked to bring "energetic systems theory to ecology." Their ecological theories, which, according to historian Peder Anker, were geared to "bring human activities into balance with the ecosystem through natural, social, and technological engineering," fit perfectly within the constellation of ideas about ecology and technology promoted by *Whole Earth* and *CQ*.[73] Along with O'Neill, the Odums lent scientific legitimacy to the discussion of space colonies. Environmental historian Donald Worster has argued that the Odums "did more than anyone else to define" the ecological "science of the postwar period" and brought to the science of ecology "a strong admiration for the achievements of technological, Apollo-making man."[74] For Worster, like many contemporary critics, the Odums' simultaneous fascination with space technology and ecosystem preservation on Earth risked "contradictory policies" that might support "the old dream of the conquest of the earth, now ironically sought in the name of environmentalism."[75] Worster insightfully pointed out that the Odums fit ecology scholar Peter Taylor's definition of "eco-technocrats," who shared with earlier technocrats a desire to use a top-down model to "protect Spaceship Earth from any explosion."[76] Within the deep history of ecological ideas Worster so beautifully chronicles in his classic *Nature's Economy* (1977), space colonies certainly could seem contradictory or even heretical. But within the context of the counterculture environmentalism of *Whole Earth*, they made some sense.

Brand's advocacy of space colonies in the 1970s expanded on the countercultural vision for life in space best captured by Gene Roddenberry's 1966 cult-classic television series *Star Trek*. A western set in space, *Star Trek* episodes focused on social issues of the 1960s and featured pioneering ethnic actors in lead roles. The role of technology in undermining or fostering social justice and environmental quality of life was a persistent theme of the show, with a particular emphasis on how these issues gained significance on the "final frontier." *Star Trek* brought the speculations and concerns of the science fiction genre to a wide audience and tied together scientific fact, technological development, and literature for a generation hungry for the type of boldly speculative research that Brand promoted through *Whole Earth* and more directly in *CQ*. Although space exploration was a logical topic for Brand, many of his readers were unconvinced.

The space colonies debate in *CQ* clearly demonstrated the limits to the techno-ecology Brand had promoted over the years. After *Space Colonies*, the divide between nature purists like visionaries Gary Snyder and Wendell Berry and technophiles like O'Neill seemed greater than it had been when Brand first breathed fresh air into the environmental culture of the 1960s.[77] No single topic dominated the ten-year run of *CQ* like the space colonies dispute. Brand seemed to revel in the ruckus he caused and gave more than equal space to his critics. The debate that played out on the pages of the quarterly was interesting less for the technical discussions about whether these colonies were feasible and more for the deeper concerns within the environmental community about what constituted an appropriate technology in opposition to ecotechnocracy. Many of the environmentally minded readers of *CQ* and longtime supporters of *Whole Earth* considered space exploration and colonization completely inappropriate. Notable participants in the debate included Lynn Margulis, Paul and Anne Ehrlich, E. F. Schumacher, Lewis Mumford, Buckminster Fuller, Ernest Callenbach, David Brower, Carl Sagan, Gary Snyder, Hazel Henderson, Wendell Berry, and astronaut Russell Schweickart, among others. The response was mostly angry or confused, with many writers drafting both public responses and private letters to Brand in search of explanations for what many perceived as a significant shift away from the traditions of *Whole Earth*. Even many who supported the AT vision of *Whole Earth* thought that space colonies were too technocratic, dependent on hierarchical centralization of scientific authority and massive federal support. Those who had followed the evolving discussions of ecologically minded technology through the various *Whole Earth* publications understandably thought the support for colonization was an unwelcome tangent or regression to an older mode of thinking about technology and nature and a clear move away from the E. F. Schumacher school of small-scale ecotech.

None was angrier than Wendell Berry. Brand played a central role in promoting Berry's eloquent writings on American agriculture and the environment, but the two parted ways over space colonies. Berry's article-length response to the issue was filled with personal criticism of Brand's decision to move into such controversial territory. "Your dismissal, out of hand," Berry wrote, "of so many people's objections and doubts—solicited by you—is an alarming display of smugness."[78] Other writers were less incensed, but many echoed Berry's concerns about both the subject and its presentation. "I do have to agree with Wendell to a degree," Gurney Norman wrote in a letter to Brand, "that your advocacy of space colonies represented a change in approach as an editor; that it broke the consistency you had built from the earliest catalogs, a consistency which people had come to count on, that you were about *tools*, not political abstractions." "So I think," he concluded, "you have inadvertently turned *CoEvolution Quarterly* . . . into a magazine of *narrow political advocacy*."[79] Readers who had not blinked over

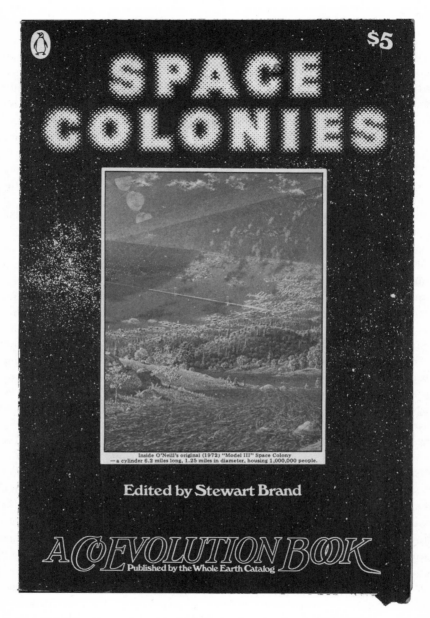

Inside O'Neill's original (1972) "Model III" Space Colony
—a cylinder 6.2 miles long, 1.25 miles in diameter, housing 1,000,000 people.

Edited by Stewart Brand

A CoEVOLUTION BOOK
Published by the Whole Earth Catalog

$5

Cover of the controversial *CoEvolution* book that collected all the articles and highlights of the contentious debates within Brand's community of contributors, readers, and supporters that demonstrated the limits to his melding of ecology and technology. (Courtesy of Department of Special Collections and University Libraries, Stanford University Libraries) contributor R. Crumb's extraordinarily politically incorrect cartoons, which appeared regularly in the quarterly over the years, had coped with articles about how to eat roadkill, and had seemed generally approving of other controversial topics were now deeply divided by the leap to space. Why? Space as the final frontier made perfect sense to Brand and tapped elegantly into the counterculture love of libertarian-leaning environmental science fiction like Robert Heinlein's *The Moon Is a Harsh Mistress* and Frank Herbert's *Dune* (1965) that reinvented the western myth and tied it to evolving technologies and expanding consciousness.[80] Space colonies also offered the best opportunity yet for Brand to explore science practice as it worked its way from fringe speculation toward possible broader acceptance, a persistent theme in *Whole Earth* and *CQ* that presaged Bruno Latour's proposition that we can only understand *Science in Action* (1987).[81]

For environmentalists like Berry, however, the space program, let alone space colonies, represented a dangerous distraction from the very immediate environmental problems on Earth. He was also deeply offended by the idea that space, and the Moon in particular, were up for grabs as new frontiers when he and others considered it a wilderness that should be preserved in perpetuity. David Brower and other wilderness proponents echoed this sentiment. This divide is not surprising, and in fact nicely reflects the real divide within the environmental movement of the 1970s over the proper path toward an environmentally healthy future and the role of technology in that process. *Whole Earth* had always represented the human-centered, pro-technology view, but not to such a degree that wilderness advocates and back-to-the-landers were offended. The technology position of *Whole Earth* through the *Epilog* drew most directly from the work of E. F. Schumacher and his "small is beautiful" economics. Berry's most pointed criticism of the space colonies debate focused on the ways that space colonies and support of space exploration subverted Schumacher's conception of AT that had provided such a welcome bridge between wilderness supporters and environmentalists who focused on the human side of the equation.[82] Likewise, R. Crumb was dumbfounded by *CQ*'s support for space research. Brand had sent Crumb to a Space Day celebration, and Crumb chronicled his profound disillusionment with the whole idea of space technology in an amazing four-page cartoon lampooning all the major players and graphically admonishing readers not to be "duped by foolish Buck Rogers dreams of glorious adventures among the planets!!" "The saddest part," he concluded with a drawing of himself slumped over and marching away from the celebration, "is that a lot of otherwise intelligent people are falling for the space hype . . . hook, line and sinker!!"[83]

Undaunted by Crumb and Berry's withering critique, Brand seemed pleased that his interest in space had spawned such an intense debate. In hindsight, it was an important debate and one that brought to light deep divisions within the

Whole Earth community of readers and writers over the very issue that had spawned the publication in the first place. The passion of the space colonies debate clearly indicated that Brand's call to recognize the preeminence of humans and human agency when thinking of environmental problems that had been accepted by millions of readers of the catalog could not stand up to the type of in-depth exposure that *CQ* afforded. By advocating moving off the Earth and building something new and perhaps better, Brand was simply following his own advice to be "as gods" to its logical extreme. Even after a decade of significant research on ATs and some productive integration of those who placed nature at the center of their advocacy with those who placed humans at the center, the environmental movement was still deeply divided. Environmentalists like Berry, Brower, and Crumb were relatively comfortable with ATs that seemed to come from a better and more Jeffersonian tradition of self-sufficiency. When the technology was high tech, which space technologies obviously were, there was much less understanding and acceptance.

In 1977, while he was immersed in the space colonies issue, Brand had an opportunity to take his iconoclastic ideas to the California governor's office, where they spurred more debate and reached a much wider audience. Governor Jerry Brown offered Brand a loosely defined appointment as the "Consultant to the Governor," which he served as between 1977 and 1979. Brand's official tenure with Brown did not begin until 1977, but Brand had influence in the administration dating back to 1974 and the early issues of *CQ*. In fact, the hybrid politics of Jerry Brown and the often iconoclastic activities of his administration might as well have been a section heading in *CQ*, so large was the role that Brown's tenure in California played in the magazine. Ultimately several key *Whole Earth* contributors joined Brown's cabinet and exerted a clear influence on the path of his administration. J. Baldwin worked in Brown's California Office of Alternative Technology. *CQ* contributor and astronaut Russell Schweickart became science advisor to Brown in 1977 and was appointed chairman of the California Energy Commission in 1979.[84] Ecological design pioneer Sim Van Der Ryn became California State Architect, and Huey Johnson had the most significant job as Brown's Secretary of Resources for California from 1978 to 1982, the state's top environmental official, with "fourteen thousand employees and a billion-dollar budget."[85]

Notable among his many accomplishments working with Brown was Johnson's Investing for Prosperity (IFP) 100-year planning initiative that provided for investment of proceeds from the sale of public natural resources into environmental programs aimed at the future. These funds went to urban forestry, fish and wildlife programs, and parks, and significant funding went to appropriate technology development and support.[86] The IFP program provided inspiration for

Stewart Brand and Governor Jerry Brown in northern California, c.1976. (© Ilka Hartmann 2007)

Johnson's founding of the Resource Renewal Institute and his internationally significant Green Plan initiatives.[87] More than any other person, Johnson was able to weave together his experiences as an environmental activist, Point board member, and key Brown cabinet member into a coherent set of environmental plans and policies that shaped international environmentalism from the mid-1970s through the year 2000. In 2001, his remarkably consistent innovation won him the prestigious United Nations Environment Programme (UNEP) Sasakawa Environment Prize.[88]

Johnson, Van Der Ryn and, to a lesser but still significant degree, Brand influenced Brown's environmental policies and ideas.[89] This was especially true in the area of space science. Space technology played a critical role in California's economy in the 1970s and, as an advisor to Brown, Brand played an important role in shaping the governor's enthusiasm for exploring new frontiers. At the height of the space colonies debate, Brown sponsored Space Day on August 11, 1977, and convened an important conference and invited preeminent space scientists to California to debate space colonies and the environmental prospects and consequences of space exploration.[90]

It was the summer that *Star Wars* hit theaters and reenergized interest in space travel. Russell Schweickart coordinated the event and chaired the panels, with Brand moderating sessions that featured luminaries like Carl Sagan, Gerard O'Neill, Jacques Cousteau, National Aeronautics and Space Administration (NASA) Robert Frosch, Johnson Space Center director Christopher Kraft, and Robert Anderson, president of Rockwell International, which was the builder of the new space shuttle *Enterprise* (named after *Star Trek*'s famous vessel) that was scheduled for its first atmospheric flyover of Edwards Air Force Base the next day.[91] Brown introduced the meeting with a statement on the value of space for California and the West: "Ecology and technology find a unity in Space." He went on to place the event firmly in a Turnerian western context: Going to space is an investment. It's not a waste of money . . . As long as there is a safety valve of unexplored frontiers, then the creative, the aggressive, the exploitive urges of human beings can be channeled into long term possibilities and benefits."[92] As governor of California, Brown had good reason to celebrate space research in all its forms. Brand pointed out to *CQ* readers that California was the American leader in space research and support industries with 51 percent of NASA's contracts going to the state.[93] But, on that first Space Day, only one significant environmentalist signed on without reservations.

Jacques Cousteau was one of the few prominent environmentalists to support the U.S. space program. "My major interest is that Cousteau is an environmentalist and is almost alone in seeing the environmental benefits of the Space Program," Brand wrote about his ally.[94] It is not surprising that a scientist who spent

(Left to right) California Governor Jerry Brown, Judge Ronald Robie, Huey Johnson, Brown's secretary of resources for California, and Dr. Priscilla Grew, February 25, 1978. Of all the *Whole Earth* and Point contributors who worked with Brown none had more influence or responsibility than Johnson, who shaped California environmental policy during a critical period of change. (Courtesy Huey Johnson)

a remarkable percentage of his life wholly reliant on technology while exploring the inner space of the Earth's oceans would have an optimistic and adventuresome perspective on space technology. In an interview, Brand asked Cousteau to explain his position. "Since the beginning," he responded, "I was convinced that our duty was to explore outer space as much as or more than inner space."[95] Cousteau responded, with prescient insight, that for anyone concerned with the health of the massive oceans or any other part of the Earth for that matter, space science offered the best hope for developing monitoring systems capable of analyzing planetary health and shaping pragmatic responses to global environmental problems. Had Brand focused on Cousteau's perspective on space exploration instead of O'Neill's, there might have been less controversy. All space exploration was dependent on centralized high science and technocratic bureaucracies, but this might have been more palatable to *CQ* readers if not linked with the rhetoric of utopian colonialism attached to the colony idea. Still, in hindsight, it is safe to say that even with this most provocative environmental heresy, Brand was on to something of profound importance for environmental science that most environmentalists of the period missed.[96]

Space technology has completely changed the environmental sciences and our perception of the human impact on the Earth. Our understanding of global warming, changing weather patterns, ocean temperatures, and deforestation depends on space technologies. If there was a true genius to *Whole Earth*, it was that LSD-inspired foundational insight that viewing the Earth from space was critical to sustaining life in the future.[97] Our ability to understand global warming owes much to *Whole Earth* and the environmental heretics like Brand, Cousteau, and O'Neill, who recognized that the technological enthusiasm that caused our twentieth-century environmental crisis could get us out of it in the twenty-first century. Once the enthusiasm for space colonies faded, space-based environmental research slowly regained legitimacy as a tool for ecological understanding that might foster environmentalism from a variety of perspectives. One of the first concrete environmental tools to emerge from space research was the *California Water Atlas* (1978). The atlas was a beautiful giant-format book produced by Brown's office and shaped by a large group of *Whole Earth* and *CQ* alums, including Brand, who served as chair of the publications advisory group, Huey Johnson, Peter Warshall, Russell Schweickart, and Raymond Dasmann. Schweickart championed the Brown administration's efforts to use satellite imagery for environmental education. The atlas made good use of satellite photos to show watershed relationships and bioregional connections and illuminate this most important resource for any western state. Peter Warshall recalled the experience with some regret for the failure of the editors to take on the most controversial aspect of water and power in California but concluded that the book perfectly captured the "hydraulic civilization of the twentieth century" in brilliant pictures accessible to a popular audience.[98]

The space colonies issue only served as an exaggerated proxy for the longer debate between environmental pragmatists and nature purists while highlighting the persistence of a divide within the environmental movement that centered on humans and the products of their labor. Space colonies may seem like the worst example of the hopeless utopian aspects of Buckminster Fuller's outlaw area, or a strange regression by Brand back to his days as a spokesman for radically disconnected communitarian escapism, but in hindsight they were really just a short detour on a road of support for pragmatic technological developments resulting in environmentally friendly technologies that offered the best hope for the next generation to understand the environmental crisis they faced and craft viable solutions.

A review of the full run of *CQ* reveals a remarkable list of first publications of insights and theories that went on to enjoy extensive coverage in the mainstream media. In their aptly titled book *News That Stayed News,* editors Art Kleiner and Brand demonstrated the extent to which the publication had served as a springboard for discussions that remained relevant in the 1990s, when their collection was published, and continued to command headlines in 2007.[99] A key difference between *Whole Earth* and *CQ* was the way the two publications dealt with politics. In *Whole Earth,* Brand followed Buckminster Fuller's advice to avoid politics, and overt politics were not featured. In *CQ*, politics were fair game and often front and center. Clearly, *CQ* had a very different position than the catalogs on integrating direct political discourse into the magazine. It seems unlikely, however, that readers of the catalog, who tended to be an exceptionally thoughtful group, never noticed the politics that were embedded in the publication from the beginning. *CQ* only brought the politics to the center in a way that demanded attention and raised very interesting questions about the ways that the counterculture built on and reconfigured the landscape of western American politics. Brand was engaged in political life during this period in a way he never had been before, and *CQ* became a thoughtful voice for an emerging hybrid politics that came from the western counterculture and drew inspiration from the pragmatic environmentalism promoted by *Whole Earth.*

CHAPTER SIX

Free Minds, Free Markets

The consumer has more power for good or ill than the voter.
Stewart Brand, The Last Whole Earth Catalog, *1972*

In the summer of 2000, the *Whole Earth Review* devoted an issue to the theme of "Beyond Left and Right." Guest editor and longtime contributor Jay Kinney, best known for his remarkable illustrations for underground comic publications like *Arcade* and *CoEvolution Quarterly* (*CQ*) and as the publisher of the spiritual magazine *Gnosis,* launched the issue with an insightful overview of oppositional politics in America.[1] Among other things, Kinney argued for a "Third Way" in American politics, a hybrid of Left and Right that supported ecology, decentralization, and regionalism.[2] By the year 2000, reevaluations of traditional political configurations were common in the wake of President Bill Clinton's category-bending centrist politics, as were thoughtful, and often bitter, reconsiderations of the political legacies of the 1960s. The summer *Review* featured authors who worked in or advocated the middle landscape of alternative politics that were not clearly linked to either Left or Right as traditionally defined. Most interesting were two excellent summations of converging trends in ecological thinking and hybrid politics by Green Party cofounder Charlene Spretnak and the always insightful Stephanie Mills, who wrote about the Green Party and bioregionalism, respectively. Both authors were frustrated by how long it was taking for their ideas to make an impact on broader party politics and worried about the persistence of issues they had discussed for decades. These essays also stand out for their precise summation of how hybrid politics intersected with environmentalism leading up to the new millennium.[3]

Spretnak was particularly discouraged that "a savvy magazine like *Whole Earth*" was asking contributors to ponder the idea of moving beyond traditional political frameworks when she and her Green Party cohorts had spent two decades formulating a platform that, she argued, did just that. "Beyond Left and Right" was, in fact, the slogan of the Green Party.[4] The Green Party had some excellent showings in states like New Mexico in the 1990s, and Ralph Nader's 2000 presidential bid spotlighted the movement. Still, the U.S. Greens failed to capture the imagination of the voting public in the way the European Green Parties had. The obvious hurdle for the Greens in the United States was convincing liberals that a Green vote was not a wasted vote, that voting Green did not have to split the liberal vote. A less obvious reason that the Greens had failed to provide a Third Way of broad appeal was that their primary agenda rejected, for thoughtful and compelling reasons, the "ideologies of modernity" and particularly the centralizing tendencies of technologically driven capitalism. Likewise, Mills unpacked the political problems that she and her "Luddite cohorts" attributed to the "political bias" of the technologies "so pervasive that we never think to debate it, any more than fish debate water."[5]

Both Mills and Spretnak made compelling cases that few environmentalists would contest. Still, both of these arguments fell squarely within the framework of traditional environmentalism and were too closely aligned with the Left to capture the middle ground of pragmatic environmental politics fostered by *Whole Earth* over the years. The more iconoclastic perspective on Third Way, or hybrid political configurations, is the hip-Right "reversionary-technophiliac synthesis," as Theodore Roszak dubbed the fusion of counterculture enthusiasm with technology, environment, and market economics.[6]

No one better captures this world of hybrid politics and fusion of technophilia, environmentalism, and western regionalism than former Wyoming cattle rancher, Grateful Dead lyricist, and *Whole Earth* contributor John Perry Barlow. Except by Deadheads, Barlow is best known for his work as a pioneer for electronic freedom and his association with the Electronic Frontier Foundation.[7] He famously referred to cyberspace as the "Electronic Frontier" and was only one of many who framed the new world of Web-based economy and culture in terms of western history and the frontier mythology.[8] Barlow penned the classic libertarian statement on cyberspace—"A Declaration of the Independence of Cyberspace" (1996)—which was widely circulated on the Web and became a manifesto for free information and free markets.[9] A classically western libertarian, Barlow nonetheless passionately avoided traditional politics, arguing that "to engage in the political process was to sully oneself to such a degree that whatever came out wasn't worth the trouble put in."[10]

As cyberspace grew into a significant force, it spawned a particularly western political philosophy characterized by a "left-right fusion of free minds and free markets" melded with a strong environmental ethic that critics have called the "California ideology" or "cyberlibertarianism."[11] The term *California ideology* appears most often as a pejorative label applied by new and old Leftists who were deeply troubled by the libertarian turn to the Right by many in the computer world and particularly disturbed that this turn was so directly connected to the counterculture.[12] A more neutral and accurate label for the politics represented by Barlow and his cohort of western new frontier technophiles is "counterculture libertarians" or maybe even the "hip Right."[13] This new generation of western libertarians, unlike anarchists, recognized the paramount role of the state in certain spheres, but emphasized the invisible hand of free-market economics whether the product was drugs, computers, or land to be preserved.

The hip Right blended the individualism and liberal social values of the counterculture with a traditionally western distrust of big government and centralized authority. In so doing, they tapped into the persistent legacy of "Bull Moose conservationism," as Barlow put it, a legacy of western conservatism that Teddy Roosevelt embodied in his western life, if not in his Washington, D.C., policies.[14] This cohort exhibited an embrace of technology unique to their generation while rejecting the national orientation and emphasis on collective achievement that characterized the Right and Left. They valued individual agency over communal action and championed the free flow of information and access to tools as the best means of empowerment and change. They also incorporated a strong environmental ethic in their philosophy, with a special emphasis on the possibilities of technologically facilitated green living. Barlow's environmentalism resembles the position of free-market environmentalists who draw inspiration from Garrett Hardin's essay "The Tragedy of the Commons" (1968), in which he argued that common ownership of resources led to their decline. Free-market environmentalists believe that individual property rights provide better protection in the long run. The counterculture libertarians tended to lean to the Right of many in their generation but were hip in a way that the budding leaders of the western-based New Right could never comprehend or achieve. This countercultural evolution of western politics blurred traditional party lines and created some unusual political coalitions. Try, for example, to envision a relationship between the Grateful Dead and Vice President Dick Cheney and you start to see how weird these connections can be.[15]

Barlow had known Stewart Brand since the acid-test days in San Francisco, but his roots were deep in the ranchlands of windswept Wyoming, and his environmentalism came directly out of his experiences working his family ranch. Like many ranchers across the Rocky Mountain West, Barlow took conservation for granted; it was a way of life. "I was a rancher and also the head of the Wyoming Outdoor Council (WOC)," Wyoming's leading statewide environmental group. He was also a member of the Stock Growers Association, which was "constantly at war" with the WOC; "it was tough being a central member of both," he later recalled.[16] If ever there was someone who didn't fit within the confining boxes of Left/Right politics, it was Barlow. When he did participate in politics, his actions blended sensibilities that are often depicted as oppositional, even irreconcilable. Within the deeper context of western history, however, Barlow's politics made sense. Barlow was one of historian Michael Allen's "Cosmic Cowboys," who linked the counterculture with western traditions in ways that complicate our understanding of what separates liberal and conservative.[17]

John Perry Barlow at Finn Barn on his Wyoming ranch, 1975. Well known as a lyricist for the Grateful Dead and as a founder of the Electronic Frontier Foundation and an early information-age free speech advocate, Barlow was also an environmentalist with a long track record of grassroots activism in his home state, including four years as president of the Wyoming Outdoor Council. (Courtesy John Barlow)

For most of his life, Barlow was "quite comfortable being Republican." It seemed like the right thing for him, given the Republican Party's strong traditions of Progressive conservation. If one looks at a picture of a young Barlow dressed in his Levis and chaps, covered with dust from a hard day at the ranch, and slumped up against a weathered barn, his conservative politics fit the image. It is only when one remembers that Barlow was also a key contributor to the Grateful Dead, whose counterculture credentials rivaled Brand's, that the picture clouds our assumptions about identity and politics during the 1960s and 1970s. A look at Barlow's environmentalism further complicates the picture. Known for his music and as a cyberphilosopher, Barlow was also a significant western environmentalist who linked ranchers with hippies, providing another example of how often concern for regional environments has been a bridge between generations, political factions, and cultural groups in the American West.

The counterculture appears political in western history usually only when cast as a foil for conservatives like Ronald Reagan, Barry Goldwater, and Richard Nixon.[18] The stoned hippies of Haight-Ashbury seemed to have almost as little in common with serious New Left activists as they did with Richard Nixon and his crew-cut cabinet. These characterizations gloss over the intense, if stealthy, politics associated with adherents of the counterculture.[19] A more cutting critique of the counterculture comes from those who assume that their sensibility was nothing more than a sad sellout to savvy marketers who quickly co-opted the counterculture lifestyle and philosophy and turned it into a tool to get their hooks into the expanding youth market. In this telling, the counterculture is a frivolous false consciousness on the part of spoiled middle-class white kids and a distraction from real political contribution that helped cement the failure of the New Left and ultimately led to the rise of a powerful new western conservatism that enabled the careers of Nixon, Goldwater, Reagan, and the Bush dynasty. Dismissing the counterculture as the apolitical sellouts of the 1960s and 1970s misses the rich contributions this cultural mode made to politics and culture and leaves no room for protagonists like Brand and Barlow.

CQ exposed savvy readers to the false dichotomy between culture and politics on the Left, and, by presenting thinkers who leaned to the Right, raised important questions about who's in and who's not when describing the counterculture.[20] Brand's conservatism was often so subtle, and so counterintuitive, that it snuck up on people until he pressed their hot button with some issue that seemed out of synch with what many readers assumed was the liberal spirit of the *Whole Earth* publications but was in perfect keeping with Brand's personal philosophy. In one particularly eloquent diatribe, artist, reader, and contributor R. Crumb asked, "Which side are you on? Where do you stand politically?" Crumb's letter captured the dismay of some readers at Brand's willingness to take on conservative issues that flew directly in the face of cherished liberal tenants.[21] But the politics of *CQ*, as was the case with *Whole Earth*, were never clearly informed by a traditionally liberal sensibility.[22] Notwithstanding some harsh criticism from time to time that often came from friends, Brand never seemed overly concerned about the vagueness of his politics or the conservative viewpoints he highlighted much to the distaste of liberal readers. When the political discussions hinged on money and markets, Brand was completely consistent in his enthusiasm for business and unfettered capitalism as the best means for those like himself to make a meaningful contribution to American culture. When critics like R. Crumb were enraged by articles by Michael Phillips on topics like investing overseas, Brand could point back to the earliest issues of *Whole Earth* where he argued, "The consumer has more power for good or ill than the voter," and markets, investment, and commerce were always at the heart of the catalogs.[23]

Particularly in the West, the argument can be made that the counterculture entrepreneurs who skipped the protest movement and the polls were the ones who made the most lasting contribution to the politics of the last decades of the century.[24] Their libertarian hybrid sensibility was not a fringe movement and this pragmatic philosophy, as historian Patricia Nelson Limerick reminds us, captured the spirit of western myth and updated it for a new generation who searched for individualism and community reinvention through the electronic frontier of cyberspace, the promising world of alternative technology, the freedom of small business, and the individualistic everyday environmentalism enabled by thoughtful green consumption.[25]

In his role as adviser to California Governor Jerry Brown, Brand was able to provide political access for many of the influential environmental thinkers—those acting from within the counterculture or those who were exerting influence on the movement from outside it—whose work was published in *Whole Earth* and *CQ*. Brand arranged meetings between Brown and creative intellectuals like Herman Kahn and technologically enthusiastic environmentalists like Amory Lovins, as well as Gregory Bateson, Thomas Szasz, Marshall McLuhan, and Ken Kesey.[26] Brand sent a steady stream of iconoclastic intellectuals to Brown's office, and Brown shaped California environmental and social policy partly according to their recommendations. By the late 1970s, Brand's political cache enabled him to move from the fringes of western politics to the center of influence, or off-center, given the politics and policies of Brown's tenure. Brand was valuable as a political adviser because he was not a traditional politician or supporter of the traditional political process; he remembered, "I was able to work directly with Jerry Brown because I was out getting experience and not marching." He added that there were "lots of examples of counterculture businesspeople who became very successful and have influence in many different ways."[27]

California Governor Jerry Brown talks with iconoclastic intellectual Herman Kahn after a marathon late-night meeting in the governor's office of Brown, Kahn, Brand, and Amory Lovins. (Photo by Stewart Brand. Courtesy of Department of Special Collections and University Libraries, Stanford University Libraries)

To understand the political landscape of the twenty-first century and the American West in particular, you must look closer at the politics that simmered beneath the psychedelic veneer of the counterculture. These politics, as the activities of Brand, Barlow, and Brown illustrate, don't necessarily take a traditional form, and many of the individuals who made significant contributions to the evolution of politics in the West did so through alternative means. These westerners are hard to trace in politics because they often didn't vote or participate in political activism in any direct way. For some, they may not even fit within standard definitions of political, but behind the scenes they were building on many deep western political tendencies: the search for new frontiers, utopian desire for a new beginning, individualism, escape, and distrust of the federal government as problem solver.[28] They embraced individual agency and inventiveness above all and worked to open markets to innovative models of production and consumption. Figures like Brand and Barlow are examples of westerners who don't fit the mold of traditional political history, yet played a role in shaping the political landscape of the modern West. Likewise, innovative western entrepreneurs like Patagonia founder Yvon Chouinard and Point Foundation board member Paul Hawken expanded the reach of the Third Way in western politics as they created a new and extremely marketable vision of western authenticity that reconfigured the use of western mythology in American politics and culture and challenged some long-standing political alliances. Through their actions, not their advocacy, they brought a countercultural version of environmentalism and social responsibility into the boardroom and the marketplace.

Much of what has been written about counterculture libertarian thought focuses on the role of libertarian politics in the business and culture of the Internet; thus the term *cyberlibertarians.*[29] Counterculture libertarian thinking, however, was also evident in the contentious world of environmental politics. There was a grassroots libertarian strain of environmentalism that differed dramatically from the guardian model of government-legislated reform that so changed the landscape of the American West. Counterculture libertarian environmentalists like John Barlow focused their energies on grassroots and individual action and technologically sophisticated entrepreneurship to move environmentalism out of the wilderness and into the market and the home. Like Brand, these environmentalists were more likely to find inspiration from Frank Herbert's *Dune* or Robert Heinlein's *The Moon Is a Harsh Mistress* than from John Muir or Aldo Leopold.[30] As a rancher in Wyoming, Barlow could certainly identify with Heinlein's western themes of dependence and distant control so compellingly presented in his counterculture version of lunar colonies at the mercy of a distant government that undermined the rights of the individual and the property owner.

Libertarian-leaning environmentalism arose in part as a response to the guardian model of environmentalism that worked very well for protecting land but not as well as a modifier of consumption and regulation of quality of life. There is only so much centralized government can do to convince people to recognize their own self-interest in protecting the environment. Barlow worried about the influence of groups like the Sierra Club on government environmental legislation. He felt that the mainstream of the environmental movement was a fundamentally biased "urban professional class versus the rural working class," and, worse, he was convinced that environmentalism in the 1970s failed to "realize the extent to which human beings are a part of nature."[31] Echoing sentiments voiced by Brand and many contributors to *Whole Earth* and emerging insights from the nascent ecological design movement, Barlow also felt alienated from the preservation and wilderness movements.[32] Like previous generations of conservationists, Barlow was perfectly comfortable with the weaving of consumption and recreational use of the land into the grazing and extractive uses he practiced. As someone who knew nature through work and play, both closely linked to market economics, Barlow distrusted environmental advocates who failed to take into account the ways that their use of nature was also enmeshed in markets, technological development, and culturally defined recreational practices.[33]

The fundamental conundrum for American environmentalism has always been the tension between capitalism, consumption, and the desire for environmental health. Despite the remarkable successes of mainstream environmentalism, it has never succeeded in resolving the problems with American materialism; environmentalism requires government regulation but also personal responsibility. Environmentalists have succeeded to a remarkable degree in generating support for federal regulation on behalf of the environment. Despite this, pragmatic solutions that enable individuals to reconcile, but not abandon, their desires for technologically enhanced living with their wishes for a sustainable and ecologically viable future have been few and far between. For technologically minded counterculture innovators like Barlow and Brand, much of what traditional environmentalists advocated just did not make sense, leading them to seek a Third Way.

Exploring the Third Way in politics, environmentalism, and technological development was one of the things that *CQ* did exceedingly well. *CQ* consistently shed light on gray areas and the innovators who lurked there, often unrecognized by the mainstream media or academic world. The debates about technology, politics, and markets in *CQ* presaged the emergence of the concept of *sustainability* that entered the larger environmental discourse in 1987 through the United Nations–sponsored Brundlandt Commission that focused on reconciling economic growth and environmental health.[34] Like that of *Whole Earth*, *CQ*'s sharp focus on design and designers showed how sustainability and "right livelihood" was not just a possibility for the future but a reality for individuals living and working within the mainstream in the 1970s. Some of the most subtle thinking about the nexus of environmentalism, technology, design, markets, and culture emerged from the quarterly's "Nomadics" section. Nomadics, along with soft tech, were J. Baldwin's territory, the zone where politics took a backseat to tools and the most practical information about how to use them. Unlike Brand, Baldwin was steadfastly nonpolitical even as *CQ* took a political turn. Baldwin preferred to focus all of his prodigious energies on design problems. He was particularly interested in recreational design and pushing and promoting the outlaw edges of technologies that facilitated outdoor recreation. One of the most frequent criticisms of the counterculture was its obsession with leisure and play. But J. Baldwin was not countercultural in any sense other than as a fellow traveler, and a close look at the "Nomadics" section reveals that even beneath the quest for leisure, recreation, and "New Games" lurked a promising effort to provide a framework for pragmatic ecological living. As Hal Rothman pointed out in his detailed Bay Area recreation study, *The New Urban Park: Golden Gate National Recreation Area and Civic Environmentalism,* San Francisco and its remarkable collection of urban open spaces exemplified changing perceptions about the relationship between recreation and public lands during a period when "the nation grappled with urban uprisings, and empowered constituencies," searching for positive outlets for urban aggression, and the *Whole Earth* crew were in the vanguard of this change.[35]

Nomadics

The history of American outdoor recreation in the twentieth century is, to a certain degree, a history of leisure-tool enthusiasm. Outdoor sports have always wielded influence inversely proportionate to their purity; in other words, sports with lots of cool toys tended to gain power because of the wider appeal and the marketing clout of outfitters. The tools of recreation linked constituents to the market and gave certain groups greater access to political power than the purists enjoyed. The mainstream environmental movement has been so influential that we tend to assume that most Americans must have envisioned a disconnect between nature and technology; however, in reality, the natural and technical worlds were very much entwined in American thinking, and *Whole Earth*, despite its seemingly eclectic mix, appealed to that latent connectedness—it struck a chord precisely because the incongruity championed by the wilderness crowd was not actually the mainstream of American thought. Because we assume that only environmentalists (and especially wilderness advocates) spend a great deal of time thinking about nature, we tend to make assumptions about the role wilderness proponents played in shaping American thinking about the relationship between nature (as sublime and natural) and technology (as intrusive and artificial); but, actually, most Americans probably think about nature a good deal and especially about ways to avoid being killed by it, enjoy it, adjust to it, and master it, and *Whole Earth* was a needle entering a rich vein of ubiquitous everyday nature thinking.

American tourists and outdoor recreationalists, in particular, have long embraced the very ambiguities and contradictions between technology, nature, and consumption that the history of *Whole Earth* reveals.[36] And yet, in 1968, the same year that *Whole Earth* was founded to foster connections between technology, culture, and nature, the American environmental movement was beginning to distance itself from the recreationalists and alternative technologists who constituted a key constituency of the movement throughout the twentieth century. The beginnings of the break with recreation coincided with the culmination of the most dramatic expansion of recreational tourism in American history and the rise of the outdoor sports technology industry.[37]

In keeping with the well-documented counterculture proclivity to use consumer culture as a means of resistance to mass culture, *Whole Earth* celebrated the symbiosis between technologically enhanced recreation, thoughtful consumption, and nature appreciation through its "Nomadics" section.[38] Dick Raymond, *Whole Earth*'s patron saint, started his career as a recreation economics consultant at the Stanford Research Institute, and "Nomadics" section editor J. Baldwin began his career in industrial design building tents for Bill Moss.[39] Recreation and leisure loom large in the history of the counterculture and play a significant role in shaping the cultural context for environmentalism in the 1960s and 1970s.[40] It was no coincidence that Baldwin, the publication's longtime technology expert and editor, was also the editor of the "Nomadics" section. As an early recreational designer, Baldwin knew firsthand about the relationship between recreation tech and environmental concerns. A close look at the "Nomadics" section over the years reveals strong support for a diversity of recreational activities geared more toward less mobile urban populations who were unlikely to be participants in the Sierra High Trip school of Bay Area recreation or have the time, money, or inclination to spend weeks hiking the John Muir Trail.

Historian Joseph Taylor has observed "an emerging tension within environmental culture" in 1968.[41] The tension manifested in several ways, but centered on the growing split between those who embraced the ambiguous area of American culture that accepted that human technology, human culture, and ecology were inexorably linked and those who wished to draw the line between nature and culture more sharply. For the environmental movement in 1968, this debate emerged most clearly in the discussion of new trends in outdoor recreation and tourism, which were supplanting resource extraction as the primary interaction between the American public and the public lands. Some in the environmental movement, like Sierra Club director David Brower concluded that the outdoor recreation culture of the late 1960s, "in which marketing and play go hand in hand," no longer fit with the politically energized environmental movement.[42] Stewart Brand provided a counterargument to this view through all of his publishing ventures but even more directly with his work on the New Games Tournaments.

Unlike the Bay Area recreational model of the Sierra Club that encouraged members to leave the city to commune with and recreate in the nature of the Sierras or elsewhere, *Whole Earth* provided a forum for those ecologically minded westerners who wanted to find nature closer to home within the city or its immediate hinterlands.

In the spring of 1973, when Brand was preparing to revive the *Whole Earth* with the *Epilog*, he also began planning a major event exploring new forms of counterculture-inspired outdoor recreation.[43] The "Nomadics" section of *Whole Earth* had always reflected Brand's intense interest in recreation. Prior to the publication of the first *Whole Earth*, Brand designed an event titled "World War IV," a scenario game for "the peaceniks I was dealing with" who, according to Brand, "seemed very much out of touch with their bodies in an unhealthy way."[44] "World War IV" reflected Brand's long interest in games and his desire to explore how creative play might be used to release aggression without real conflict. Later, his work chronicling the rise of the early computer game *Spacewar* for *Rolling*

Stone elicited a long run of thoughtful pieces on the role that recreation and games play in shaping culture and quality of life. Both of these experiences convinced Brand to explore alternative recreation further as a means of working on cultural differences, increasing awareness of environmental issues, bringing together diverse groups of actors, and, of course, having fun.[45]

Brand spent a considerable amount of his Point Foundation time and money exploring this evolving notion that he called "Softwar," a "conflict which is regionalized (to prevent injury to the uninterested), refereed (to permit fairness and certainty of a win-lose outcome), and cushioned (weaponry regulated for maximum contact and minimum permanent disability)."[46] He discussed these ideas with friend George Leonard, author of the 1970s classic *The Ultimate Athlete* (1974).[47] Leonard was a leading advocate of alternative sports and an enthusiastic supporter of Brand's idea for an alternative recreation event in the Bay Area. All of this work on recreation led to the idea for the New Games Tournament, an event intended to explore new forms of outdoor recreation, with a few video games thrown in, and encourage the Bay Area community "to relate to its natural environment in a new and creative way."[48] More than any other single event, New Games exemplified the nexus of thinking about economy, sport, pragmatic environmentalism, and social activism that was emerging from Northern California in the early 1970s and the granting efforts of the Point Foundation and the network of *Whole Earth* publications that supported it.[49]

The New Games Tournament was held on two consecutive weekends in October 1973 at the Nature Conservancy's 2,200-acre Gerbode Preserve, just north of the Golden Gate Bridge. The event was sponsored by the Point Foundation, with funds coming from Brand and Huey Johnson.[50] Brand had encountered some resistance to proposals to hold events, like a series of *Whole Earth* jamborees that became a fondly remembered Bay Area tradition, in Golden Gate Park from the newly formed Golden Gate National Recreation Area Citizens Council and turned to Huey Johnson's Trust for Public Land to provide a large open space within easy striking distance of the city.

On those two weekends of the first New Games Tournament, over six thousand Bay Area adventure seekers paid $2.50 to gather in the beautiful Gerbode Valley for a remarkable celebration of alternative nature recreation. Onlookers perched on the rolling hillsides could watch hang gliders silently swooping over crowds of players intently balancing on beams while being pummeled with gunnysacks in a game of Tweezli-Whop, or head-banded and bearded participants diving for a high-flying Schmerltz. Large groups, sometimes numbering in the hundreds, played Slaughter, Caterpiller, Vampire, Hagoo, Planet Pass, Orbit, and Siamese Soccer. Parachutes fluttered in the wind around smaller groups working on inventing their own new games or quietly playing the Native American–inspired

Stewart Brand meditates on the Earth Ball while attempting to stay aloft for as long as possible at the New Games Tournament, October 1973. (Ted Streshinsky/CORBIS)

Bone Game. Nothing drew more of a crowd than the game that became the icon of the New Games movement: Earth Ball.[51]

At six feet in diameter, painted as a globe by prankster artist Roy Sebern, the giant rubber and canvas Earth Ball was the focal point of the New Games Tournament. Participants used the ball for many games, from balancing on it to crawling under it. But mostly the ball was the object of rushing crowds who hoisted and threw it, chased it, rolled it, and piled on it. The primary Earth Ball activity was a loosely structured game where two sides worked to push the giant ball across a boundary marker with the goal of keeping the game going as long as possible; there were no winners in the traditional sense. Participant William deBuys remembered having the time of his life while wallowing in the mud at the bottom of a pile of jubilant New Gamers under the Earth Ball.[52] Prior to the event, the ball drew curious onlookers as it was slowly inflated at a San Francisco gas station. Then, large crowds chased the bounding ball down San Francisco streets, where onlookers kept joining the parade, a trend that came to be known as *snowballing*, building a crowd like one builds a snowball. Kids in particular loved the Earth Ball, and versions of the ball and the noncompetitive sports ideal it represented found their way to gym classes all over America in the late 1970s.

The New Games Tournaments were a major success and received significant national news coverage, including a live broadcast with Walter Cronkite and a feature article in *Sports Illustrated*.[53] The sight of a large bunch of hippies, grandmotherly types, children, and a nice multiracial mix of Bay Area citizens playing weird new games in a beautiful Bay Area setting was a no-lose proposition for the media. The coverage, for the most part, however, did not parody and succeeded in capturing the thoughtful ideas that drove the event. *Sports Illustrated* reporter Keith Power carefully explained the unusual event, Brand's motives, and the countercultural context for a reevaluation of sport and urban open-space use. Although he concluded, "This is obviously not the kind of constructed diversion apt to catch on with the average American in pursuit of New Leisure," he nonetheless captured the spirit of an event that was more than a stunt.

New Games exemplified the convergence of alternative sport, environmentalism, and urban social activism that drove the granting efforts of the Point Foundation and provided the intellectual framework for a generation of Bay Area counterculture environmentalists who used recreation as a spearhead for a pragmatic everyday environmentalism. Community organizer and New Games planner Pat Farrington made it clear that "one of our major goals is to make people aware of public lands and promote using them in an ecologically sound manner."[54] The New Games and Point Foundation efforts in the Bay Area during the mid-1970s highlighted the ways that technologically enthusiastic environmentalists were building new coalitions of recreationalists and conservationists while

Stewart Brand (in top hat and buckskin shirt) lays out sticks for the "Bone Game" at the New Games Tournament, October 1973. Sponsored by the Point Foundation, the New Games Tournament was widely covered in the national media and showcased the innovative environmentalism of Point. (Ted Streshinsky/CORBIS)

reaching out to new urban constituencies long ignored by the mainstream environmental movement.

Not all approved of the New Games scope, and some worried about the enthusiasm for technologically oriented new sports that were changing the tone of open-space use in the Bay Area during the 1970s. For those accustomed to viewing the Marin hills free of crowds, or quietly walking or hiking the valleys, New Games might have seemed like an invasion. New Games also brought with it lots of toys and tools, from hang gliders to the Earth Ball and big colorful parachutes, causing the "bare hills of Gerbode, covered only with the parched grasses of summer," to "burst into color."[55] Most of the technologies deployed during New Games, however, were simple and left little lasting impact on the land. Huey Johnson thought the potential for New Games to link the preservation of wild and scenic lands surrounding San Francisco with innovative community building addressing pressing urban social and environmental issues of the early 1970s easily outweighed concerns about new and expanded use. Further, the hang gliders and other simple tools for New Games were in perfect keeping with long-established trends in technologically facilitated nature use and were only more obvious because they were being deployed in the new urban parks, where their use was visible to millions.

While David Brower and the Sierra Club were worried about the impact of dramatically expanding interest in outdoor recreation and the sports-marketing revolution it fueled, federal land-management agencies like the National Park Service were working, as they had since 1916, to reconcile preservation and recreation in keeping with the trends of the day. Under the leadership of Secretary of the Interior Stewart Udall, the National Park Service shifted responsibility for recreation management to a new agency, the Bureau of Outdoor Recreation (BOR), created in 1962 as part of a larger federal effort to rethink outdoor recreation and the public lands with a special emphasis on providing urban recreational opportunities.[56] The newly formed BOR played a particularly significant role in the Bay Area where Point Reyes National Seashore had been recently established and the groundwork for the National Park Service concept of the urban park was coalescing around the spectacular Golden Gate open spaces that formed the heart of San Francisco and Marin County. The BOR expressed early interest in helping promote the New Games Tournament as an example of the type of innovative urban recreation the agency was created to foster.[57] The involvement of the BOR in New Games and favorable press in the agency's publication *Outdoor Recreation Action* helped legitimize the event and its countercultural version of urban outdoor recreation.

The subtle conflicts surrounding New Games and other similar events like the *Whole Earth* jamborees demonstrated the persistence of conflicting ideas about what constituted environmentalism even among groups with similar concerns about the same place. Specifically, the New Games Tournaments highlighted differences between the urban-centered and human-centered environmentalism of *Whole Earth*, the Point Foundation, and the national environmental groups that were late in recognizing the importance of making outdoor recreation meaningful and viable for diverse urban populations.[58] It also demonstrated class and generational conflicts within the Bay Area community of outdoor enthusiasts. Supporters of hiking, walking, and traditional forms of urban and suburban leisure often reacted against emerging forms of alternative recreation like mountain biking, hang gliding, and mass participatory events, like the giant tug-of-war or Earth Ball mayhem that New Games facilitated. Although traditionalists might have viewed New Games as nothing more than a weird distraction from nature communion, the event was a harbinger of critical changes in the recreational economy and the culture of outdoor sports.

The second New Games Tournament, held a year later, and again sponsored with Point Foundation funds from Huey Johnson and Stewart Brand, advertised in Spanish and Chinese and actively sought to move far beyond the normal constituency for outdoor recreation and bring together players "across diverse social, ethnic and economic backgrounds" so that they might better relate "socially through play."[59]

Huey Johnson articulated the environmental importance of the event nicely. "As a professional environmentalist," he wrote, "I feel it is urgent that we reach the urban center. Our origins are WASP in funding, energy and direction. If there is one lesson we have learned in the last few years, it is that inherent in the meaning of 'ecology' and 'human ecology' is the awareness that environmental relationships do not stop at 'boundaries,' be they nations, wilderness areas, cities or neighborhoods."[60] Johnson was one of the most consistent and eloquent voices for this view of his generation of environmentalists. He refashioned the earlier efforts of conservationists such as Arthur Carhart and Edward Hilliard, who worked against the trend in environmental advocacy to compartmentalize land and activities and sought instead to craft an environmental ethic that could appeal to a broader spectrum of American society.[61] In keeping with the mission of his new organization, the Trust for Public Land, Johnson emphasized the importance of shifting the focus of the environmental movement to the urban center, which he thought would "give a much-needed diversity" and "modernize a movement still based on the early conservation philosophy of Thoreau and Leopold."[62]

Although never a specific goal of the New Games Tournament, marketing of outdoor recreation to a new constituency and reconfiguring long-standing trends of recreational consumption had great potential for expanding environmentalism and providing avenues for individual agency through conscientious recreational consumption. J. Baldwin made this point consistently through his series of articles on recreational design in *CQ* over the years. Johnson articulated the argument from inside the environmental movement and understood, ahead of many, that the "city dweller feels removed from the land," and traditional environmental issues like wilderness were too often an abstraction for the bulk of urban Americans. "What is needed is awareness," he argued, and building on the long-standing links between technologically enhanced recreation, consumption, and ecological awareness was one way to accomplish this goal. The New Games movement peaked in the Bay Area with the second tournament but spread quickly across the nation and, by 1976, became an international trend spawning several popular books and significant changes in the way recreation was perceived by a new generation of outdoor enthusiasts. Interest in New Games faded in the 1980s but not before leaving a legacy of enthusiasm for new forms of urban outdoor recreation, many of which endure into the 2000s and can be seen daily in urban parks across the nation and in the recreational economy.

New Games was a significant factor in the changing landscape of outdoor recreation in the 1970s. The work of emerging companies like Patagonia, presented to readers of the *Last Whole Earth*, was even more significant. In the winter 2004

Patagonia company catalog, sandwiched between a colorful layout of women's Capilene high-performance Tech Panties and a page dedicated to the new PCR Get Down Jacket made from long-filament polyester fabric from recycled plastic bottles, you can find a thoughtful article by Yvon Chouinard, the cofounder and owner of Patagonia, "On Corporate Responsibility for Planet Earth." Beginning with a quote from the late environmentalist and longtime Sierra Club president David Brower that sensibly warns, "There is no business to be done on a dead planet," Chouinard outlines his philosophy as an alpinist, surfer, fly fisherman, environmentalist, and wildly successful business man.[63] Chouinard's first catalog came out in 1972, the same year as *Last Whole Earth Catalog* and was clearly modeled on Brand's most popular publication.[64] The *Last Catalog* featured Chouinard and Tom Frost's young company in the "Nomadics" section as a source for high-quality, high-tech mountain gear but also a source of history and environmental philosophy.[65] Like *Whole Earth*, it was much more than a listing of products; it was a work of art and an articulation of an emerging philosophy of outdoor recreation and environmentalism.[66]

Chouinard in particular, along with several business-savvy partners, including climbing pioneer Tom Frost, linked extreme sports, environmental advocacy, and consumption in a manner similar to *Whole Earth* but aimed at a very different audience. Together they founded the wildly successful outdoor apparel company Patagonia, and changed the dress code for the new West.[67] Slightly older than Stewart Brand, Chouinard spent a good part of the fifties and sixties living a bohemian dropout life in Yosemite National Park's legendary climbers hangout, Camp 4. During this period, Camp 4 was full of young men and a few women who had given up on materialism and headed to the mountains in search of "authentic experiences." What separated this generation of wilderness truth seekers was their decidedly entrepreneurial genius.[68]

It was while he was living in Camp 4 during the 1950s and into the early 1960s that Chouinard got his start in business. The dusty picnic tables of Camp 4 produced no less than three founders of internationally successful corporations during the fifteen-year period between 1958 and 1973 alone. These legendary pioneers of rock climbing, who were the first to scale the seemingly impossibly vertical granite of Yosemite's El Capitan, are significant for their contributions to the worldwide evolution of rock climbing as a sport, as well as for their technical innovations and contributions to a major economic revolution in outdoor equipment and apparel.[69] Chouinard, Frost, and Royal Robbins all founded companies that went on to great success and helped create the multi–billion dollar outdoor sports industry. Rock climbers like Royal Robbins and Chouinard are examples of bohemian extreme-sports enthusiasts who turned their passion into successful businesses and set the model for the counterculture entrepreneurs who followed.

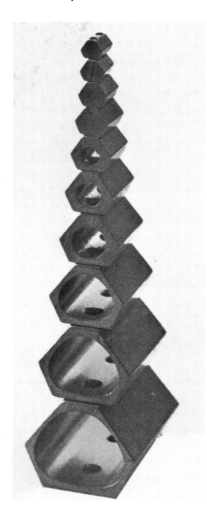

Artfully displayed lightweight Hexentrics in the iconic 1972 Chouinard Equipment Company catalog. Hexentrics replaced rock-damaging pitons and enabled the clean climbing revolution that Chouinard and Frost celebrated in their classic statement on environmentally responsible recreation, "A Word," which prefaced the beautiful photography and innovative gear that followed. Inspired by *Whole Earth,* the Chouinard catalog set the standard for the outdoor industry. (Photo by Tom Frost)

Patagonia, which started as the Great Pacific Iron Works with $600 Chouinard borrowed from his mother, was the quintessential garage business running, even after great success, out of Brand's beloved "low road" buildings.[70] Dissatisfied with the quality of European pitons (metal spikes that climbers use), Chouinard started making his own high-quality "chromolly" units in his garage for his own use. Word spread and soon the demand grew and a business was born. Chouinard expanded his operation to include clothing and formed two companies: Chouinard Equipment and Patagonia. Both of these companies were successful almost from the start, in part because Chouinard found a real need and filled it, but more importantly because he built his businesses around a powerful set of political and social concerns. Early on, it was "clean climbing," the idea that rock climbing and other outdoor activities had to take care of the resources they used and do as little damage to the rock as possible. Clean climbing was a revolution that reshaped the sport worldwide and opened the door for the mass marketing of what had been up to that time a fringe sport for eccentrics and dropouts.

The clean climbing revolution had implications beyond the climbing community. Chouinard developed a host of thoughtful and environmentally sensitive products for climbers. More significantly, he was among a pioneering group of American businessmen who, in the 1970s, built a business philosophy that united environmentalism, outdoor sports, social responsibility, libertarian-leaning views on government, and huge profits. Like Brand and the counterculture entrepreneurs who followed them, these climbers were socially liberal, distrusted the government, and had very little desire to work for any traditional political movement.[71] The politics of the climber/entrepreneurs, like that of the ecological designers, was embedded in their engineering and marketing. Chouinard, for example, worried that the successful technological developments of the postwar period had made access to the rarefied cliffs of Yosemite a little too easy and environmentally harmful. He argued that "no longer can we assume the Earth's resources are limitless; that there are ranges of unclimbed peaks extending endlessly beyond the horizon. Mountains are finite, and despite their massive appearance, they are fragile."[72] He helped reinvent his sport and in the following decades infused his hybrid environmental politics into the apparel industry by transmitting his message to a large and willing audience through his artful catalogs. Chouinard's is just one example of a green business model that shaped American consumerism in the late twentieth century.[73]

By the 1990s, Camp 4 alumni Frost, Chouinard, and Robbins wielded considerable political power and used their influence to help preserve the park they had grown to love as disheveled climbing bums. Like many of their generation, they had used their disengagement from politics very productively and found themselves moving to positions of political power from the most unlikely of trajectories. Like Brand, it was their lack of participation in traditional politics and their disengagement from the traditional political process during key periods in their lives that gave them political power and influence later in their careers.

Central to Chouinard's notion of corporate responsibility was a faith in the value of individual and corporate agency over centralized hierarchy. "I don't trust my government," Chouinard argued. "I support the front-line activist, the river keepers and tree sitters who save a single patch of land or stretch of water." As the owner of a successful corporation, Chouinard could put his philosophy

Chuck Pratt (left) and Yvon Chouinard, testing a prototype hauling bag in Yosemite National Park's Camp 4, 1969. (Photo by Glen Denny)

into action, most notably with his "1% for the Environment" campaign that encouraged corporations to contribute tens of millions of dollars to environmental causes in the 1990s and early years of the new century. This independent thinking and desire to focus on grassroots rather than institutional or government reform is in keeping with Patagonia's history as a maverick western company that never fit any standard business model. From the beginning, Patagonia used cutting-edge appropriate technologies wedded to liberal social values and a strong environmental ethic to create a fabulously successful market niche and loyal consumer base. According to cofounder Tom Frost, "Our business activities mirrored our rock climbing philosophy that emphasized the style and purity of the activity. Keeping the products simple and pure with designs that came from nature and worked with nature drove our business model."[74] While building their company from a tiny garage operation that produced high-quality specialty hardware for rock climbers into an international success and flagship of green consumption, Chouinard and Frost blazed a path for an eclectic assortment of counterculture entrepreneurs who followed them.

Patagonia's philosophy, evolving during the 1970s, provides a good example of a distinctly countercultural model of consumption, right livelihood, and business activism that is not often discussed as a part of western history. Although Chouinard and his generation of outdoor entrepreneurs probably would not consider themselves countercultural (they came of age mostly in the bohemian 1950s), his politics mesh nicely with Barlow and his counterculture libertarian cohort and with the model of ecologically thoughtful and market-savvy recreation that Brand was promoting through his New Games Tournaments. These efforts helped push trends in fashion and culture, creating a new iconography of western regional authenticity.

Just after Lyndon Baines Johnson rode his ranching roots into the White House and while Ronald Reagan was gearing up his conservative western political imagery, a growing segment of the western population was putting on their first Patagonia Synchilla jacket made from recycled plastic soda bottles. By the mid-1980s, for many westerners putting on the *Pat* took the place of putting on the cowboy hat. Yvon Chouinard added a new wardrobe to the closet of western politics in the process, creating a new western iconography that rivaled the success of the cowboy hat–wearing traditional politicians. Chouinard's high-tech clothing, tested in the harshest environments on Earth, became a political statement for millions who never intended to use the clothes for anything more than a trip to the store. The burgeoning outdoor industry began clothing a legion of new westerners in outfits as laden with political and cultural symbolism as the western wear that preceded them.

The Money Tool

The outdoor sports industry exploded in the late 1970s, and the American West was the focal point for much of the recreation and the business.[75] The new western imagery of Patagonia tapped into western mainstream liberalism in a powerful way. At the same time, these trends pointed to a continued blurring of the lines between conservative and liberal. The Left/Right politics of counterculture libertarians coupled with thoughtful consumerism, innovation, and business acumen found its finest expression on the Technicolor pages of *Wired* magazine.[76] In *Wired*, the libertarian new western politics reached its zenith of influence during the dot-com boom of the 1990s when flip-flop–wearing Silicon Valley CEOs (chief executive officers) crashed the gates of the corporate world, average Americans felt empowered by purchasing business machines, and a powerful new voting block of what conservative critic David Brooks called Bobos (bourgeois bohemians) ushered in a new era of mass consumption.[77] Brooks viewed the orgy of technonatural consumption of the 1990s as further proof that the counterculture was a fraud and that its adherents were dupes who didn't understand that consumption was consumption whether it was BMWs or bamboo floors for your home yoga gym. But this view missed the entrepreneurial spirit built into the fabric of the counterculture from the beginning and overlooked that the consumption trends of the 1980s and 1990s were not evidence of a liberal sellout as much as an example of the extent that hybrid Left/Right counterculture politics had always played in the countercultural sensibility.[78]

The marketing savvy and business genius of the counterculture have been fruitful topics for discussion among historians and cultural critics. Much of this literature is critical of the counterculture "sellouts" who traded in their souls for a buck or critical of the cynical marketers who stole the soul of the counterculture and used it to hawk Nike sportswear to poor kids and yuppies.[79] Of these, Thomas Frank's *The Conquest of Cool* (1997) stands out as the most thoughtful. Frank deftly highlights the ways that counterculture from its earliest days mixed perfectly with capitalism. There were many in the counterculture who forcefully asserted their business interests and forged an alternative business network that built a strong foundation for future political activism. Bohemian pacesetters like Yvon Chouinard and a generation of counterculture businesspeople who followed were very consistent in their drive to make products and deliver services that they believed in and that they thought could make a difference in the world.[80] As Stewart Brand related later, "As they followed the mantra 'Turn on, tune in and drop out,' college students of the '60s also dropped academia's traditional disdain for business. 'Do your own thing' easily translated into 'start your own business.'"[81] These new entrepreneurs stuck with their ideals as they moved away from traditional politics and toward business ventures built on values that enabled them to work toward political goals through alternative means. Young entrepreneurs, Steward Brand insisted, "brought an honesty and a dedication to service that was attractive to vendors and customers alike. Success in business made them disinclined to 'grow out of' their countercultural values, and it made a number of them wealthy and powerful at a young age."[82] This history suggests that in western politics dropping out was perhaps more effective than tuning in.

The particular brand of western green consumerism and entrepreneurial enthusiasm epitomized by the *Whole Earth Catalog* was very effective in enabling its participants to reinvent powerful western symbols and negotiate or advance a hybrid Left/Right political agenda via the market economy. The political realism of green consumption and the emerging rhetoric of sustainability provided an easily digestible version of western Left/Right politics—individualistic, rugged, cool, hip, antiestablishment, and yet still closely linked with the traditions of the region. The focus on lifestyle and intellectual exploration by Stewart Brand and his generation of creative entrepreneurs and thinkers provided an alternative model for activism far different than the contentious union liberalism and polarizing protest movements that characterized much of liberal politics in the West during the twentieth century and sent many middle-of-the-road westerners scurrying toward Ronald Reagan and the New Right.

Convincing American producers and consumers that there was a middle ground between capitalism and environmentalism was no easy task. Some good examples of those working toward this reconciliation include Paul Hawken, Amory Lovins, and Hunter Lovins, who provided a captivatingly simple model for this reconciliation in their influential book *Natural Capitalism* (1999).[83] Building on Hawken's earlier work in *The Ecology of Commerce* (1993) and *Growing a Business* (1987), they argued that an environmental ethic based on realistic use of existing appropriate technologies was the key not only to the health of the planet but also to the future of corporate success and profitability. Of course, what these contemporary authors proposed is not without precedent. In some ways, their model of natural capitalism harkens back to the "gospel of efficiency" of Gifford Pinchot, Teddy Roosevelt, and the Progressive conservationists. Natural capitalism and the pragmatic ideal of sustainability it supported harkened even further back to Thomas Jefferson and, in particular, Ralph Waldo Emerson. Emerson, as Richard White explains, "reconciled nature with the busy, manipulative world of American capitalism. He reconciled utilitarianism with idealism."[84] Unlike earlier proponents of the "Machine in the Garden," however, natural capitalists argue more stridently that environmentalism is best left to individuals and corporations who will use the free market to correct environmental waste and abuse.[85] Much of the political debate about how to best reconcile commerce and environmentalism had played out in the pages of the *Whole Earth Catalog*s.

While *Whole Earth*'s readers learned about social, cultural, and technological alternatives, they also got an implicit and explicit lesson about green capitalism and green consumerism. Starting in 1968, *Whole Earth* and spin-off publications made significant contributions to the reevaluation of capitalism, consumerism, technology, and the environment. In the pages of the *Whole Earth Catalog,* appropriate technology, advice on business, and counterculture politics happily and effortlessly commingled. Had Daniel Bell studied the ideal of capitalism espoused in the catalogs or, more importantly, embodied by the project of making and selling the catalogs, he might have found a rough model for resolving the *cultural contradictions of capitalism.*[86] In the Left/Right world of counterculture libertarianism, there were no contradictions of capitalism; it was all a part of the same sensibility.

With hindsight, *Whole Earth* was clearly at the forefront of a new trend in American business, a trend that resonates in our own time. Brand was a pioneer in the greening of American business, and his business ventures were a harbinger of a new political calculus at least two decades ahead of its time. On the pages of *Whole Earth* and in his own books, Brand articulated a world of counterculture capitalism while many of his contemporaries were still advocating a socialist revolution. Brand was not alone in business or philosophy. The founders of Ben & Jerry's, Apple Computers, Smith & Hawken, Williams-Sonoma, and Patagonia all shared parts of this vision. Those who created these successful companies shared some direct connection or general affinity for the countercultural sensibility and used their knowledge of cultural trends to create powerful corporations, find new market niches, and reshape the American economy and American business.[87]

In the 1970s, scholars like Daniel Bell argued that capitalism was on the verge of collapse, doomed to destroy itself because the tenuous balance between consumption and production was becoming unsteady as the Protestant ethic faded in the hedonistic climate of the day.[88] In hindsight, it is easy to see how one might have failed to perceive a future where capitalism would be reconciled with liberal social values, environmental concerns, and hedonistic self-expression. Similarly, it was pretty clear by the late 1970s that the age of the organization man was on the wane but not clear to most what would take his place. While many liberals wrung their hands over the hedonism and seeming lack of political engagement of the counterculture, Brand, Chouinard, and a host of innovative entrepreneurs had already created a model for integrating their social politics within the existing framework of the capitalist market.

Reading the first few *Whole Earth Catalog*s, you will find nothing less than an attempted reconciliation of nature and capitalism, freedom and safety, technology and environment, rural and urban, adventure and domestication, and the holy and the profane. Central to this encompassing reconciliation was a casual acceptance of capitalism. Long before many in his cohort, Brand was willing to utter, in a positive way, the most profane word of his generation: *money*. "So along with shit, fuck . . . and the rest, I wanted to say among my friends money, not to swear but to honor its function."[89] This willingness to be open about the business and capital side of his endeavor provided a powerful, and often emulated, counterculture business model. Brand and *Whole Earth* helped invent the "weird hybrid zone where creativity and commerce intersect" that epitomized the dot-com boom of the 1990s and the parallel greening of American business.[90] While they worked without shirts and played volleyball every noon, the crew at *Whole Earth*, like their Silicon Valley colleagues who followed, were deadly serious about achieving their goals and thought nothing about putting in eighty-hour workweeks.

Whole Earth and its business model provided inspiration for a new generation of business leaders. The influence of *Whole Earth* on significant business trends is pervasive enough to argue that it is in the realm of business where *Whole Earth* has left its most lasting mark. Brand was well aware of this significance, commenting, "The effects of this sub-economy could become even more consequential than the subculture has been. Aquarius and its independent life-support system saving the crew when the command module malfunctions."[91]

Green consumption and the desire for environmentally friendly alternatives drove successful national efforts to change American business practices without altering the fundamental economic system. One of the most significant of these efforts was the 1989 grassroots "McToxics" campaign aimed at McDonald's restaurants' use of Styrofoam packaging and organized by Love Canal activist Lois Gibbs's Citizens' Clearinghouse for Hazardous Waste.[92] Gibbs and her organization convinced American consumers that it was unacceptable that a corporation like McDonald's would dump 1.3 billion cubic feet of CFC (critical flocculation concentration)-laden styrene foam into landfills each year. This effort soon went national with consumers engaging in boycotts and "send-it-back" efforts against McDonald's franchises that refused to switch to cardboard and paper packaging. Ultimately McDonald's capitulated, and the foam was replaced with paper wrappers. The McToxics effort illustrated the degree to which environmentalism had successfully infiltrated consumer culture and provided a model of an alternative type of consumer-based political action. It also demonstrated that although Americans were willing to use their economic power to advocate environmental issues, they were unwilling to challenge the basic economic system: They still wanted the hamburger, and they were willing to buy it from an enormous corporation, but they preferred that it not be wrapped in Styrofoam.

The influence of the counterculture libertarians peaked with the dot-com boom of the 1990s. Their techno-utopian rhetoric and enthusiasm for the New Economy and the electronic frontier lost some of their luster in the stock crash and scandals of the early 2000s. The environmental version of the counterculture libertarian sensibility faded from view as the California ideology of the cyber-economy rose to prominence. The election of George W. Bush and rise of the so-cial conservatives severed many of the tenuous ties between the hip Right and the Republican Party. John Perry Barlow had worked on Dick Cheney's Wyoming campaign when Cheney ran as a small government, pragmatic environmentalist (yes, he really did consider himself an environmentalist) and free-market fiscal conservative. Barlow especially appreciated Cheney's environmental views and policies and considered him an ally in his Wyoming environmental activism.[93] Their alliance was troubled at times, most famously leading to the lyrics for the Grateful Dead song "Throwing Stones" (first performed in 1982). By the early 2000s, Barlow could not stomach the Bush administration's "very authoritarian, assertive form of government . . . in the guise of libertarianism."[94] For counter-culture libertarians like Barlow, the Bush administration was a dangerous failure on two counts. First and foremost, the administration demonstrated "an unwill-ingness to engage in any kind of mitigation of the free market," and, second, they were intensely adversarial to the liberal social values that characterized the "hippie-mystic strain" of libertarianism that had contributed to the Left/Right fusion of the cyberlibertarians and the environmental pragmatists.[95] It is a fasci-nating moment in American history when hippies are saddened by the lack of core conservative values evidenced by fundamentalist Republicans.

The criticism runs both ways. While Barlow and others on the hip Right be-came disenchanted with the New Right, they received much condemnation, espe-cially from the Left. For liberal critics, the blurring of political lines risked a dilu-tion of liberal claims to social responsibility in return for a hybrid libertarianism that seemed unlikely to appeal to social conservatives or traditional liberals. Fur-ther, the wave of social conservatism that swept the nation in the aftermath of 9/11 demonstrated some of the limitations of thoughtful consumption and renegade business models as a political strategy.

Western liberals might wonder whether the hybridization of Left and Right politics of the counterculture libertarians contributed to the rise of the radical Right while diluting liberal constituencies by moving pro-business moderates to the Right. A more positive view could argue that the hybrid movement represents the best bet for a popular consensus. By holding Republicans to the conservative traditions of fiscal responsibility and protection of the free market while standing firm on their liberal social agenda, counterculture libertarians offered an avenue for a different type of centrist politics that could appeal to politically moderate westerners from the Left or the Right. Moreover, the natural capitalism of com-panies like Patagonia, captured perfectly in Chouinard's recent *Let My People Go Surfing* (2005), still represents a means of integrating social responsibility into daily life that even during periods of intense political polarization appeals to a wide spectrum of Americans from both sides of the political fence.

In fact, the counterculture's model of politically realistic, consumer-friendly environmentalism might provide the best hope for a future meeting of political minds in the West. Conservation and preservation evolved into environmentalism because of a collective realization that protecting the environment was a personal choice that influenced quality of life. Millions of Americans love the outdoors and go there as often as they can. A decent percentage of these outdoor enthusiasts support environmental protection and give money to groups that lobby on behalf of the environment and work toward Progressive legislation. Many average outdoor fans may even vote for candidates who have some sense of an environmental ethic, but are those actions more or less important than when they walk out of their way to recycle a can or read the label of their new jacket to find out what it was made of, or give a few seconds of thought to how their consumption fits into the chain of ecology that we are a part of despite of how divorced we are from the production side of the capitalist equation?

Conservatives have done a good job of recognizing that personal choices and preferences for quality of life and values can, and often do, supersede American interests in policy plans and decisions. The counterculture libertarians recognized this also and helped shape a model of indirect political response based on individual agency. Reconciliation and meeting on middle grounds is always a laudable goal: Could this hybrid philosophy of politics be a model? Or is it just another utopian dream that played out on the well-used western stage? Poet William Carlos Williams famously said that "the pure products of America go crazy." Western environmental politics generally proves this true, which makes it likely that the future of western politics will be some hybrid of Left and Right. Understanding the pragmatic mode of environmental thinking that emerged from our very recent past offers a hope, if only a hope, for a practical new human- and community-centered environmental culture for the twenty-first century—an environmentalism reconciled with the market economy and distanced from the contentious political debates that characterized the movement in the past century.

What Happened to Appropriate Technology?

When he wrote his introduction to the 1994 *Millennium Whole Earth Catalog,* J. Baldwin preemptively answered a question that many longtime readers might have asked after perusing this slick new version of the catalog that was full of the latest information technology and the utopian cyberculture enthusiasm in the air at the beginning of the dot-com boom of the early 1990s. "But where are the windmills?" he began. *Whole Earth* "once featured pages of bold experiments and make-it-yourself plans along with commercial windmills and where to get them." The new catalog, readers would quickly discover, contained much less of this environmental pragmatism. J. Baldwin was quite aware of this significant change. "Aren't windmills important any more?" he asked both himself and the pragmatic readers who always turned to his sections first. "Has it turned out that they don't work? Has *Whole Earth* sold out to Big Power? Was wind power just a fad?" What, he was really asking, happened to the appropriate technology (AT) movement, in which he himself figured so prominently, which promised an alternative pragmatic environmentalism that wedded design and technological innovation, marketing savvy, and individual agency?

Baldwin understood that history had dealt his movement some cruel and unanticipated blows: ATs, like all technologies, had unintended consequences; promising research led to dead ends; corporations took much longer to grasp the significance and benefits of natural capitalism than anticipated; and AT was sucked into the cycles of American politics despite the desires of many of the movement's leaders. Nonetheless, he was still confident in the mid-1990s that while there had been setbacks, the movement represented by the iconic windmill was still alive and to a certain degree had simply become common sense. "Renewable energy," he argued, "has become accepted as a concept; grade-school kids know the term. Manufacturers, regulators and bankers are attentive now." He concluded, "Not being much interested in common knowledge or nostalgia, we no longer need to do windmills and other well-developed renewables."[1] Baldwin could take pride in *Whole Earth*'s role in this transformation in American thinking that had moved environmentalism from the wilderness into the home, classroom, and boardroom, but his optimism about the degree to which this sensibility had become mainstream may have been premature. During the subsequent five years, SUV (sports utility vehicle) sales exploded, so-called McMansions sprang up in massive subdivisions, and modest sized homes that had served several generations well were "scraped off" in neighborhoods across the country during the explosive housing boom of the 1990s and early 2000s. The average American home grew from 1,300 square feet in 1970 to over 2,200 by 2005, while the percentage of these homes that were built using green principles actually declined.[2] Much of this excess was fueled by the cyberrevolution that was the primary subject of the *Millennium Whole Earth*. Where was AT and the pragmatic environmentalism so effectively promoted by *Whole Earth* during the most significant housing boom since the post–World War II phenomenon that played such a significant role in shaping the environmental movement in the 1950s?

Despite its successes, the AT movement was not without its ironic consequences or its critics. In 1980, political scientist Langdon Winner published a landmark essay on AT: "Building the Better Mousetrap." His account remains one of the most subtle, concise versions of the rise and fall of the movement. Written as Ronald Reagan began his small-government revolution that threw a bucket of water on the nicely caught fire of AT, Winner's essay was cynical about the prospects of the movement and critical of the contributions of *Whole Earth* in particular. He especially disdained the "catalogue-browsing consciousness of the New Age," which, he argued, "was not one that wanted to be bothered by well-reasoned arguments."[3] Further, Winner rightfully pointed out that the key to any success the AT movement had prior to 1980 was due to the ways that the new technologies tapped into deep conservative traditions of craft and thrift with an eye toward the market. This argument echoed many critics of *Whole Earth* and lumped the catalog into the dismissive category of New Age mainly because the vision of AT presented in the catalogs lent itself so well to lifestyle reimaginings that Winner found a pale shadow of pragmatic political organization. Winner failed to realize that it was that very connection with a deeper conservative American tradition that made the AT world presented through *Whole Earth* so significant.

In 1980, the prospects for the movement seemed bleak, and there were many solid reasons for critiquing the utopian aspects of the movement. The idealism that drove AT often failed to account for the degree to which even small-scale and individualistic ideas, like the personal computer, could very rapidly be incorporated into, and even strengthen, the very centralizing systems they were designed to subvert. Spectacular failures like Biosphere II provided fodder for critics of the more utopian wing of the movement. Biospherians posing for pictures in uniforms that looked strikingly similar to those from the original *Star Trek* television series and proposals for a Biosphere connected to the Sands Hotel on the Las Vegas Strip did little to counter arguments from thoughtful critics like Winner.[4] Behind some of the more farcical aspects of this well-funded project were excellent scientists and the best-funded attempt to reconcile nature and the machine to date. Biosphere II's Director of Space Applications, Mark Nelson, called it the perfect "marriage of ecology and technology."[5] And *Whole Earth*'s Kevin Kelly highlighted the "symbiosis of nature and technology" that Biosphere II represented as a real attempt to learn how "to live within nature and with our machines."[6] Biosphere II was "a very futuristic idea," according to Peter Warshall, "because obviously that's how the whole world was going to work from now on— a combination of the ecological infrastructure and the human-constructed infrastructure . . . the ideal of *Whole Earth*."[7] Biosphere II represented both the hard work and innovation that were emerging from the ecological design community at the same time it became an unfortunate example of how far the dreams were from the reality.

Behind the media circus, Biosphere II hosted the first Shell Oil Learning Conference, sponsored by Shell and Volvo, and "brought together what would become the Global Business Network (GBN)," which, according to Biosphere naturalist Peter Warshall, "was a way of bridging the corporate world, business world, and environmental world."[8] GBN became Stewart Brand's next big venture, a potent collaboration with the potential to work behind the scenes to greatly expedite the shift toward natural capitalism in the 1990s and new century, while pushing once again the limits of Brand's environmental heresies.[9] Brand's other contemporary venture, The Long Now Foundation, dedicated to innovative research in "Time and Responsibility," highlights the tensions in Brand's environmentalism and all efforts to unite technological enthusiasm with ecological awareness. Established in 1996 to build a giant 10,000-year clock and library to foster long-term and holistic thinking about time the same way the picture of the whole Earth inspired holistic thinking about space, the foundation's lectures and pod-casts have become a significant forum for the types of iconoclastic innovators *Whole Earth* once featured. The foundation is building its remarkable clock and hope to house it in a rock chamber high on the flanks of Nevada's Mt. Washington, overlooking the Spring Valley, nestled among one of the only privately controlled groves of ancient Bristlecone Pines in the West.[10] About one hundred miles southwest of the beautiful Long Now site is another project concerned with very long-term thinking; the Yucca Mountain Nuclear Waste Repository. Brand's most controversial current heresy is his support for nuclear power. It seems somehow fitting that both the Long Now clock, which may inspire people to think hard about the future, and the waste product of nuclear power, which will compel people to think hard about future environmental impacts of our technology, will live so close to each other. Both of these technologically driven projects have and will continue to inspire debate about the relationship between nature and technology and highlight the profound consequences we face if we don't find a way to reconcile our machines with nature. The middle ground seems to be sustainable appropriate technologies.

The prototype 10,000–year "Long Now" clock designed by computer scientist Danny Hillis. The nonprofit Long Now Foundation plans to build the full–scale version and place it in a chamber high on Nevada's remote Mt. Washington. (Photo by Rolfe Horn, courtesy of The Long Now Foundation)

Appropriate technology as a movement did fade from the public prominence it enjoyed during the 1970s and was certainly seriously damaged by the open hostility of the Reagan administration in the early 1980s. But, as historian of technology Carroll Pursell has observed, "although the appropriate technology 'movement' of the 1970s disappeared, the technologies themselves persisted and developed."[11] Moreover, the organizations and individuals represented on the pages of *Whole Earth* continued to work on perfecting new technologies despite reductions in federal funding.

Although the AT revolution may not have played out the way postscarcity theorists expected, the majority of the AT initiatives have had an overwhelmingly positive impact on American culture and American environmentalism and offer a suggestion for how to move environmentalism out of the wilderness. The promotion of renewable energy resources and energy conservation through technological invention is one example of success. Energy-efficient houses, thermal windows, solar power, and high-efficiency electrical devices have become widely accepted to the point of becoming standard features of American culture, just as Baldwin argued. Curbside recycling and the proliferation of postconsumer waste recycling have also gained wide approval and to a certain extent become a part of daily life. Many of these technologies and services that seem so obvious and sensible that they go unnoticed today resulted from the radical innovation of counterculture environmentalists. Whether they went back to the land or into the laboratory, they infused environmentalism with an optimistic hope that one day the nagging question of how to reconcile the tension between the modernist desire to exploit the progressive potential of technological innovation with the antimodernist desire to preserve the natural world might be resolved through enlightened technical innovation.

Events of the dawning years of the new millennium lent a new, and recently frantic, urgency to the search for viable models for an ecologically sustainable future. The American public, faced with a resurgent energy crisis and alarmingly accelerating cycles of global warming and its cataclysmic effects, have once again demanded action from politicians and environmental leaders. The century of environmental advocacy that seemed to reach a satisfying conclusion with the significant gains of the 1960s forward suddenly seemed an inadequate model for dealing with the new realities of the twenty-first-century environmental crisis. Global warming vividly illustrates the necessity of whole-systems thinking. Now more than ever, we need to have the picture of the whole Earth in our minds as we think about how to respond to current environmental realities. We also need to remember that only through the careful application of technology will we extricate ourselves from our astonishingly foolish and heavy-handed application of technocratic solutions to the environmental and economic problems caused by the Industrial Revolution. In his startling Oscar-winning documentary *An Inconvenient Truth* (2006), Al Gore points out that Americans have too often moved from denial to despair in one step without pausing to ask how we might use our capacity for innovation to solve the problem. Gore's ability to illustrate global warming owes much to the *Whole Earth* network of ecological collaborators, who recognized that the same remarkable brains and opposable thumbs that caused our twentieth-century environmental crisis could get us out of it in the twenty-first century. This insight was the difference that made a difference at *Whole Earth*.

By 2007, public sentiment was leaning hard toward a simple acceptance of the fact of human-caused global warming. Pictures of mountains from around the globe denuded of the glaciers that captivated mountaineers for two centuries make for compelling evidence easily grasped by an American public reluctant to accept the reality of our ability to cause environmental changes so sweeping in scope and swift in coming. The issue remained contentious, but there was a dawning realization that something was happening and that something ought to be done to mitigate the problem. Even Wal-Mart got into the game when it hired AT visionary and longtime *Whole Earth* contributor Amory Lovins to help its efforts to create a greener business model and building program.[12]

Although the political players are different, the rhetoric and propositions for change are remarkably familiar. The effort to create a viable appropriate technology dates to the 1960s, with more ad hoc efforts going back at least into the late 1800s. The environmental philosophy articulated by Stewart Brand, J. Baldwin, Peter Warshall, and Huey Johnson in particular, and the programmatic environmental activities of the Point Foundation, seem more important and more relevant to any thoughtful analysis of ongoing efforts to create a harmonious and sustainable relationship between people and nature in America than they did in the 1970s.

The story of the Point Foundation's debates about appropriate uses of funds and discussions about the type of environmental activities that they wanted to support provides insights into the concerns that were shaping one model for a commerce-path environmental culture aimed at the future we now inhabit. Now more than ever, it is important to understand past experiences in alternative environmentalism and antecedents for current efforts in AT sustainability and ecological design. One answer to the question of what happened to the AT movement is that it became dependent on the guardian model of federal funding and thus wedded to the ups and downs of the American political cycle. In 1980, when Ronald Reagan came into office, federal funding for AT dried up quickly, and organizations and agencies previously flush with government cash found themselves without a funding source. This bleak situation supports Jane Jacobs's thesis about monstrous hybrids of commerce and government wedded to the benefit of neither.

Throughout this dark period, *Whole Earth* remained the voice and center of community for the grassroots AT and ecological design movements. Stewart Brand was one of the most astute observers of trends in these areas, and as early as 1974 saw trouble on the horizon. The funding crisis in the early 1980s confirmed his worst fear of the AT movement's direction in the 1970s: "We've been led to believe that invention comes either from lone men in their basements or gaudy corporation and university laboratories, and nowhere between. Crash Government funding for energy relief is being dispensed accordingly. What a waste."[13] Brand continued to advocate individual agency and autonomy while exploring the possibilities for corporate environmentalism as an antidote to monstrous hybrids. The story of the Point Foundation, however, demonstrates that even a steadfast proponent of individual agency and self-education like Brand was willing to intervene directly when he thought the cause and time were right. Point was designed to provide a loosely structured facilitation system for more direct advocacy while avoiding the politics inherent in the foundation world. For a brief moment in the early 1970s, Point was one of the most significant efforts to fund and provide a more solid foundation for the grassroots AT movement while fostering a broader set of related social and cultural concerns. Point's philosophy of counterculture finance and simple living has become mainstream, if only as a secondary trend, in a culture that still gauges success by material progress toward the bigger and better.[14]

America at the dawn of the twenty-first century is a country polarized and divided by bitter political fights that have made consensus an endangered species. For many thoughtful Americans, the terms *liberal* and *conservative* have lost their historical meanings and become nothing more than verbal clubs to bludgeon opponents during dogmatic fights. Causes that could have united Americans across political lines have suffered setbacks over the past quarter-century. Environmental concerns in particular have at times in our past been meeting grounds where social, political, religious, and cultural differences were put aside during coalition building aimed at protecting the natural world that sustains us all, regardless of our beliefs. Disenchantment with the polarized status quo runs deep these days while hope for the future runs low and political pessimism appears to have eclipsed even the dark days after the Vietnam War and Watergate. The techno-ecological utopia many hoped would flower in the new century has been delayed by the political polarization that has made the project of environmental coalition building more difficult than at any time in the previous century.

At the same time, recent events have raised critical questions about how the nation might use the best insights from ecological science and American technological genius to work toward a survivable future for the nation and the fragile Earth it sits on. The war in Iraq, if it accomplished nothing else, raised awareness in America about the perils of our energy addiction and dependence. Once again,

Americans are asking questions about technology, commerce, and environment. Is there a system that unites ecology and technology? Is there a model for how to be an *environmentalist* free from the political baggage that term has carried for much of the last quarter-century? Is there a viable intellectual tradition of a cultural/political pragmatism that lies somewhere between conservative and liberal that might offer a means toward a less divisive future? The environmental activities of the counterculture as expressed on the pages of *Whole Earth* may have been on the fringes of environmental politics in the 1960s and 1970s, but they had moved squarely to the center of American environmental culture by the beginning of the new millennium.

At the turn of the new century, America was on the brink of a second energy crisis. Californians, faced with 500 percent increases in power bills, scrambled for energy alternatives. Fears of foreign oil dependence in the wake of the September 11, 2001, terrorist attacks breathed new life into federally supported AT research dormant since the Reagan administrations. Soaring fuel prices and astronomical electric bills again raised the specter of scarcity and spawned new questions about green technologies and renewable energy solutions. Future environmental advocates will need to capitalize on recurring patterns of scarcity and the connection of energy to national security to push the American public toward deeper changes in practices of production and consumption while supporting the development of new technologies that provide for a satisfactory quality of life without jeopardizing environmental health.

What the history of *Whole Earth* finally suggests is that there have always been more than two options for environmentalism, that we need not choose between nature or culture, that we can embrace our tools and use them to create a sustainable future. The problem in the past has not been the creation of new, environmentally friendly technology—innovative individuals and companies have been doing that for a century. *Whole Earth* made this information easily accessible for a generation. The problem has been convincing American producers and consumers that there was common ground between capitalism and environmentalism, nature and culture. The challenge for environmental advocates in the next century will be to move forward with the efforts of the past half-century and reconcile American notions of technological progress, consumerism, and economy with environmentalism to create a sustainable economics. The *Whole Earth Catalog*, which ended, for now, its long publication run in the spring of 2003, was the best attempt yet to integrate the insights of pragmatic environmentalism into American popular culture. In the early 2000s, the environmental issues explored in the catalogs seemed remarkably contemporary. In 1968, when a young Stewart Brand launched his venture after being struck with inspiration while flying home from his father's funeral, the *Whole Earth Catalog* was nothing less than revolutionary.

Abbreviations

HJP   Huey Johnson Papers, Resource Renewal Institute, San Francisco

PFR   Point Foundation Records, Department of Special Collections, Stanford University Libraries

SBP   Stewart Brand Papers, Department of Special Collections, Stanford University Libraries

WECR   *Whole Earth Catalog* Records, Department of Special Collections, Stanford University Libraries

Preface and Acknowledgments

1. Peter Braunstein and Michael William Doyle, eds., *Imagine Nation: The American Counterculture of the 1960s and '70s* (New York: Routledge, 2002), 5, 10; and Michael William Doyle, "Debating the Counterculture: Ecstasy and Anxiety over the Hip Alternative," in *The Columbia Guide to America in the 1960s*, ed. David Farber and Beth Bailey (New York: Columbia University Press, 2001), 143–156.

Introduction: One Highly Evolved Tool Box

1. Barbara Ward, *Spaceship Earth: The Impact of Science on Society* (New York: Columbia University Press, 1966). The toolbox reference in the title of this chapter comes from J. Baldwin, "One Highly Evolved Toolbox," *CoEvolution Quarterly* 5 (Spring 1975): 80–85.

2. J. Baldwin, ed., *The Essential Whole Earth Catalog* (New York: Doubleday, 1986), 402.

3. "Missal for Mammals," *Time*, November 21, 1969, 74–76; "Windmill Power," *Time*, December 2, 1974, 12; and Stewart Brand, ed., *The Last Whole Earth Catalog* (Portola Institute and Random House, 1971), 439. Brand also retells this story with a little more depth in the introduction to *The Essential Whole Earth Catalog*, ed. J. Baldwin (New York: Doubleday, 1986), 4.

4. J. Baldwin, ed., *Essential Whole Earth Catalog*, 402.

5. Stewart Brand's scrawled notes from that plane ride, reprinted in *Last Whole Earth Catalog*, 439.

6. J. Baldwin, ed., *Essential Whole Earth Catalog*, 402.

7. There are many fine sources on the development of appropriate technology. For example, see David Dickson, *Alternative Technology and the Politics of Technical Change* (Glasgow: Fontana/Collins, 1974); Nicolas Jéquier, ed., *Appropriate Technology: Problems and Promises* (Paris: Organization for Economic Co-operation and Development, 1976); Franklin A. Long and Alexandra Oleson, eds., *Appropriate Technology and Social Values: A*

Critical Appraisal (Cambridge, MA: Ballinger, 1980); Witold Rybczynski, *Taming the Tiger: The Struggle to Control Technology* (New York: Penguin, 1985); Mathew J. Betz, Pat McGowan, and Rolf T. Wigand, eds., *Appropriate Technology: Choice and Development* (Durham: Duke University Press, 1984); Ron Westrum, *Technologies and Society: The Shaping of People and Things* (Belmont, CA: Wadsworth, 1991); and Theodore Roszak, *Where the Wasteland Ends: Politics and Transcendence in Postindustrial Society* (Garden City, NY: Anchor, 1973). Two more recent works shed new light on the history of alternative technology within the context of environmental politics: Martin W. Lewis, *Green Delusions: An Environmentalist Critique of Radical Environmentalism* (Durham, NC: Duke University Press, 1992); and Charles T. Rubin, ed., *Conservation Reconsidered: Nature, Virtue, and American Liberal Democracy* (Lanham, MD: Rowman & Littlefield, 2000). The best history of alternative technology as a political movement is Jordan Kleiman's excellent doctoral dissertation, "The Appropriate Technology Movement in American Political Culture" (University of Rochester), 2000.

8. The quote is from Peter Warshall, interview by the author, Tucson, Arizona, November 11, 2006. One of the guiding principles of the *Whole Earth* was the desire to present information in as unmediated a fashion as possible. "Except for occasional self-dubious asides," Brand wrote, "I think we should deny our contemporaries and ourselves all this interpretation. Such indulgence leads rapidly to advertising language such as 'lifestyle,' where we are the commodity. We are far better employed as journalists or field anthropologists, immersed in story. Let other poor drabs do the critiquing" (Stewart Brand to David Shelzline, March 1, 1976, *Whole Earth Catalog* Records [hereafter cited as WECR], M1045, box 3:5, Department of Special Collections, Stanford University Libraries).

9. The *Whole Earth Catalog* has had many incarnations. Because of the editor's iconoclastic style and alternative publishing methodology, *Whole Earth* can be difficult to cite consistently. The first edition was published in 1968 as the *Whole Earth Catalog: Access to Tools,* edited by Stewart Brand and published by the Portola Institute with distribution provided by Random House. Several revised versions followed between 1969 and 1971, all with Brand as the lead editor, when *The Last Whole Earth Catalog* (Portola Institute & Random House, 1971) appeared. *The Last Whole Earth* won the prestigious National Book Award in 1972. All of the *Whole Earth*s were reprinted many times, and often there were supplemental editions. Between 1972 and 1999, there were several notable editions for anyone interested in the history of the publication. See especially Stewart Brand, ed., *The Next Whole Earth Catalog: Access to Tools* (Point Foundation with distribution by Rand McNally in the United States and by Random House in Canada, 1980). This particular edition is notable for shear size (608 oversized pages) and breadth of coverage. There were also several *Whole Earth* companion volumes that focused on particular themes or issues, such as J. Baldwin and Stewart Brand, eds., *Soft-Tech* (New York: Penguin, 1978); Howard Rheingold, ed., *The Millennium Whole Earth Catalog* (San Francisco: Harper San Francisco, 1994); and Peter Warshall, ed., *30th Anniversary Celebration: Whole Earth Catalog* (San Rafael, CA: Point Foundation, 1999). The thirtieth-anniversary edition includes a wonderful collection of alternative technology and counterculture essays by leaders from the 1960s to the 1990s. For more on this theme, see Kevin Kelly, ed., *Signal: Communication Tools for the Information Age* (New York: Harmony, 1988).

10. "Report on 1984 Hacker's Conference," *Whole Earth Review*, May 1985, 49.

11. Brand to Shelzline, March 1, 1976.

12. The various *Whole Earth Catalogs*, *CoEvolution Quarterly*s, and *Whole Earth Reviews* were collected and shared like *National Geographic* and *Mad* magazine. This meant that copies of *Whole Earth* in its various forms, not unlike an academic book, had a wider subsidiary ad hoc distribution and readership than print numbers alone imply. The extent to which this happened is hard to chart, but Editor Peter Warshall points to the regular offers he received during his tenure to take control of vast *Whole Earth* archives assembled by individuals and organizations, extensive library holdings, and the ready availability of even small run issues for sale on the internet (Warshall, interview).

13. Kevin Starr, "*Sunset Magazine* and the Phenomenon of the Far West," http://sunset-magazine.stanford.edu/html/influences_1.html. The Web site is Stanford University's wonderful *Sunset* online archive.

14. Beth Bailey, "Sex as a Weapon: Underground Comix and the Paradox of Liberation," in Braunstein and Doyle, *Imagine Nation*, 307.

15. Robert Gottlieb, *Forcing the Spring: The Transformation of the American Environmental Movement* (Washington, DC: Island, 1993), 100.

16. James Livingston, *Pragmatism and the Political Economy of Cultural Revolution, 1850–1940* (Chapel Hill: University of North Carolina Press, 1997), xix.

17. It is admittedly difficult to generalize about the readers of the catalog and would be a mistake to claim a monolithic set of shared beliefs. Extensive letters to the editor and careful correspondence collected in WECR (M1045, Reader Correspondence Files) do, however, indicate trends in thinking. Of particular interest is a series of letters from readers focused on their perceived relationship with the publication and how they acted on ideas they encountered in the catalogs (box 4:9).

18. Mark Levenson to Stewart Brand, WECR, M1045, box 6:5.

19. Over the past decade, interest in various aspects of *Whole Earth*'s history has produced some remarkable multidisciplinary scholarship. Notable recent contributions include English professor Bruce Clark's work on whole systems, art historian Linda Henderson's work on Buckminster Fuller and the "Fourth Dimension," architectural historian Simon Sadler's studies of *Whole Earth*'s role in promoting ad hoc architecture, Fred Turner's remarkable study of how *Whole Earth* contributed to the rise of digital utopianism, French historian Caroline Maniaque's wonderful comparative studies of American and French countercultural underground publishing, Jordan Kleiman's comprehensive study of the appropriate technology movement, and Sam Binkley's work on countercultural lifestyle publishing efforts. Certainly others are hard at work on *Whole Earth* even now that I am unaware of. For further reading, see Bruce Clark and Linda Henderson, eds., *From Energy to Information: Representation in Science and Technology, Art, and Literature* (Stanford: Stanford University Press, 2002); Linda Dalrymple Henderson, *The Fourth Dimension and Non-Euclidean Geometry in Modern Art* (Princeton, NJ: Princeton University Press, 1983); and Jonathan Hughes and Simon Sadler, eds., *Non-plan: Essays on Freedom, Participation and Change in Modern Architecture and Urbanism* (Oxford: Architectural, 2000). Sadler greatly expands his coverage of *Whole Earth* in his unpublished essay "Selling Buildings through the *Whole Earth Catalog*," from "The Whole Earth: Parts Thereof," a symposium held at the University of California–Davis on

May 8, 2006. Fred Turner, "Where the Counterculture Met the New Economy: The Well and the Origins of Virtual Community," *Technology and Culture* 46 (July 2005): 485–512, and *From Counterculture to Cyberculture: Stewart Brand, the Whole Earth Network, and the Rise of Digital Utopianism* (Chicago: University of Chicago Press, 2006); Caroline Maniaque, "Searching for Energy," in *Ant Farm 1968–1978*, ed. Constance Lewallen and Steve Seid (Berkeley: University of California Press, 2004), 14–17; Jordan Kleiman, "The Appropriate Technology Movement in American Political Culture" (Ph.D. diss., University of Rochester, 2000); and Sam Binkley, "Consuming Aquarius: Markets and the Moral Boundaries of the New Class, 1968–1980" (Ph.D. diss., New School University, 2001).

20. Leo Marx, *The Machine in the Garden: Technology and the Pastoral Ideal in America* (New York: Oxford University Press, 1964), 23.

21. David Nye, *American Technological Sublime* (Cambridge, MA: MIT Press, 1996), and *America as Second Creation: Technology and Narratives of New Beginnings* (Cambridge, MA: MIT Press, 2004). Nye points out that, before Marx, Perry Miller explored the technological sublime in his Pulitzer Prize–winning *The Life of the Mind in America* (New York: Harvest, 1965). Nye provides a nice concise historiography of this analytical thread in *Technological Sublime*, xv.

22. Matt Wray, "A Blast from the Past: Preserving and Interpreting the Atomic Age," *American Quarterly* 58 (June 2006): 467–483. *Whole Earth*'s Stephanie Mills was a consistently insightful critic of a persistent technological enthusiasm that failed to recognize the lasting effects and unintended consequences of technology. Mills's views on technology are most clearly revealed in two edited collections: Robert Theobald and Stephanie Mills, eds., *The Failure of Success: Ecological Values vs Economic Myths* (Indianapolis, IN: Bobbs-Merrill, 1973); and Stephanie Mills, ed., *Turning Away from Technology: A New Vision for the 21st Century* (San Francisco: Sierra Club, 1997). For changing perceptions of technology at the *Whole Earth Catalog*, see *Whole Earth Review*, Winter 1991, no. 73. Dedicated to the theme of "Questioning Technology," this significant issue features articles by J. Baldwin, Langdon Winner, and Ivan Illich, and provides a nice point of comparison with presentations of technology in earlier issues of the catalog.

23. Thomas P. Hughes, *American Genesis: A Century of Invention and Technological Enthusiasm, 1870–1970* (New York: Penguin, 1989), 443. Anyone studying the history of technology owes a debt to Hughes for this monumental and insightful overview and his remarkable collected works. For his views on the counterculture role in shaping the history of technology, see 443–472.

24. Stewart Brand, "Environmental Heresies," *MIT Technology Review* online, May 2005, 1, http://www.technologyreview.com/Energy/14406/; John Tierney, "An Early Environmentalist, Embracing New 'Heresies,'" *New York Times*, February 27, 2007, D1, D 3; and Stewart Brand, "Environmentalists Stifle New Things," *Environmental Action Visions*, May/June 1985, 15. A nice discussion about tools versus environmentalism is contained in J. Baldwin's "What's an Ecolog?" in *Whole Earth Ecolog: The Best of Environmental Tools and Ideas*, ed. J. Baldwin and Stewart Brand (New York: Harmony, 1990), 3. Siva Vaidhyanathan, "Rewiring the 'Nation': The Place of Technology in American Studies," *American Quarterly* 58 (September 2006): 559, a special issue of the *Quarterly* that explores a wide variety of multidisciplinary approaches to analyzing technology from a humanities perspective. For

a thoughtful and passionate refutation of Brand's viewpoint and the general trend in the conservative media to dismiss environmentalists as gluttons for bad news see David W. Orr, *The Nature of Design: Ecology, Culture and Human Intention* (New York: Oxford University Press, 2002), 86–89, in particular. Orr presents a completely uncompromising view of what ecological design could be and what little comfort we should take with, what he views, the small steps we've taken to integrate the insights of ecology into our culture and economy.

25. Thomas Kuhn, *The Structure of Scientific Revolutions*, 2nd ed. (Chicago: University of Chicago Press, 1970). For an excellent concise overview of science and technology in the 1960s, see Timothy Moy, "The End of Enthusiasm: Science and Technology," in *The Columbia Guide to America in the 1960s*, ed. David Farber and Beth Bailey (New York: Columbia University Press, 2001), 305–311 (quote, 310).

26. Cynthia Hamilton, "Industrial Racism, the Environmental Crisis, and the Denial of Social Justice," in *Cultural Politics and Social Movements*, ed. Marcy Darnovsky, Barbara Epstein, and Richard Flacks (Philadelphia: Temple University Press, 1995), 190.

27. Louis Menand, *The Metaphysical Club: A Story of Ideas in America* (New York: Farrar Straus Giroux, 2001), xi.

28. Bruce Braun and Noel Castree, eds., *Remaking Reality: Nature at the Millennium* (New York: Routledge, 1998), 32.

29. Menand, *Metaphysical Club*, xi.

30. For David Wrobel's views on exceptionalist thinking and implications for American culture, see his *The End of American Exceptionalism: Frontier Anxiety from the Old West to the New Deal* (Lawrence: University Press of Kansas, 1993).

31. More recently, a generation of environmental historians have argued that "there is no clear line between us and nature," and that environmentalism must value the connectedness between humans and nature that is so obvious to those who live, labor, and play in close contact with their environment (Richard White, *The Organic Machine: The Remaking of the Columbia River* [New York: Hill & Wang, 1995], 109). This remarkable little book uses the Columbia River as a metaphor for the contested territory where nature and culture, "the mechanical and the organic[,] meet" (110). For the most insightful proponents of a more historically grounded pragmatic environmental culture, see Patricia Nelson Limerick's "Mission to the Environmentalists," in her *Something in the Soil: Legacies and Reckonings in the New West* (New York: W. W. Norton, 2000), 171–185; William Cronon's "The Trouble with Wilderness; or, Getting Back to the Wrong Nature," and Richard White's "Are You an Environmentalist or Do You Work for a Living?" both in *Uncommon Ground: Toward Reinventing Nature*, ed. William Cronon (New York: W. W. Norton, 1995), 69–90; and 171–185. Braden Allenby, *Reconstructing Earth: Technology and Environment in the Age of Humans* (Washington, DC: Island, 2005).

32. Buckminster Fuller's design science revolution grew out of the technocracy movement of the 1930s of which he was a fellow traveler. Those of the technocracy movement, especially during the early years of the Great Depression, placed their hopes in a thorough reordering of society that would place engineers and technical experts at the top of the social pyramid for the purpose of effecting a dramatic top-down restructuring of American society according to the principles of technical efficiency, and the eradication of scarcity through technical innovation.

33. "Pangs and Prizes," *Time*, April 24, 1972, 88, does a good job explaining the controversy that attended the awarding of the prize to *Whole Earth*. When the catalog won the vote in the newly inaugurated contemporary-affairs category, dissenting judge Garry Wills walked out on the proceedings. "The winner," he complained, "was a non-book, and the product not of a writer but of a large group of collaborators." See also Henry Raymont, "Juror Quits Book Panel over 'Whole Earth Catalog,'" *New York Times*, April 5, 1972, 35.

34. Cindi Katz, "Under a Falling Sky: Apocalyptic Environmentalism and the Production of Nature," in *Marxism in the Postmodern Age*, ed. Antonio Callari, Stephen Cullenberg, and Carole Biewener (New York: Guilford, 1995), 276–282.

35. Stewart Brand, *The Clock of the Long Now: Time and Responsibility* (New York: Basic, 1999), 108.

36. Al Gore, *An Inconvenient Truth: The Planetary Emergency of Global Warming and What We Can Do about It* (New York: Rodale, 2006); Alex Steffen, ed., *World Changing: A User's Guide for the 21st Century* (New York: Abrams, 2006); and the World Changing Web site, http://www.worldchanging.org.

1. Environmental Heresies

1. Stewart Brand described his early twenty-first-century environmental views as heresies in an essay of the same name ("Environmental Heresies," *MIT Technology Review* online, May 2005, 1, http://www.technologyreview.com/Energy/14406/. For an excellent concise statement of Brand's views on environmentalism, see his "Review of *The Resourceful Earth*," *Whole Earth Review*, March 1985, no. 45:15. Brand favorably reviews this conservative rebuttal to the apocalyptic Global 2000 Report in *The Resourceful Earth: A Response to Global 2000*, ed. Julian L. Simon and Herman Kahn (Hagerstown, MD: Basic Blackwell, 1984).

2. In this chapter, I use the term *antimodernism* to group individuals and organizations who defined themselves in opposition to the prevailing twentieth-century belief in progress through technological innovation. Antimodernists in the conservation and preservation movements rarely rejected the modernist/Progressive ideal that societies are improvable; they simply rejected the notion that improvement required looking forward to new technologies to solve old problems.

3. Bruce Braun and Noel Castree, "The Construction of Nature and the Nature of Construction: Analytical and Political Tools for Building Survivable Futures," in *Remaking Reality: Nature at the Millennium*, ed. Braun and Castree (New York: Routledge, 1998), 3–42; and David Louter, *Windshield Wilderness: Cars, Roads, and Nature in Washington's National Parks* (Seattle: University of Washington Press, 2006).

4. Michael McClosky, "Wilderness Movement at the Crossroads, 1945–1970," *Pacific Historical Review* 41 (August 1972): 346–361; Samuel P. Hays, "From Conservation to Environment: Environmental Politics since World War Two," *Environmental Review* 6 (Fall 1982): 14–41; and Mark W. T. Harvey, "Echo Park, Glen Canyon, and the Postwar Wilderness Movement," *Pacific Historical Review* 60 (February 1991): 43–67.

5. Peter Braunstein, "Forever Young: Insurgent Youth and the Sixties Culture of Rejuvenation," in *Imagine Nation: The American Counterculture of the 1960s and 1970s*, ed. Peter Braunstein and Michael William Doyle (New York: Routledge, 2002), 258.

6. Jesse Kornbluth, ed., *Notes from the New Underground* (New York: Viking, 1968), 180.

7. Arthur Carhart is the example I know best. Universally considered a leading activist in the 1940s and 1950s, he has been dismissed by environmental historians primarily because his wilderness philosophy was not pure enough. For a corrective to these tendencies, see Charles T. Rubin, *Conservation Reconsidered: Nature, Virtue, and American Liberal Democracy* (Lanham, MD: Rowman & Littlefield, 2000). This excellent collection of essays takes on the tendencies of historians to depict conservation, preservation, and environmentalism as oppositional movements. Particularly useful is Bob Pepperman Taylor's afterword.

8. William Cronon, ed., *Uncommon Ground: Toward Reinventing Nature* (New York: W. W. Norton, 1995), 69.

9. For a remarkably similar argument against elevating a mythically pristine wilderness at the expense of the rest of the environment, see Arthur Carhart, *Planning for America's Wildlands* (Harrisburg, PA: Telegraph, 1961). Carhart has often been criticized for his failure to support the wilderness bill at a time when his influence and access to a national audience were at a high point. Carhart argues, convincingly, that wilderness as defined by the Wilderness Society did not really exist in any pure state, but was an "experience," a construct that lived "within your mind" (19) rather than in a particular place. Carhart refused to support the wilderness bill in 1964 because he felt that arguing for wilderness purity would be a de facto concession to those who sought to develop lands not considered pristine.

10. Euell Gibbons, *Stalking the Wild Asparagus* (1962; rept., Chambersburg, PA: Alan C. Hood, 2005).

11. This section on counterculture environmentalism and the *Whole Earth Catalog* owes a great deal to an essay I wrote for an edited collection on the counterculture: Andrew G. Kirk, "Machines of Loving Grace: Appropriate Technology, Environment, and the Counterculture," in Braunstein and Doyle, *Imagine Nation*, 353–378.

12. Murray Bookchin, *Post-scarcity Anarchism* (Berkeley, CA: Ramparts, 1971).

13. Ibid., 12.

14. Ibid., 11.

15. Steven Levy, *Hackers: Heroes of the Computer Revolution* (New York: Penguin, 1994).

16. Jennifer L. Burns, "O Libertarian, Where Is Thy Sting?" *Journal of Policy History* 19, 4 (Fall 2007): 453–471.

17. To see Brand's libertarian sensibility applied to environmental issues, see his remarkable letter to John Brademas, chairman of the U.S. House Subcommittee on Education, where he relates the "involvement of Government with mass communications," even in the cause of ecological education, as "dangerous as re-joining Church and State." He then outlines his alternatives, which emphasize individual or private initiative with government reward of good efforts at most (Stewart Brand, ed., *Whole Earth Catalog* Supplement, July 1970, 1).

18. The classic study of the conservation movement is Samuel P. Hays's *Conservation and the Gospel of Efficiency: The Progressive Conservation Movement, 1890–1920* (Cambridge, MA: Harvard University Press, 1959). Also useful is Stephen Fox's *The American Conservation Movement: John Muir and His Legacy* (Madison: University of Wisconsin Press, 1981). See also Char Miller's excellent *Gifford Pinchot and the Making of Modern Environmentalism* (Washington, DC: Island, 2001) and his *Ground Work: Conservation in American Culture* (Durham, NC: Forest History Society, 2007).

19. William Vogt, *Road to Survival* (New York: William Sloan, 1948), 37–38.

20. See Paul Sutter, *Driven Wild: How the Fight against Automobiles Launched the Modern Wilderness Movement* (Seattle: University of Washington Press, 2002).

21. Aldo Leopold, *A Sand County Almanac: And Sketches Here and There* (New York: Oxford University Press, 1949).

22. Robert Gottlieb's remarkable *Forcing the Spring: The Transformation of the American Environmental Movement* (Washington, DC: Island, 1993) provides the best contemporary overview of the Clean Air Act and National Environmental Policy Act, and how all of these important milestones fit within a complicated set of social concerns and cultural trends.

23. Hal K. Rothman, *Saving the Planet: The American Response to the Environment in the Twentieth Century* (Chicago: Ivan R. Dee, 2000), 86–87; and Mark W. T. Harvey, *A Symbol of Wilderness: Echo Park and the American Conservation Movement* (Albuquerque: University of New Mexico Press, 1994), 203–204.

24. Arthur Carhart wrote numerous articles against the dam. Two of the best are "State Must Not Toss Away Scenic Dinosaur Park," *Denver Post*, April 7, 1954, 41; and "The Menaced Dinosaur National Monument," *National Parks Magazine* 26 (January/March 1952): 19–30. Mark W. T. Harvey's *A Symbol of Wilderness* is an excellent study of Echo Park as a critical tipping point in American environmental values.

25. Adam Rome, *The Bulldozer in the Countryside: Suburban Sprawl and the Rise of American Environmentalism* (Cambridge: Cambridge University Press, 2001), a wonderful history of American culture and the rise of everyday environmentalism; Kenneth Jackson, *Crabgrass Frontier: The Suburbanization of the United States* (New York: Oxford University Press, 1985); and Dolores Haden, *Building Suburbia: Green Fields and Urban Growth, 1820–2000* (New York: Vintage, 2003).

26. Haden, *Building Suburbia*, 132–137.

27. As quoted in Rome, *Bulldozer in the Countryside*, 43.

28. The story of Bolinas is told well by Orville Schell in *The Town That Fought to Save Itself* (New York: Pantheon, 1976). Peter Warshall, interview by the author, Tucson, Arizona, November 11, 2006.

29. Fairfield Osborn, *Our Plundered Planet* (Boston: Little, Brown, 1948).

30. For an excellent overview of the effect of atomic technology on American culture, see Paul Boyer, *By the Bomb's Early Light: American Thought and Culture at the Dawn of the Atomic Age* (New York: Pantheon, 1985). For sci-fi and environmentalism, see Carl Abbott, *Frontiers Past and Future: Science Fiction and the American West* (Lawrence: University Press of Kansas, 2006), 68–69.

31. John Eastlick, "Proposed Collection of Conservation of Natural Resources," FF-51, box 4, Conservation Library Collection archive.

32. Fox, *American Conservation Movement*. Fox highlights Muir's antimodernist rhetoric as evidence that the conservation movement had, from the beginning, two distinct strains of thought: one, progressive and modern, focused on efficiency and reform; whereas the other, antimodernist, focused on the esthetic and spiritual values of wilderness. A further discussion of these ideas can be found in Max Oelschlaeger's *The Idea of Wilderness: From Prehistory to the Age of Ecology* (New Haven, CT: Yale University Press, 1991).

33. Oelschlaeger, *Idea of Wilderness*, 2.

34. Rachel Carson, *Silent Spring* (Boston: Houghton Mifflin, 1962).

35. Barry Commoner, *The Closing Circle: Nature, Man, and Technology* (New York: Alfred A. Knopf, 1971).

36. Jacques Ellul, *The Technological Society*, trans. Joachim Neugroschel (New York: Continuum, 1980), first published in French in 1954 and in English in 1964. The quote is from Thomas P. Hughes, *American Genesis: A Century of Invention and Technological Enthusiasm* (New York: Penguin, 1989), 450.

37. The quote is from Langdon Winner, "Building the Better Mousetrap: Appropriate Technology as a Social Movement," in *Appropriate Technology and Social Values: A Critical Appraisal*, ed. Franklin A. Long and Alexandra Oleson (Cambridge, MA: Ballinger, 1980), 33.

38. Herbert Marcuse, *One-Dimensional Man: Studies in the Ideology of Advanced Industrial Society* (Boston: Beacon, 1964). For more on Marcuse and how he popularized the insights of the Frankfurt school of Marxian philosophers and sociologists, see Hughes, *American Genesis*, 445–446; and Gottlieb, *Forcing the Spring*, 91–93.

39. Lewis Mumford, *Technics and Civilization* (New York: Harcourt, Brace & World, 1963). For the technocracy movement, see Charles Alexander, *Nationalism in American Thought, 1930–1945* (Chicago: Rand McNally, 1969). Richard White also has a concise and insightful discussion of Mumford in *The Organic Machine: The Remaking of the Columbia River* (New York: Hill & Wang, 1995), 109–110, pointing out why thinkers like Mumford should not be dismissed by environmental historians.

40. Hughes, *American Genesis*, 446–450; and Lewis Mumford, *The Myth of the Machine: The Pentagon of Power* (New York: Harcourt Brace Jovanovich, 1970).

41. For an in-depth look at the machine in American culture, see Leo Marx, *The Machine in the Garden: Technology and the Pastoral Ideal in America* (New York: Oxford University Press, 1964).

42. Theodore Roszak, *The Making of a Counter Culture: Reflections on the Technocratic Society and Its Youthful Opposition* (New York: Doubleday, 1969).

43. Ibid., 8.

44. Charles A. Reich, *The Greening of America: How the Youth Revolution Is Trying to Make America Livable* (New York: Random House, 1970).

45. E. F. Schumacher, *Small Is Beautiful: Economics as if People Mattered* (New York: Harper & Row, 1973).

46. Ibid., 124.

47. A useful taxonomy of technologies can be found in Marilyn Carr, ed., *The AT Reader: Theory and Practice in Appropriate Technology* (New York: Intermediate Technology Development Group of North America, 1985), 6–11.

48. Witold Rybczynski, *Paper Heroes: A Review of Appropriate Technology* (Garden City, NY: Anchor, 1980), 1–4.

49. David Dickson, *Alternative Technology and the Politics of Technical Change* (Glasgow: Fontana/Collins, 1974), 148–173.

50. Samuel Hays, *Beauty, Health, and Permanence: Environmental Politics in the United States, 1955–1985* (Cambridge: Cambridge University Press, 1987), 262.

51. Lewis Herber [Murray Bookchin], *Our Synthetic Environment* (New York: Alfred A. Knopf, 1962); and Bookchin, *Post-scarcity Anarchism* (quote, 22). Also, see Ulrike Heider, *Anarchism: Left, Right, and Green* (San Francisco: City Lights, 1994); and Arthur Lothstein, ed., *"All We Are Saying . . .": The Philosophy of the New Left* (New York: Capricorn, 1970).

52. Herber, *Our Synthetic Environment*, 22.

53. Bookchin, *Post-scarcity Anarchism*, 21.

54. The best overview of the New Left, the counterculture, and environmentalism can be found in Robert Gottlieb's excellent *Forcing the Spring: The Transformation of the American Environmental Movement* (Washington, DC: Island, 1993), 81–114. See also Martin Lewis, *Green Delusions: An Environmentalist Critique of Radical Environmentalism* (Durham, NC: Duke University Press, 1992). For a very different viewpoint from Gottlieb's and from that in this essay, see Hays, *Beauty, Health, and Permanence*, 259–265. Hays argues that there were only superficial similarities between the "negative" counter-culture and the "positive" environmental alternative lifestyle movement.

55. For a subtle analysis of the relationship between the New Left, environment, and social movements, see Gottlieb, *Forcing the Spring*, 93–98.

56. Stewart Brand, phone interview by the author, September 9, 2004.

57. Carr, *AT Reader*, 9.

58. Tom Wolfe, *The Electric Kool-Aid Acid Test* (New York: Bantam, 1969), 2.

59. Stewart Brand journals, 1955–1956 notebooks, Stewart Brand Papers (SBP), M1237.

60. Warshall, interview.

61. Ed McClanahan and Gurney Norman, "The Whole Earth Catalog," *Esquire*, July 1970, 118.

62. Dick Raymond, phone interview by the author, July 25, 2005.

63. Biographical information on Stewart Brand: Brand Résumé, *Whole Earth Catalog* Records, M1045, Department of Special Collections, Stanford University Libraries. Stewart Brand journals, 1955–1959, SBP, cover high school through Stanford. Brand also compiled biographical materials on his sections of the Long Now Foundation and the Global Business Network (*The Last Whole Earth Catalog* [New York: Portola Institute & Random House, 1971], 239–240). Brand's life and work were covered extensively in the press. One of the best concise biographical articles is Andrew Brown's interview with Brand ("Whole Earth Visionary," *Guardian Online*, August 4, 2001).

64. Paul Ehrlich biography, Center for Conservation Biology, Stanford University, 2004.

65. For an overview of the rise of population as an environmental concern, see Hays, *Beauty, Health, and Permanence*, 207–216. See also David Wrobel, "Malthus Revisited," in *The End of American Exceptionalism: Frontier Anxiety from the Old West to the New Deal* (Lawrence: University Press of Kansas, 1993), 112–121.

66. Paul R. Ehrlich, *The Population Bomb* (New York: Ballantine, 1968).

67. Gottlieb (*Forcing the Spring*, 97–98) has a concise discussion of leftists' arguments against politics of population control.

68. Brand, phone interview. See also the population discussion in Brand, "Environmental Heresies," 1. To see just how far the these two have diverged in their environmental views about the role of consumption and technology in shaping a viable future, compare Brand's aforementioned article with Ehrlich's most recent book (Paul R. Ehrlich and Anne H. Ehrlich, *One with Nineveh: Politics, Consumption, and the Human Future* [Washington, DC: Island, 2005 paperback edition]).

69. Stewart Brand journals, 1955–1959, demonstrate Brand's growing interest in biology and conservation. In addition to Ehrlich, early environmental influences included

Fairfield Osborn's *Our Plundered Planet* (Boston: Little, Brown, 1948) and an ecology talk by Aldous Huxley at Stanford (April 16, 1957).

70. The best source on the history of ecology is Donald Worster, *Nature's Economy: A History of Ecological Ideas*, 2nd ed. (Cambridge: Cambridge University Press, 1994).

71. Stewart Brand, *The Clock of the Long Now: Time and Responsibility* (New York: Basic, 1999), 108.

72. Stewart Brand, "CoEvolution and the Biology of Communities," in *News That Stayed News: Ten Years of CoEvolution Quarterly*, ed. Art Kleiner and Stewart Brand (San Francisco: North Point, 1986), 3.

73. Brand, *Clock of the Long Now*, 134. Al Gore's documentary *An Inconvenient Truth* does a fantastic job of graphically illustrating another version of Ehrlich's formulation with technology in the place of population (*An Inconvenient Truth: The Planetary Emergency of Global Warming and What We Can Do about It* [New York: Rodale, 2006], 232–233).

74. This was the philosophy voiced by many of the leaders of the American environmental organizations. Organizations like the Wilderness Society consistently pointed to population as the crux of the environmental crisis (for example, see Olaus Murie to John Spencer, April 2, 1953, Bancroft Library, Sierra Club Records, 71/103c, box 63:18). Especially influential were Vogt, *Road to Survival*, and Ehrlich, *Population Bomb*. For a thorough discussion of population and environmental politics, see Bob Pepperman Taylor, *Our Limits Transgressed: Environmental Political Thought in America* (Lawrence: University Press of Kansas, 1992), 27–50.

75. Joni Seager, *Earth Follies: Coming to Feminist Terms with the Global Environmental Crisis* (New York: Routledge, 1993), 213–219.

76. Brand, phone interview.

77. Raymond, phone interview.

78. Stewart Brand journals, 1955–1959.

79. Brown, "Whole Earth Visionary," 1–3

80. Brand, phone interview.

81. For libertarian influences and reading, see Ayn Rand, 1966–1969 Notebooks, 88–92, SBP.

82. John Markoff, *What the Dormouse Said: How the 60s Counterculture Shaped the Personal Computer Industry* (New York: Viking, 2005), 156–157. Markoff dates this thinking to Brand's time at Phillips Exeter Academy, but Brand's journals suggest that it was actually during his first year at Stanford that he began to think specifically about these issues while writing a journal for a class assignment.

83. Ibid., 58–60. In *From Counterculture to Cyberculture*, Fred Turner expands and complicates this trajectory of Brand's career and *Whole Earth*'s history and provides the critical context for how Brand, a biologist, could loom so large in the rise of the cyberculture (*From Counterculture to Cyberculture: Stewart Brand, the Whole Earth Network, and the Rise of Digital Utopianism* [Chicago: University of Chicago Press, 2006]).

84. David Farber, "The Intoxicated State/Illegal Nation: Drugs in the Sixties Counterculture" (in Braunstein and Doyle, *Imagine Nation*, 18), one of the best articles on any countercultural topic. See also Jay Stevens, *Storming Heaven: LSD and the American Dream* (New York: Grove, 1987).

85. Markoff, *What the Dormouse Said*, 60. Markoff has an intensely detailed description of Brand's initial experience with LSD.

86. Brand's journals indicate that, even though a voracious consumer of a variety of drugs, he was very thoughtful about his own drug use and well aware of the vicious downsides to chemical experimentation. The 1966–1969 Notebooks (100–120) outline pros and serious cons of drug use during the creation of the first *Whole Earth*.

87. Brand, Résumé. The Trips Festival was the West Coast version of Andy Warhol's Dom event that most chroniclers point to as "the beginning" of the counterculture and first true multimedia happening.

88. Stewart Brand, notebooks no. 4, "Notions," box 6:1–15, SBP.

89. Brand, 1955–1956 notebooks.

90. Brand, notebooks no. 4, "Notions," 13.

91. Brown, "Whole Earth Visionary," 2.

92. Barry Miles, *Hippie* (New York: Sterling, 2004), 46. This is not an analytical book, and far too celebratory, but it is a fantastic collection of rare images.

93. Brand, as quoted in Brown, "Whole Earth Visionary." Brand's journals display an early and remarkably sophisticated appreciation of Native cultures dating back to his early teens and even earlier childhood. See also Stewart Brand, "Indians and the Counterculture," in *History of Indian-White Relations*, ed. Wilcomb E. Washburn, vol. 4 of *The Smithsonian Handbook of North American Indians* (Washington, DC: Smithsonian Institution, 1988), 570–572.

94. David Farber, *The Age of Great Dreams: America in the 1960s* (New York: Hill & Wang, 1994), 181; and Stewart Brand, 1964–1966 notebooks, SBP.

95. Peter Braunstein and Michael William Doyle, "Historicizing the American Counterculture of the 1960s and 1970s," in the same authors' *Imagine Nation*, 7. On Braunstein and Doyle's list of overhyped big moments, the Acid Tests are number four.

96. Charles Perry, *A History of the Haight-Ashbury* (New York: Vintage, 1985), 46–47. Perry does the best job chronicling the weird events leading up to the Trips Festival and the remarkable cast of characters who made it happen.

97. Ibid., 47.

98. The Trips Festival was one of the first, if not *the* first, significant counterculture events to charge admission. Some point to Trips as the first sellout of the San Francisco counterculture and marketing of the spirit of the period.

99. Roszak, *Making of a Counter Culture*.

100. Timothy Miller, "The Sixties-Era Communes," in Braunstein and Doyle, *Imagine Nation*, 327–352.

101. Theodore Roszak, *From Satori to Silicon Valley: San Francisco and the American Counterculture* (San Francisco: Don't Call It Frisco, 1986), 22–26.

102. The best account of this campaign comes from Stewart Brand himself: "Why Haven't We Seen the Whole Earth?" in *The Sixties: The Decade Remembered Now, by the People Who Lived It Then*, ed. Linda Obst (New York: Random House/Rolling Stone, 1977), 168–169 (quote, 168).

103. Bruce Mau, *Massive Change* (New York: Phaidon, 2004), 104.

104. Ibid., and Brand, "Why Haven't We Seen," 168.

105. Glen Deny, interview by the author, San Francisco, February 14, 2006.

106. Brand, phone interview; Brand, "Why Haven't We Seen"; Stewart Brand, "The Earth from Space," *Rolling Stone*, May 15, 2003, 124; Neil Maher, "Shooting the Moon," *Environmental History* 9 (July 2004): 526–531; and "Stewart Brand on the Long View," on-line interview by Jennifer Leonard, in Mau, *Massive Change*, 104–105. For Tom Wolfe's version of the story, see his *Electric Kool-Aid Acid Test*, 224.

107. Maher, "Shooting the Moon," 529.

108. Ibid., 529–530.

109. Brand, phone interview.

2. Thing-Makers, Tool Freaks, and Prototypers

1. Stewart Brand journals, 1966–1969 notebooks, 47, SBP, M1237. The phrase "Thing-makers, Tool Freaks, and Prototypers" comes from J. Baldwin and Stewart Brand, eds., *Soft-Tech* (New York: Penguin, 1978), 8.

2. Joan Didion, *Slouching towards Bethlehem* (New York: Farrar, Straus & Giroux, 1961), 84.

3. Ed McClanahan and Gurney Norman, "The Whole Earth Catalog," *Esquire*, July 1970, 124. Of all the popular media stories on *Whole Earth*, the long *Esquire* article is the best historical source. Stewart Brand usually got the spotlight in the popular media, but the 1970 *Esquire* article gave Dick Raymond very good coverage, with both authors clearly understanding how critical Raymond was to the success of *Whole Earth* (ibid.).

4. Stewart Brand, ed., *Whole Earth Catalog*, Fall 1969, no. 129; and Richard Raymond to Donald Pritzker, June 21, 1968, Point Foundation Records (PFR), M1441, box 1:2. In a series of letters, Raymond lays out Portola Institute idea and solicits advice about forming a new foundation.

5. Richard Raymond to Mr. John E. Booth, October 11, 1966, PFR, M1441, box 1:1.

6. Two important early letters outline Dick Raymond's thinking about foundation work and philanthropy: August Heckscher to Mrs. Lois Chale, August 13, 1964, PFR, M 1441, box 1:2; and Dick Raymond to Mr. John E. Booth, October 11, 1966, PFR, M1441, box 1:2.

7. For an excellent overview of the significance of the Stanford Research Institute during this time, see Art Kleiner, *The Age of Heretics: Heroes, Outlaws, and the Forerunners of Corporate Change* (New York: Doubleday, 1996), 394.

8. McClanahan and Norman, "Whole Earth Catalog," 124.

9. Dick Raymond, phone interview by the author, July 25, 2005. There is a nice, concise Portola Institute history in the introduction to Michael Phillips's *The Briarpatch Book* (San Francisco: New Glide, 1978), vii–ix.

10. For more on pre–*Whole Earth* work at Portola, see Ronald Gross, review of the Portola Big Rock Candy Mountain project, *New York Times*, May 21, 1972, BR44.

11. McClanahan and Norman, "Whole Earth Catalog," 124.

12. Michael Phillips, phone interview by the author, March 2, 2006.

13. Michael Phillips, *The Seven Laws of Money* (Boston: Shambhala, 1997).

14. Good background on Glide Church may be found on the church's Web site, http://www.glide.org.

15. McClanahan and Norman, "Whole Earth Catalog," 95; and Raymond, phone interview.

16. McClanahan and Norman, "Whole Earth Catalog," 124. Information on the educational fair is from Stewart Brand journals, 1967 Notebooks, 47, SBP, M1237.

17. Stewart Brand, "History," in *The Last Whole Earth Catalog*, ed. Stewart Brand (New York: Portola Institute & Random House, 1971), 439. One of the wonderful things about researching *Whole Earth* is that virtually every *Whole Earth* publication included a substantial introspective section on the history to date, with design, publication, and distribution details. Anyone who wants to know the history of *Whole Earth*, written by its founders in concise and compelling prose, need only to turn to its back pages.

18. Ibid.

19. Phillips, *Seven Laws of Money*, 18.

20. Stewart Brand journals, 1966–1969 Notebooks, 88, SBP, M1237.

21. Peter Warshall, interview by the author, Tucson, Arizona, November 11, 2006.

22. Doug Finley to Stewart Brand, *Whole Earth Catalog* Records (WECR), M1045, box 4:9, Department of Special Collections, Stanford University Libraries.

23. Brand, "History," 439.

24. Stewart Brand, "Genesis," in *The Essential Whole Earth Catalog*, ed. J. Baldwin (New York: Doubleday, 1986), 402.

25. Brand, "History," 439.

26. Ibid.

27. Brand, "Genesis," 402.

28. Brand, "History," 439.

29. Brand, "Genesis," 402.

30. "Missal for Mammals," review of the *Whole Earth Catalog*, ed. Stewart Brand, *Time*, November 21, 1969; 74–76.

31. J. Baldwin and Stewart Brand, eds., *Whole Earth Ecolog: The Best of Environmental Tools and Ideas* (New York: Harmony, 1990), 128.

32. Brand, "Genesis," 402.

33. Brand, "History," 441.

34. Baldwin, *Essential Whole Earth Catalog*, 402.

35. "Missal for Mammals," 76.

36. James J. Farrell, *The Spirit of the Sixties: Making Postwar Radicalism* (New York: Routledge, 1997), 11.

37. Peter Warshall, *Septic Tank Practices: A Guide to the Conservation and Re-use of Household Wastewater* (New York: Anchor reprint edition, 1979).

38. Helen and Scott Nearing, *Living the Good Life: How to Live Sanely and Simply in a Troubled World*, 2nd ed. (New York: Galahad, 1970); Keith Melville, *Communes in the Counterculture: Origins, Theories, Styles of Life* (New York: William Morrow, 1972); Bennett M. Berger, *The Survival of a Counterculture: Ideological Work and Everyday Life among Rural Communards* (Berkeley: University of California Press, 1981); and Timothy Miller, *The Hippies and American Values* (Knoxville: University of Tennessee Press, 1991), and *The 60s Communes: Hippies and Beyond* (Syracuse, NY: Syracuse University Press, 1999).

39. Robert V. Hine, *California Utopianism: Contemplations of Eden* (San Francisco: Boyd & Fraser, 1981).

40. Tom Wolfe, *The Electric Kool-Aid Acid Test* (New York: Bantam, 1997), 191–200.

41. Baldwin and Brand, *Soft-Tech,* 5.

42. Peter Coyote, *Sleeping Where I Fall* (Washington, DC: Counterpoint, 1998), 298. This amazing autobiography reads like fiction and should not be missed by anyone who wants to understand the counterculture.

43. Stewart Brand, "The Commune Lie," in Brand, *Last Whole Earth Catalog,* 181. There is more on negative views about communes and their prospects in McClanahan and Norman, "Whole Earth Catalog," 119–120.

44. Langdon Winner, "Building the Better Mousetrap: Appropriate Technology as a Social Movement," in *Appropriate Technology and Social Values: A Critical Appraisal,* ed. Franklin A. Long and Alexandra Oleson (Cambridge, MA: Ballinger, 1980), 32 (quote from Winner).

45. It is tempting to use the term *lifestyle* to describe the trend, but the term was so loathed by Brand and others involved in the *Whole Earth Catalog* project that it cannot accurately represent their goals. Clearly an emphasis on lifestyle choices shaped the counterculture of the 1970s, but the term itself, so often casually tossed about in superficial media stories about the counterculture, is too laden with the connotation of frivolousness to be useful for representing the culture of the *Whole Earth Catalog.*

46. "Technocracy's Children" is the title of chapter 1 of Theodore Roszak's *Making of a Counter Culture: Reflections on the Technocratic Society and Its Youthful Opposition* (New York: Doubleday, 1969), 1. Witold Rybczynski, *Paper Heroes: A Review of Appropriate Technology* (Garden City, NY: Anchor, 1980), 94.

47. Gareth Branwyn, "Whole Earth Review," in StreetTech, http://www.streettech. com/bcp/BCPgraf/CyberCulture/WholeEarthReview.html (accessed September 21, 1998).

48. Stewart Brand, ed., *The Next Whole Earth Catalog: Access to Tools* (New York: Random House, 1980), 6.

49. Al Gore, *An Inconvenient Truth: The Planetary Emergency of Global Warming and What We Can Do about It* (New York: Rodale, 2006); and Alex Steffen, *World Changing: A User's Guide for the 21st Century* (New York: Abrams, 2006); and the World Changing Web site, http://www.worldchanging.org.

50. "Windmill Power," *Time,* December 2, 1974, 12.

51. The quote is from R. Buckminster Fuller and Robert Marks, *The Dymaxion World of Buckminster Fuller* (Garden City, NY: Anchor, 1973), 12 (among the many books that chronicle Fuller's life and work, *Dymaxion World* is one of the most accessible and concise); Robert Marks, ed., *Buckminster Fuller: Ideas and Integrities* (Englewood Cliffs, NJ: Prentice-Hall, 1963); Robert Snyder, ed., *Buckminster Fuller: Autobiographical Monologue/ Scenario* (New York: St. Martin's, 1980); E. J. Applewhite, *Cosmic Fishing* (New York: Macmillan, 1985); J. Baldwin, *Bucky Works: Buckminster Fuller's Ideas for Today* (New York: John Wiley & Sons, 1996); and Thomas T. K. Zung, ed., *Buckminster Fuller: Anthology for the New Millennium* (New York: St. Martin's, 2001). Fuller's astounding personal record of his life as "Guinea pig B," what he called his "Chronofile," is housed at Stanford University Department of Special Collections, M1090. Fuller kept almost every scrap of paper he generated or collected, providing a truly impressive personal record of one remarkable life.

52. This view of R. Buckminster Fuller is nicely presented by E. J. Applewhite in notes from an address in Zurich celebrating the opening of a major Fuller exhibit "Your Private Sky" in 1999 (Zung, *Buckminster Fuller*, 360–362).

53. Stewart Brand, ed., *Whole Earth Catalog*, Fall 1968, 2.

54. Snyder, *Buckminster Fuller*, 38.

55. Ibid., 54–55.

56. J. Baldwin does the best job concisely explaining Fuller's idea of "comprehensive anticipatory design science" as an alternative to politics (*Bucky Works*, 62–67).

57. Clark Secrest, "'No Right to be Poor': Colorado's Drop City," *Colorado Heritage*, Winter 1998, 14–21.

58. Paolo Soleri's vision of an alternative world created through revolutionary architecture was even more iconoclastic than Buckminster Fuller's. Soleri's radical design ideas were popularized in *Arcology: The City in the Image of Man* (Cambridge: MIT Press, 1969) and epitomized by his still-unfinished life project: Arcosanti in the Arizona Desert. Like Soleri, Moshe Safdie focused on alternative designs for communal living (see his *Beyond Habitat* [Cambridge, MA: MIT Press, 1970]).

59. Theodore Roszak, *From Satori to Silicon Valley: San Francisco and the American Counterculture* (San Francisco: Don't Call It Frisco, 1986),18.

60. Ibid., 17.

61. Ibid., 19.

62. Ibid., 19–20.

63. Baldwin, *Bucky Works*, 63.

64. Roszak, *From Satori to Silicon Valley*, 22.

65. Fuller and Marks, *Dymaxion World*, 196–198.

66. Joachim Krausse and Claude Lichtenstein, eds., *Your Private Sky, R. Buckminster Fuller: The Art of Design Science* (Baden, Switzerland: Lars Muller, 1999), 354.

67. Baldwin, *Bucky Works*, 62.

68. Buckminster Fuller, World Design Science Decade, 1965–1975, document 5, as quoted in Stewart Brand, ed., *The Whole Earth Catalog*, 1968; first edition reprint in *Whole Earth Review*, Winter 1999, no. 95:4.

69. J. Baldwin is as good at explaining complicated ideas as he is discussing the proper screwdriver for the job. The quote is from J. Baldwin, "Stuff & Nonsense," *Whole Earth Review*, Winter 1995, no. 88:108.

70. Review of *On Growth and Form*, by D'Arcy Wentworth Thompson, in *Whole Earth Catalog*, 1968 reprint, 9.

71. For examples of whole-systems thinking in practice, see Amory Lovins, L. Hunter Lovins, and Paul Hawken, "A Road Map for Natural Capitalism," in *Harvard Business Review on Business and the Environment* (Boston: Harvard Business School Press, 2000), 7–10.

72. Lars Skyttner, *General Systems Theory: Ideas and Applications* (River Edge, NJ: World Scientific, 2001); and Fritjof Capra, *The Web of Life: A New Scientific Understanding of Living Systems* (New York: Anchor, 1997).

73. Norbert Weiner, *Cybernetics: Or Control and Communications in the Animal and the Machine*, 2nd ed. (Boston: MIT Press, 1965); and Nick Heffernan, *Projecting Post-Fordism: Capital, Class, and Technology in Contemporary American Culture* (London: Pluto, 2000). The latter has an interesting brief discussion of Stewart Brand, the *Whole Earth Catalog*, and cybernetics.

74. Brand, *Whole Earth*, 1968 reprint, 34.

75. J. Baldwin, interview by the author, Penngrove, California, February 13, 2006.

76. The quote is from an interview in Chris Zelov and Phil Cousineau, eds., *Design Outlaws on the Ecological Frontier* (Easton, PA: Knossus, 1997), 44.

77. Ibid., 45.

78. Brand, "Buckminster Fuller," *Whole Earth*, 1968 reprint, 3.

79. Baldwin, interview.

80. Baldwin and Brand, *Soft-Tech*, 6; and J. Baldwin, "Encounters with the Mentor," in Zelov and Cousineau, *Design Outlaws*, 44–49. *Design Outlaws on the Ecological Frontier* (1997) is the companion to the film *Ecological Design: Inventing the Future* (2002). Both the film and the book provide excellent interviews of key figures in the ecological design movement. The book is basically a collection of oral histories.

81. Zelov and Cousineau, *Design Outlaws*, 47.

82. Baldwin, *Bucky Works*, 197–198.

83. Baldwin, interview.

84. Ibid.

85. J. Baldwin, "Making Things Right," *Whole Earth Review*, Fall 1995, no. 86:6.

86. Ibid., 4–7.

87. Ibid., 6.

88. One of the best sources on Baldwin's tool/technology philosophy and views on alternative technology is contained in a wonderful day-long interview of Stewart Brand and J. Baldwin by longtime *Whole Earth* supporter Marlon Brando. Brando asks great practical alternative technology questions, and Baldwin and Brand are forced to really articulate their ideas ("Marlon Brando Plans a Whole Earth TV Special" (a conversation with Stewart Brand and J. Baldwin) *CoEvolution Quarterly*, Winter 1975, 60–69.

89. Baldwin, interview. Baldwin recently nicely revisited his Moss experience in his article "The Case for Long Shots," *Strategy + Business* online, Autumn 2006, http://www.strategy-business.com.

90. Baldwin, "Making Things Right," 6.

91. "Moss the Tentmaker," *Time*, July 26, 1976, 66. See also the Moss Web site, http://www.mossinc.com/whoweare/history.aspx.

92. Ibid., 7.

93. Edward Hilliard's Colorado-based Redfield Rifle Site Company is another good example of convergence of industrial design and outdoor recreation technology and how these trends were shaping an alternative environmentalism designed to avoid traditional politics. For more on Hilliard, see Andrew Kirk, *Collecting Nature: The American Environmental Movement and the Conservation Library* (Lawrence: University Press of Kansas, 2001), 82–92.

94. Baldwin, interview.

95. Baldwin, "Stuff & Nonsense," 109.

96. For information on the Econoline, see the Ford Web site, http://media.ford.com/newsroom/release_display.cfm?release=7168.

97. Baldwin, "Making Things Right," 7.

98. The term *outlaw designers* appeared in Brand, *Last Whole Earth Catalog*, 112. The idea of the "outlaw edge" of design comes from Buckminster Fuller and was featured in the first *Whole Earth* in 1968.

99. Paul Feyerabend is best known for his *Against Method: Outline of an Anarchistic Theory of Knowledge* (London: Verso, 1979).

100. Baldwin, interview; and Lloyd Kahn, *Domebook 2* (Bolinas, CA: Shelter, 1972). Kahn followed up with the amazing *Shelter*, ed. Kahn (Bolinas, CA: Shelter, 1973). This classic study of hand-built homes from around the world is as fresh today as it was in the early 1970s.

101. Baldwin, interview.

102. J. Baldwin, "Designing Designers," *Whole Earth Review*, Summer 1995, no. 86:14–16. Baldwin also explains his design teaching strategies and philosophy on becoming a working designer in *Bucky Works*, 222–225.

103. Lloyd Kahn, "Pacific Domes," in *DomeBook 2*, 20–34. For fantastic pictures of the construction of the domes by the students, see Lloyd Kahn, "Pacific High School Revisited," and J. Baldwin, "Sun and Wind in New Mexico," both in Kahn's *Shelter*, 119, 164.

104. Baldwin, interview.

3. Baling Wire Hippies

1. J. D. Smith to Michael Phillips, WERC, M1045, box 7:7.

2. Stewart Brand, "History," in *The Last Whole Earth Catalog*, ed. Stewart Brand (New York: Portola Institute & Random House, 1971), 439; and Stewart Brand interview in Chris Zelov and Phil Cousineau, eds., *Design Outlaws on the Ecological Frontier* (Easton, PA: Knossus, 1997), 76.

3. Stewart Brand, "Alloy," in Brand, *Last Whole Earth Catalog*, 111.

4. Ibid., 112.

5. Ibid., 111.

6. Ibid., 112.

7. From J. Baldwin and Stewart Brand, eds., *Soft-Tech* (New York: Penguin, 1978), 8.

8. Peter Rabbit, *Drop City* (New York: Olympia, 1971); Timothy Miller, *The Hippies and American Values* (Nashville: University of Tennessee Press, 1991); Bill Voyd, "Funk Architecture," in *Shelter and Society*, ed. Paul Oliver (New York: Praeger, 1969); Steve Baer, "Zomes," in *Shelter*, ed. Lloyd Kahn (Bolinas, CA: Shelter, 1973), 134–135, and *Sunspots: Collected Facts and Solar Fiction*, 2nd ed. (Albuquerque, NM: Zomeworks, 1977); and Simon Sadler, "Drop City Revisited," *Journal of Architectural Education* 59 (February 2006): 5–14. Although not about the actual Drop City commune, T. C. Boyle really captures the madness and magic of this era and the fine line that separates liberal from conservative utopian visions in his novel, *Drop City* (New York: Viking, 2003). The quote is from the Farm Web site, http://www.thefarm.org/lifestyle/root2.html.

9. Roberta Price, *Huefano: A Memoir of Life in the Counterculture* (Amherst: University of Massachusetts Press, 2004). Price provides a mesmerizing and beautiful account of commune life. Libre was one of the communes that Stewart Brand visited on his early tours with his Whole Earth Truck Store (see 108–109).

10. "The Plowboy Interview: Steve and Holly Baer," *Mother Earth News* 22 (July/August 1973): 8; and Steve Baer, *Dome Cookbook* (Corrales, NM: Lama Fund, 1968).

11. Baer, *Sunspots;* Steve Baer, "The Clothesline Paradox," in Baldwin and Brand, *Soft-Tech*, 50–51; and "Plowboy Interview," 6–15.

12. Steve Baer, *Dome Cookbook*. Brand credits Steve Baer for inspiration in Brand, *Last Whole Earth Catalog*, 435.

13. The Manhattan Project designers and Alloy participants make for an interesting comparison regarding changes in thinking about science, technology, and environment in one short generation. For more on Trinity, see Ferenc Morton Szasz, *The Day the Sun Rose Twice: The Story of the Trinity Site Nuclear Explosion, July 16, 1945* (Albuquerque: University of New Mexico Press, 1995).

14. J. Baldwin, interview by the author, Penngrove, California, February 13, 2006.

15. Sadler, "Drop City Revisited."

16. Baldwin, interview.

17. Brand, *Last Whole Earth*, 113.

18. The quote is from Bruce Braun and Noel Castree, *Remaking Reality: Nature at the Millenium* (New York: Routledge, 1998), 33.

19. Brand, *Last Whole Earth Catalog*, 111.

20. Robert Frank, *The Americans* (New York: Aperture, 1978); originally published as *Les Americains* (Paris: R. Delpire, 1958). The Robert Frank Collection, in the Houston Museum of Fine Arts, houses Frank's remarkable documentary film collections, including *Pull My Daisy* (1959) and *Life-Raft Earth* (1969), but unfortunately not Alloy.

21. J. Baldwin, "One Highly Evolved Toolbox," *CoEvolution Quarterly* 5 (Spring 1975): 80–85; and Scott Landis, *The Workshop Book: A Craftsman's Guide to Making the Most of Any Work Space* (Newton, CT: Taunton, 1987), 140–144.

22. Baldwin, "One Highly Evolved Toolbox," 8.

23. Ibid. 152. A very nice recap of J. Baldwin's thinking on tools and their role in a fast-changing world is his "The Ultimate Swiss Omni Knife," *Whole Earth Review*, 30th anniversary issue, Winter 1998, 20.

24. Stewart Brand quoting Baldwin in the introduction to J. Baldwin, ed., *The Essential Whole Earth Catalog* (New York: Doubleday, 1989), 4.

25. Baldwin, interview.

26. Stewart Brand, "Introductions," in Baldwin and Brand, *Soft-Tech*. Brand provides a nice, concise overview of Baldwin's remarkable AT experiences and explains in brief how soft tech became a central feature of the *CoEvolution Quarterly*, 4.

27. Sim Van Der Ryn, "Abstracted in Sacramento: The Late Great State Architect Tells What He Learned about Power," *CoEvolution Quarterly* 22 (Summer 1979): 16–21. Van Der Ryn taught a course on outlaw building at the University of California–Berkeley in the fall of 1971.

28. For more on his work and philosophy of design, see Sim Van Der Ryn and Stuart Cowan, *Ecological Design* (Washington, DC: Island, 1996); and Sim Van Der Ryn, *The Toilet Papers: Recycling Waste and Conserving Water*, rept. ed. (Sausalito, CA: Ecological Design Press, 1999).

29. Van Der Ryn, "Abstracted in Sacramento," 16–21. This is the best discussion of how the alternative sensibility of *CoEvolution* translated into the halls of power and western politics of the 1970s. Carroll Pursell, *The Machine in America: A Social History of Technology* (Baltimore: Johns Hopkins University Press, 1995), 305.

30. Baldwin, interview.

31. Brand, "History," 752.

32. Patricia Nelson Limerick, *The Legacy of Conquest: The Unbroken Past of the American West* (New York: Norton, 1988), 94; and Jay Kinney, phone interview by the author, November 10, 2005.

33. Jonathan Hughes and Simon Sadler, eds., *Non-plan: Essays on Freedom, Participation and Change in Modern Architecture and Urbanism* (Oxford: Architectural, 2000), viii. Especially useful are the Hughes and Sadler essays, which both relate larger trends in urbanism and architecture to counterculture efforts in the American West and briefly to the *Whole Earth Catalog*. For concise essays on key issues in the encounter between professional architecture and emerging ideas of ecological design, see various authors, *Sustainable Architecture White Papers*, ed. David E. Brown, Mindy Fox, and Leslie Hoffman (2001; rept. New York: Earth Pledge, 2005), and James Steele, *Ecological Architecture: A Critical History* (New York: Thames & Hudson, 2005), 184–192.

34. Constance M. Lewallen and Steve Seid, *Ant Farm 1968–1978* (Berkeley: University of California Press, 2004). See also Simon Sadler, *The Situationist City* (Cambridge, MA: MIT Press, 1998).

35. Arthur Carhart, *How to Plan the Home Landscape* (New York: Doubleday, 1935); Chris Wilson, *The Myth of Santa Fe: Creating a Modern Regional Tradition* (Albuquerque, NM: University of New Mexico Press, 1997), 292–293; and Peter Van Dresser, *Development on a Human Scale: Potentials for Ecologically Guided Growth in Northern New Mexico* (New York: Praeger, 1972), and *Homegrown Sundwellings* (Santa Fe, NM: Lightning Tree, 1977).

36. Van Dresser, *Development*.

37. Wilson, *Myth of Santa Fe*, 293. For an earlier discussion of Decentrists, see Jane Jacobs, *The Death and Life of Great American Cities* (New York: Vintage, 1992 edition), 17–21.

38. For more on counterculture trends in self-build, see Jonathan Hughes, "After Non-plan: Retrenchment and Reassertion," in Hughes and Sadler, *Non-plan*, 178–180.

39. "The Plowboy Interview: John Shuttleworth," *Mother Earth News* 31 (January/February 1975): 6–14.

40. Lloyd Kahn, *Domebook 2* (Bolinas, CA: Shelter, 1972). Kahn followed up with the amazing *Shelter* (Bolinas, CA: Shelter, 1973). For an updated version, see Lloyd Kahn, *Home Work: Handbuilt Shelter* (Bolinas, CA: Shelter, 2004).

41. James Marston Fitch includes a very nice, concise discussion of prototypes versus replicas and the cultural implications of the democratization of American architecture in his *Historic Preservation: Curatorial Management of the Built World* (Charlottesville: University Press of Virginia, 1990), 8.

42. Lloyd Kahn, *Refried Domes* (Bolinas, CA: Shelter, 1989). This newsprint broadside has a nice, concise history of Kahn's publishing efforts.

43. This view is represented in Lloyd Kahn's Domebooks and the alternative technology bible, *Rainbook: Resources for Appropriate Technology* (New York: Schocken, 1977), to name just a couple of examples.

44. Joseph J. Corn and Brian Horrigan, *Yesterday's Tomorrows: Past Visions of the American Future* (Baltimore: Johns Hopkins University Press, 1984), 62, 67.

45. For Lloyd Kahn's views on the failure of the geodesic dome concept as a home, see his *Refried Domes*.

46. Stewart Brand, *How Buildings Learn: What Happens after They're Built* (New York: Penguin, 1994), 59.

47. Baldwin, interview.

48. Steele, *Ecological Architecture*, 184–192.

49. J. Baldwin, "Sun and Wind in New Mexico," in Kahn, *Shelter*, 164.

50. Helga Olkowski, Bill Olkowski, Tom Javits, and the Farallones Institute staff, *The Integral Urban House: Self-reliant Living in the City* (San Francisco: Sierra Club, 1979); and Grant to Bill and Helga Olkowski, for urban-oriented ecological design/farming, in Huey Johnson Papers (HJP), Resource Renewal Institute, Point Grants vol. 2. There's a nice discussion of the Olkowskis and the integral urban house concept in Christine Macy and Sarah Bonnemaison, *Architecture and Nature: Creating the American Landscape* (New York: Routledge, 2003), 333–338.

51. This is not to say that Buckminster Fuller's design principles were unnatural. Over the years, many similarities between Fuller's geodesics and microorganisms were uncovered.

52. For an excellent in-depth analysis of counterculture architecture and its relationship to the rise of the ecology movement, see Macy and Bonnemaison, *Architecture and Nature*. Their chapter "Closing the Circle: The Geodesic Domes and a New Ecological Consciousness, 1967" provides a detailed analysis of the geodesic dome's role in shaping and reflecting changing perceptions of architecture and nature at a critical turning point in American history. The quote is from 339.

53. Two good books on the idea of home in world history and America culture are Tracy Kidder, *House* (Boston: Houghton Mifflin, 1985), for the American view elegantly told, and Witold Rybczynski, *Home: A Short History of an Idea* (New York: Penguin, 1986), for the world perspective. For an engaging look at the changing view of the role of technology in shaping home design from the 1920s to the 1980s, see Corn and Horrigan, *Yesterday's Tomorrows*, 61–85.

54. Brand, *How Buildings Learn*, 156.

55. Ibid., 159.

56. Dennis Parks, *Living in the Country Growing Weird: A Deep Rural Adventure* (Reno: University of Nevada Press, 2001), 1.

57. Kevin Starr, *Material Dreams: Southern California through the 1920s* (New York: Oxford University Press, 1990), ix.

58. Sadler, "Drop City Revisited," 12.

59. Macy and Bonnemaison, *Architecture and Nature*, 338.

60. This was especially true when preservationists, both cultural and natural, found themselves united in a desire to protect adopted regional traditions. For more on one important nexus of regional/environmental preservation, see Wilson, *Myth of Santa Fe*, 293–294. Things may be changing now with even the conservative magazine of the National Trust for Historic Preservation, *Preservation*, linking cultural and environmental preservation (James Conway, "Stealing Beauty: The West's Common Lands Need Real Defenders," *Preservation* 58, no. 3 [May/June 2006]: 28–33, 70–71).

61. Mike Wallace, *Mickey Mouse History and Other Essays on American Memory* (Philadelphia: Temple University Press, 1996), 20.

62. This insight is thanks to a discussion with Simon Sadler and comments provided during an American Studies Association panel, Oakland, California, October 2006.

63. Stewart Brand, "Neighborhood Preservation Is an Ecology Issue," *CoEvolution Quarterly* 15 (Fall 1977): 40–44. For more on this episode, see "Waterfront Struggle: A Very Partial History," *Whole Earth Catalog* Records box 12:6ff. "Gate 5 Invaded," Garlic Press, vol. 3, August, 1977.

64. Jane Jacobs, *The Death and Life of Great American Cities* (New York: Vintage, 1989).

65. Brand, "Neighborhood Preservation," 40.

66. Baldwin and Brand, *Soft-Tech*, 6.

67. Ibid.

68. As quoted in Jason F. McLennan, *The Philosophy of Sustainable Design: The Future of Architecture* (Kansas City, MO: Ecotone, 2004), xiii.

69. In 1978, J. Baldwin and Stewart Brand edited *Whole Earth*'s *Soft-Tech*, a distillation of a decade of evolving thought about alternative technology at *Whole Earth* and the primary document for the history of the movement. In 1995, Baldwin would work to help create the stillborn International Ecological Design Institute: "Nobody had time to run it." An excellent discussion of the differences between environmentalists and ecological designers is in Baldwin's interview of February 13, 2006.

70. William McDonough and Michael Braungart, *Cradle to Cradle: Remaking the Way We Make Things* (New York: North Point, 2002), 60–61.

71. Jane Jacobs, *Systems of Survival: A Dialogue on the Moral Foundations of Commerce and Politics* (New York: Random House, 1992). The two-syndrome approach is explained in the text (23–25 and 31) and in an appendix (215).

72. McDonough and Braungart, *Cradle to Cradle*, 59–60.

73. Jacobs, *Systems of Survival*, 93–94.

74. McDonough and Braungart, *Cradle to Cradle*, 59–60.

75. Samuel P. Hays, *Beauty, Health, and Permanence: Environmental Politics in the United States, 1955–1985* (Cambridge: Cambridge University Press, 1987), 481–482. No single book better captures the universe of environmental politics than this landmark study.

76. Hal K. Rothman, *Saving the Planet: The American Response to the Environment in the Twentieth Century* (Chicago: Ivan R. Dee, 2000), 141–145. I owe thanks to Hal for many memorable discussions of this period and NEPA and the EIS in particular.

77. Peter Warshall, interview by the author, Tucson, Arizona, November 11, 2006.

78. Ibid.

79. Ibid.

80. Ibid.

81. Stewart Brand, introduction to Peter Warshall, "Streaming Wisdom: Watershed Consciousness in the Twentieth Century," *CoEvolution Quarterly* 12 (Winter 1976/1977): 4–13 (quote, 4); and Peter Warshall, "Sociodemography of Free-Ranging Male Rhesus Macaques," Ph.D. diss., Harvard University, 1973. For a very nice, concise discussion of Levi-Strauss and the relationship between nature and culture, see Dave Hickey, "Shooting the Land," in *The Altered Landscape*, ed. Peter E. Pool (Reno: University of Nevada Press, 1999), 22–32.

82. Warshall, interview.

83. Ibid.

84. Ibid.

85. For more on Orville Schell's work for *Whole Earth,* see the major "Access to China" section he contributed to *The Whole Earth Epilog: Access to Tools,* ed. Stewart Brand (New York: Point/Penguin, 1974), 626–633.

86. Warshall, interview.

87. Ibid.

88. The Bolinas *Hearsay* newspaper is a great source for the activities during the septic tank fight and generational transition in the town.

89. Warshall, interview.

90. Peter Warshall, *Septic Tank Practices: A Guide to the Conservation and Re-use of Household Wastewater* (New York: Anchor reprint edition, 1979). The story of Bolinas and the fight for control of the Public Utility District and the town itself is well told by Orville Schell in *The Town That Fought to Save Itself* (New York: Pantheon, 1976). Warshall, interview.

91. Warshall, *Septic Tank Practices,* 8.

92. Ibid., 10–11.

93. Peter Warshall, "Natural History Comes to Whole Earth," *CoEvolution Quarterly* 2 (Summer 1974): 70–71.

94. Peter Warshall, guest ed., "Watershed Consciousness" section, with Stewart Brand's introduction of Warshall (*CoEvolution Quarterly* 12 [Winter 1976]: 4–62).

95. Peter Warshall on Raymond Dasmann in "Raymond Dasmann," *CoEvolution Quarterly* 12 (Fall 1976): 25. His comments preface two significant essays by Dasmann that explore the roots of emerging conflicts among environmental activists and scholars.

96. Hays, *Beauty, Health, and Permanence,* 460–463. Hays explains how public support and pressure from environmental organizations spurred legislative action.

97. Karen R. Merrill, *The Oil Crisis of 1973–1974* (Boston: Bedford/St. Martin's, 2007); and Daniel Yergin, *The Prize: The Epic Quest for Oil, Money and Power* (New York: Free Press, 1991).

98. In California, where the energy crisis was particularly acute, appropriate technology research would eventually gain direct support from the state government under Gov. Jerry Brown, whose administration elevated ecological design proponents to new heights of influence. Sim Van Der Ryn, Stewart Brand, J. Baldwin, and Huey Johnson were appointed to significant positions in the Brown administration. And, in 1976, Brown created a state Office of Appropriate Technology in an effort to search for viable solutions to the reliance on foreign oil.

99. Daniel Horowitz, *Jimmy Carter and the Energy Crisis of the 1970s* (Boston: Bedford/St. Martin's, 2005).

100. For a thoughtful account of Earth Day and its consequences, see Robert Gottlieb, *Forcing the Spring: The Transformation of the American Environmental Movement* (Washington, DC: Island, 1993), 105–114.

101. Statistics are from Kirkpatrick Sale, *The Green Revolution: The American Environmental Movement, 1962–1992* (New York: Hill & Wang, 1993), 33.

102. Amory Lovins, "Energy Strategy: The Road Not Taken," *Foreign Affairs* 55 (October 1976): 65–96; Hugh Nash, ed., *The Energy Controversy: Soft Path Questions and Answers* (San Francisco: Friends of the Earth, 1979); and Jim Harding, ed., *Tools for the Soft Path* (San Francisco: Friends of the Earth, 1979).

103. Lovins, "Energy Strategy," 65.

104. For good biographical coverage and context for Lovins's achievements, see Art Kleiner, *The Age of Heretics: Heroes, Outlaws, and the Forerunners of Corporate Change* (New York: Doubleday, 1996), 303–305. Also useful is Chip Brown, "High Priest of the Low-Flow Shower Heads," *Outside Magazine*, November, 1991, 58; and John R. Emshwiller, "Amory Lovins Presses Radical Alternatives for Fueling the Nation," *Wall Street Journal*, March 16, 1981, 1.

105. Lovins, "Energy Strategy," 82–83.

106. The Rocky Mountain Institute Web site, http://www.rmi.org.

107. Paul Hawken, Amory Lovins, and L. Hunter Lovins, *Natural Capitalism: Creating the Next Industrial Revolution* (Boston: Little, Brown, 1999). A very good, concise overview of his perspective on evolving trends in alternative energy is in Amory B. Lovins, "Energy Lessons Learned and to be Learned: Verities That Will Astonish Some and Delight the Rest," *Whole Earth Review*, Winter 1998, 31.

108. Hawken et al., *Natural Capitalism*, 310–313. The four-mind-set idea comes from, Donella Meadows, "Seeing the Population Issue Whole," in *Beyond the Numbers*, ed. L. A. Mazur (Washington, DC: Island, 1994), 23–33.

109. The quotes are from Hawken et al., *Natural Capitalism*, 310–312.

110. Ibid., 312.

111. Stewart Brand, "Local Dependency," in Baldwin and Brand, *Soft-Tech*, 5.

112. T. Lindsay Baker, *A Field Guide to American Windmills* (Norman: University of Oklahoma Press, 1985); and Paul Gipe, *Wind Energy Comes of Age* (New York: Wiley, 1995). The best study of wind energy is Robert W. Righter, *Wind Energy in America: A History* (Norman: University of Oklahoma Press, 1996). David Rittenhouse Inglis, *Wind Power and Other Energy Options* (Ann Arbor: University of Michigan Press, 1978); Michael Hackleman, *The Homebuilt, Wind-Generated Electricity Handbook* (Culver City, CA: Peace, 1975); and Richard L. Hills, *Power from Wind: A History of Windmill Technology* (Cambridge: Cambridge University Press, 1994). See also Nicholas P. Cheremisinoff, *Fundamentals of Wind Energy* (Ann Arbor, MI: Ann Arbor Science, 1978); and Douglas R. Coonley, *Wind: Making It Work for You* (Philadelphia: Franklin Institute Press, 1979).

113. Hills, *Power from Wind*, 265–281.

114. Baldwin and Brand, *Soft-Tech*, 5.

115. *Soft Machine*, ABC/Probe, 1968; and Richard Brautigan, *The Pill versus the Spring Hill Mine Disaster* (San Francisco: Four Seasons, 1968), 1. Brautigan's poem was distributed as a free broadside by the Communications Company in Haight-Ashbury, June 1967.

116. Kevin Kelly, ed., *Signal: Communications Tools for the Information Age: A Whole Earth Catalog* (New York: Harmony, 1988), 3.

117. Stewart Brand, "Bigger, Finer, Slower," in *Whole Earth Ecolog: The Best of Environmental Tools and Ideas*, ed. J. Baldwin and Stewart Brand (New York: Harmony, 1990), 3.

118. Ibid. For more on Jobs, Wozniak, and Apple, see Steven Levy, *Insanely Great: The Life and Times of Macintosh, the Computer That Changed Everything* (New York: Penguin, 1995); Steven Levy, *Hackers: Heroes of the Computer Revolution* (New York: Penguin, 1994); and Jeff Goodell, "The Rise and Fall of Apple Inc.," *Rolling Stone*, April 4, 1996, 51–73, and April 18, 1996, 59–88.

119. Goodell, "Rise and Fall of Apple," 52.

120. For more on the environmental issues associated with personal computers, see Alex Steffen, ed., *World Changing: A User's Guide to the 21st Century* (New York: Abrams, 2006), 134–137.

121. Theodore Roszak, *The Cult of Information: A Neo-Luddite Treatise on High-Tech, Artificial Intelligence, and the True Art of Thinking* (Berkeley: University of California Press, 1994), xiii–xv.

122. The home page of the Well is at http://www.well.com (accessed September 28, 1998). Fred Turner is the leading authority on this aspect of *Whole Earth*'s history. For his marvelous analysis of the links between *Whole Earth*'s print network and the rise of the digital age, see his "Where the Counterculture Met the New Economy: The Well and the Origins of Virtual Community," *Technology and Culture* 46 (July 2005), 485–512, and especially his *From Counterculture to Cyberculture: Stewart Brand, the Whole Earth Network, and the Rise of Digital Utopianism* (Chicago: University of Chicago Press, 2006). Some similar ground is covered by journalist John Markoff in *What the Dormouse Said: How the 60s Counterculture Shaped the Personal Computer Industry* (New York: Viking, 2005); and, of course, Theodore Roszak's indispensable book, *From Satori to Silicon Valley* (San Francisco: Don't Call It Frisco, 1986).

123. Of particular interest is Paul Krassner and Ken Kesey's turn as editors under the banner of Krassner's *Realist* presents *The Last Supplement to the Whole Earth Catalog* (March 1971), featuring the work of Krassner, Kesey, R. Crumb, and a wonderful collection of counterculture authors and artists.

124. Stewart Brand, "Liferaft Earth," in Brand, *Last Whole Earth Catalog*, 35–39.

125. For a critique of the Spaceship Earth and the life-raft metaphor, see Garrett Hardin, "Living on a Lifeboat," *BioScience* 24 (October 1974): 561–568.

126. Stewart Brand, comments on *The Population Bomb*, by Paul R. Erhlich, in Brand, *Last Whole Earth Catalog*, 34.

127. J. Baldwin, "World Game," in Baldwin and Brand, *Whole Earth Ecolog*, 7; J. Baldwin, *Bucky Works: Buckminster Fuller's Ideas for Today* (New York: John Wiley, 1996), 197–203; and Hugh Kenner, *Bucky: A Guided Tour of Buckminster Fuller* (New York: William Morrow, 1973), 282–284.

128. Baldwin, "World Game," 7.

129. Ibid.

130. Stewart Brand, "Game Design," in Brand, *Last Whole Earth Catalog*, 35.

131. Andrew Fluegelman, ed., *The New Games Book* (San Francisco: Headlands, 1976), and *More New Games* (New York: Doubleday, 1981); Pat Farrington, "New Games Tournament," *CoEvolution Quarterly* 2 (Summer 1974): 94–101; and Stewart Brand, "Fanatic Life and Symbolic Death among the Computer Bums," in *II Cybernetic Frontiers*, ed. Stewart Brand (New York: Random House, 1974), 39–91.

132. Rick Field, as quoted in Stewart Brand, "Liferaft Earth," in Brand, *Last Whole Earth Catalog*, 36.

133. Brand, *Whole Earth Catalog*, 440.

134. Stewart Brand, ed., *Whole Earth Catalog: Find Your Place in Space* (Point Foundation, July 1970), 53. This supplement also contains a reprint on page 1 of a letter Brand wrote to the U.S. House Subcommittee on Education outlining his environmental philosophy in 1970.

135. Correspondence Files, *Whole Earth Catalog* Records, M1045.

136. Gregory Groth Jacobs to *Whole Earth*, in Brand, *Last Whole Earth Catalog*, 440.

137. *Last Whole Earth*, 440–441.

138. Ibid.

139. Criteria printed on the inside front cover of all the catalogs, including *Last Whole Earth Catalog*, 1. Edward Allen, *Stone Shelters* (Cambridge, MA: MIT Press, 1969).

140. Brand, *Last Whole Earth Catalog*, 435.

141. Baldwin, *Essential Whole Earth Catalog*, 402.

4. On Point

1. Stewart Brand, ed., *The Last Whole Earth Catalog* (New York: Point Foundation and Random House, 1971), 435.

2. Ibid. Brand emphasizes the doing over telling in a wonderful interview with Marlon Brando: "All of us are much more interactively doing stuff than we were then . . . I was busy being an editor . . . so now I've built my own house, I'm sailing these little boats. I think that may be the case for a lot of people who just had it [the *Whole Earth Catalog*] on the coffee table." ("Marlon Brando Plans a Whole Earth TV Special," *CoEvolution Quarterly* 8 [Winter 1975]: 68).

3. Ibid.

4. Stewart Brand, "History," in *The Whole Earth Epilog*, ed. Stewart Brand (New York: Point/Penguin, 1974), 752.

5. Brand, *Last Whole Earth Catalog*, 441.

6. Stewart Brand to Lawrence Ferlinghetti, May 14, 1974, WECR, M1045, 1: Stewart Brand Editorial Files; and Christopher Lehmann-Haupt, "Alas, It's Dead. Long Life to It!" *New York Times*, August 19, 1971, 33.

7. There are many good sources on the demise party. These are the best: Fred Moore to Dick Raymond, February 28, 1972, PFR, M1441, box 1:7;Stewart Brand, "Destination Crisis," paper for the Point Board of Directors, December 3, 1971, PFR, M1441, box 1:4; Thomas Albright and Charles Perry, "The Last Twelve Hours of the Whole Earth," *Rolling Stone* 86 (July 8, 1971): 1, 6–8; Michael Phillips, *The Seven Laws of Money* (Boston: Shambhala, 1997); Stewart Brand, ed., *The Whole Earth Epilog* (New York: Point/Penguin, 1974), 752; and Stewart Brand, "Genesis," in *The Essential Whole Earth Catalog* (New York, NY: Doubleday, 1986), 402–403.

8. Albright and Perry, "Last Twelve Hours," 1.

9. Ibid.

10. Brand, "History," 752.

11. Moore to Raymond.

12. Albright and Perry, "Last Twelve Hours," 6.

13. Point Foundation Grant Summary, *CoEvolution Quarterly* 2 (Summer 1974): 172–173.

14. Brand, "History," 752.

15. Ibid.

16. Albright and Perry, 6.

17. Ibid.

18. Moore to Raymond.

19. Peter Warshall, interview by the author, Tucson, Arizona, November 11, 2006.

20. Stewart Brand, "Money," in Brand, *Last Whole Earth Catalog*, 438.

21. Albright and Perry, "Last Twelve Hours," 8.

22. Brand, "History," 752.

23. Phillips, *Seven Laws of Money*, 102.

24. Richard Raymond to Mr. Don C. Frisbee, June 1, 1970, PFR, M1441, box 1:1.

25. Ibid.

26. Brand, "Destination Crisis," 2; and paper for the Point Board of Directors, December 3, 1971, PFR, M1441, box 1:4.

27. Michael Phillips, "Right Livelihood," in *The Briarpatch Book*, ed. Michael Phillips (San Francisco: New Glide/Reed, 1978), 174–175.

28. Dick Raymond, phone interview by the author, July 25, 2005; and Dick Raymond to Point Board, September 11, 1972, PFR, M1441, box 1:6.

29. Brand, *Whole Earth Epilog* (Baltimore: Point/Penguin, 1974), 616.

30. Michael Phillips, "Point Arrival Paper," PFR, M1441, box 1:8.

31. Michael Phillips to Stewart Brand, May 17, 1971, WECR, M1045, box 3:4.

32. Dick Raymond to J. D. Smith, March 16, 1973, PFR, M1441, box 2.

33. Meeting of Board of Directors, November 2, 1971, PFR, M1441, box 1:6; and Michael Phillips, phone interview by the author, March 2, 2006.

34. Phillips, phone interview; Phillips, *Seven Laws of Money*, 66–67; and Brand, "History," 752.

35. Dick Raymond to Mr. Bob Kochtitzky, July 27, 1973, in PFR, M1441, box 1:2.

36. Phillips, *Seven Laws of Money*, 67.

37. Michael Phillips, "Seven Laws of Money," PFR, M1441, box 2.

38. Paul Hawken, Amory Lovins, and L. Hunter Lovins, *Natural Capitalism: Creating the Next Industrial Revolution* (Boston: Little, Brown, 1999); and Paul Hawken, *The Ecology of Commerce: A Declaration of Sustainability* (New York: HarperCollins, 1993), and *Growing a Business* (New York: Simon & Schuster, 1987).

39. Paul Hawken, "Of Briars," in Phillips, *Briarpatch Book*, 176–183. See also Andy Alpine, "The Briarpatch Network," *CoEvolution Quarterly*, 8 (Winter 1975): 8–9; and Michael Phillips, "What is the Meaning of Money," PFR, M1441, box 1:1.

40. Michael Phillips to Stewart Brand. E. F. Schumacher's quote is from the subtitle of *Small Is Beautiful* (*Small Is Beautiful: Economics as If People Mattered* [New York: Harper & Row, 1973]).

41. Huey Johnson, interview by the author, San Francisco, February 14, 2006.

42. Brand, "History," 752.

43. Huey Johnson, phone interview by the author, November 8, 2005.

44. Ibid.

45. Stewart Brand, quoted in Meeting of Board of Directors.

46. Waldemar A. Nielsen, *The Big Foundations: A Twentieth Century Fund Study* (New York: Columbia University Press, 1972), 3. See also Mark Dowie, *American Foundations: An Investigative History* (Cambridge, MA: MIT Press, 2001).

47. Dowie, *American Foundations*, x.

48. Stewart Brand, "Agent Report," January 1972, PFR, M1441, box 2:22.

49. Huey Johnson, "Point Departure Paper," HJP.

50. Dick Raymond to Mike Phillips, March 16, 1973, PFR, M1441, box 1:2.

51. Meeting of Board of Directors. A small version about the founding of Point appears in Brand, "History," 752, as does the quote "radical ad-man." Also important is the official Point legal history drafted by Point lawyer Lawrence Klein (Lawrence Klein to Mr. Richard Austin, January 29, 1973, PFR, M1441, box 1:3).

52. Brand, "History," 752.

53. Brand, "Agent Report."

54. Meeting of Board of Directors, 2.

55. Ibid., 4

56. Brand, "Destination Crisis," 2.

57. Ibid., 5.

58. J.D. [Smith] to Michael Phillips, WECR, box 7:7.

59. Brand, "History," 752.

60. Point Minutes, February 6–7, 1972, PFR, M1441, box 2:22; and Johnson, phone interview.

61. Johnson, interview.

62. Brand, "Destination Crisis," 5.

63. Ibid., 5–10.

64. Brand, "Destination Crisis," 5.

65. Stewart Brand, ed., *The Media Lab: Inventing the Future at MIT* (1987; rept. New York: Penquin, 1988), 225. For a wonderful overview of science fiction as a genre with statistics on readers and how these readers cross over with *Whole Earth* publications, see Pamela Cokeley to Stewart Brand, January 16, 1976, WECR, box 4:5.

66. Statistics from Cokeley to Stewart Brand. For another perspective on sci-fi and counterculture environmentalism, see Theodore Roszak, *From Satori to Silicon Valley: San Francisco and the American Counterculture* (San Francisco: Don't Call It Frisco, 1986), 16.

67. Bill English, "A Cottage Industry," PFR, M1441, box 1:4.

68. Ibid. John Markoff places Bill English within the context of early personal computer development in Markoff's *What the Dormouse Said: How the 60s Counterculture Shaped the Personal Computer Industry* (New York: Viking, 2005), 41–42.

69. Point Board Minutes, November 9, 1972, PFR, M1441, box 1:3, 7.

70. Huey Johnson, "Activist Chairs for an Eco-Renaissance," PFR, M1441, box 1:4; and Huey Johnson to Stewart Brand, November 29, 1971, PFR, M1441, box 1:4, discussion of Johnson's views about Point activism.

71. Board of Directors Meeting, December 12, 1971, PFR, M1441, box 1:6, 3.

72. Stewart Brand to Jan Mirikitani, March 21, 1975, PFR, M1441, box 2:9.

73. Board of Directors Meeting.

74. Johnson, phone interview.

75. Bill Birchard, *Nature's Keepers: The Remarkable Story of How the Nature Conservancy Became the Largest Environmental Organization in the World* (San Francisco: Jossey-Bass, 2005), 57; and Dennis Farney, "Land for Posterity: A Conservation Group Preserves Choice Sites by Aggressive Tactics," *Wall Street Journal*, March 11, 1970, 1. For Huey Johnson's perspectives on land banking, see his "Land Banking," *CoEvolution Quarterly* 2 (Summer 1974): 72–77.

76. Birchard, *Nature's Keepers*, 58; and Johnson, phone interview.

77. Huey D. Johnson, *No Deposit—No Return: Man and His Environment—A View toward Survival* (Reading, MA: Addison-Wesley, 1970).

78. Ibid., 316.

79. Ibid.

80. Johnson materials, PFR, M1441, box 2:19. Kay Collins, newly appointed director of the Conservation Library in Denver, also received the award that year ("Women Are Winners of Eight of Twenty American Motors Conservation Awards," American Motors Conservation Awards Records, 1943–1981, box 15, Western History/Genealogy Department, Denver Public Library, Denver, Colorado).

81. Brand made the comment in print in his lead comments to Johnson, "Land Banking," 72: "Not long ago I heard myself soothing a lady ruffled by Huey's sometimes abrupt manner, 'Huey Johnson is a thug-for-Good.'"

82. Johnson, phone interview.

83. Huey Johnson, "Activist Chairs for an Eco-Renaissance," PFR, M1441, box 1:4, 2.

84. Ibid., 5.

85. Johnson, interview.

86. Johnson, "Point Departure Paper."

87. "Huey Johnson Interview," *Whole Earth Review,* Winter 1988, 61:64.

88. Huey Johnson to Michael McCurry, December 23, 1975, WECR, box 5:1. For background on the Trust for Public Land, see the Trust for Public Land Web site, http://www.tpl.org.

89. Huey Johnson to Ivah Deering, December 11, 1973, HJP, Point Foundation Grants vol. 1. This remarkable long letter outlines Johnson's thoughts about Point, Trust for Public Land, and pragmatic environmentalism.

90. Ibid.

91. Especially interesting for this topic are a series of letters from Huey Johnson to various Point board members and environmentalists outlining his efforts in urban environmentalism (Point Foundation Report for 1973, 264–271, PFR, M1441, box 1:2, in particular, Huey Johnson to Michael Phillips, June 1, 1973; also, Point Foundation Grant to Bill and Helga Olkowski, for urban-oriented ecological design/farming, HJP, Point Foundation Grants, vol. 2).

92. Cynthia F. Lambert, "Huey D. Johnson: The Green Party," *Mother Earth News* 84 (November/December 1983), http://www.motherearthnews.com/Homesteading-and-Self-Reliance/1983-11-01/Profiles.aspx (accessed April 30, 2007).

93. Point Foundation Reports for 1972–1973, PFR, M1441, boxes 1:2 and 1:3. The quote is from the 1972 report, page 3.

94. A good overview of Jerry Mander's views on Point and records of grants is in Jerry Mander to Directors of Point, March 7, 1973, PFR, M1441, box 1:2.

95. James J. Farrell, *The Sprit of the Sixties: Making Postwar Radicalism* (New York: Routledge, 1997), 210.

96. Jerry Mander, *In the Absence of the Sacred: The Failure of Technology and the Survival of the Indian Nations* (San Francisco: Sierra Club, 1991), 3–4; and Point Foundation Report 1973, PFR, M1441, boxes 1:2 and 1:3, 285–322. Mander's views on technology and environment are carefully articulated in his reports to the board. In particular, he discusses his views on television and the environment and the role of public media in shaping the response to the environmental crisis.

97. Point Foundation Board Meeting, March 13, 1973, PFR, M1441, box 1:2, 35.

98. Diana Shugart to Stewart Brand, "Resignation," PFR, M1441, box 2:16.

99. Board of Directors Meeting, March 15, 1972, PFR, M1441, box 1:6, 3.

100. Ernest Callenbach, *Ecotopia* (New York: Bantam, 1977).

101. Point Foundation Report for 1972, PFR, M1441, box 1:3; Point Foundation Report for 1973, PFR, M1441, box 1:2; and Point Foundation Minutes, 1971–1974, PFR, M 1441, box 2:22. A good summary of Point granting activities is in *CoEvolution Quarterly* 2 (Summer 1974): 171–173. Huey Johnson kept scrupulous records of all his activities at Point, and his papers provide the most detailed picture of one member's giving over a period of years. Also, Johnson required his grantees to file reports with thorough explanations of their supported activities, providing an invaluable record of grantee work during the critical first three years of Point (HJP, "Point Grants of Huey Johnson," vols. 1 and 2).

102. Anita Sue Reed to Jerry Mander, April 5, 1973, PFR, M1441, box 1:2.

103. Point Foundation Report for 1972, PFR, M1441, box 1:3.

104. Bolinas Plan, WECR, M1045, box 7:7.

105. Richard Raymond to Dick Austin, March 13, 1973, PFR, M1441, box 2:2.

106. Brand, "History," 752.

107. HJP, Point Foundation Grants vol. 2. Project Jonah is also covered in documents from PFR, M1441, 2, and WECR, M1045, box 7:8–9.

108. Nevada Outdoor Recreation Association (NORA), "Big Book," Nevada State Museum. This remarkable, truly huge book contains the story of the organization and provides maps and documents related to wilderness surveys of Board of Land Management lands by county.

109. Huey Johnson, Report on Grant to Charles Watson, November, 1972, HJP, vol. 2.

110. The quote is from the Nevada Outdoor Recreation Association home page, http://www.nora.org.; and NORA, "Big Book," sec. 1, "Commons Ecology."

111. Brian Doherty, *This Is Burning Man: The Rise of a New American Underground* (New York: Little, Brown, 2004). Doherty celebrates the weirdness that is Burning Man and provides some brief contextual links to Stewart Brand and the San Francisco counterculture.

112. Charles Watson to Stewart Brand, 1972, HJP, Point Foundation Grants vol. 2. Charles Watson letter to Huey Johnson on September 11, 1972, outlines NORA goals and how Point funding was making them happen (HJP, Point Foundation Grants vol. 2). The grant from Point was big news in Nevada and made most of the state's papers. See, in particular, "Public Land Conservation Grant Made," *Nevada State Journal*, June 1, 1972, 1, in which Watson credits Brand and Johnson for making their efforts possible at a critical moment. See also "Preservation and the Public Lands," *National Parks Magazine*, November 1965, 20–21.

113. Charles Watson and T. H. Watkins, *The Lands No One Knows: America and the Public Domain* (San Francisco: Sierra Club, 1975).

114. William Bryan, "An Identification and Analysis of Power-Coercive Change Strategies and Techniques Utilized by Selected Environmental Change Agents" (Ph.D. diss., University of Michigan), 1971.

115. William Bryan, phone interview by the author, April 21, 2006.

116. Ibid.

117. A very thoroughly documented history of Bryan's grants and activities is in William Bryan Grant file, HJP, Point Foundation Grants, vol. 1.

118. See the Cook Center Web site at http://www.cookcenter.org.

119. Dick Raymond to Point Board, box 1:6, 5.

120. William L. Bryan Jr., "Appropriate Cultural Tourism: Can It Exist? Searching for an Answer," in *The Culture of Tourism, The Tourism of Culture: Selling the Past to the Present in the American Southwest,* ed. Hal K. Rothman (Albuquerque: University of New Mexico Press, 2003), 140–163. For information on Bryan's work in environmentally minded tourism, see Off the Beaten Path at http://www.offthebeatenpath.com and Adventure Collection at http://www.adventurecollection.com for responsible tourism guidelines.

121. The best secondary source on Auroville is Stephanie Mills's wonderful *In Service of the Wild: Restoring and Reinhabiting Damaged Land* (Boston: Beacon, 1995), 168–186. The community also chronicled its own history in two publications: *Auroville Perspective* (Pondicherry, India: Sri Aurobindo Ashram, 1973) and *Auroville: The First Six Years* (Auroville, India: Auropublication, 1974). Eric, "Report from Auroville," *CoEvolution Quarterly* 1 (Spring 1974): 72; and Alan Lithman, "Revisiting Auroville: A Twenty-Year Project That's Still Growing," in *Helping Nature Heal: An Introduction to Environmental Restoration,* ed. Richard Nilsen (Berkeley: Whole Earth Catalog/Ten Speed, 1991), 94–96, and "The Subtlest of Catastrophes: Erosion vs. Reforestation at Auroville," *CoEvolution Quarterly* 25 (Spring 1980): 72–74. A nice little section on Auroville as a model for ecotechnological living removed from the demands of consumption is in Norman Myers and Jennifer Kent, *The New Consumers: The Influence of Affluence on the Environment* (Washington, DC: Island, 2004), 137–138.

122. Barry Lopez, foreword to Nilsen, *Helping Nature Heal,"* v.

123. Nancy Jack Todd, *A Safe and Sustainable World: The Promise of Ecological Design* (Washington, DC: Island, 2005), 16; and John Todd, "Pioneering for the 21st Century: A New Alchemist's Perspective," paper presented for the Limits to Growth, 1975, Assessments of Alternatives to Growth Conference, The Woodlands, Texas, October 19–21, 1975, WECR, M1045, box 3:6. See also Jeffrey Jacob, *The New Pioneers: The Back-to-the-Land Movement and the Search for a Sustainable Future* (University Park: Penn State University Press, 1997). The best coverage of New Alchemy and the broader sustainable agriculture movement is in Jordan Benson Kleiman, "The Appropriate Technology Movement in American Political Culture" (Ph.D. diss., University of Rochester), 2000. The New Alchemy Institute Collections are housed at Iowa State University, Department of Special Collections, MS-254.

124. John Todd, "Ocean Arks," *CoEvolution Quarterly* 23 (Fall 1979): 46–55; J. Baldwin, "Trails of an Ocean Ark Model," *CoEvolution Quarterly* 24 (Winter 1979): 56–58; Nancy Jack Todd and John Todd, *From Eco-cities to Living Machines: Principles of Ecological Design* (Berkeley: North Atlantic, 1993), 33, 64–66; and Todd, *Safe and Sustainable World.* The *Journal of the New Alchemists* (1–7) provides an overview of the activities at New Alchemy and situates their work within a larger community of concerned activists, artists, designers, and fellow travelers.

125. Todd, *Safe and Sustainable World,* 10; and John Todd, Biographical Sketch, WECR, M1045, box 22:9.

126. The quote is from the John Todd interview in *Design Outlaws on the Ecological Frontier,* ed. Chris Zelov and Phil Cousineau (Easton, PA: Knossus, 1997), 40.

127. John Hess, "Farm-Grown Fish: A Triumph for the Ecologist and the Sensualist; Cabbage Problems Equally Effective," *New York Times*, September 6, 1973, 32.

128. Todd and Todd, *From Eco-cities*, 33 and 65. See also J. Baldwin, "The New Alchemists Are Neither Magicians Nor Geniuses. They Are Hard Workers," *CoEvolution Quarterly* 12 (Winter 1976/1977): 104–111.

129. Todd and Todd, *From Eco-cities*, 33.

130. Point Foundation Report for 1973, 41.

131. Point Foundation Meeting, March 13, 1973, PFR, M1441, box 2:15.

132. John Todd to Stewart Brand, February 27, 1973, PFR, M1441, box 1:6.

133. Hess, "Farm-Grown Fish," 32.

134. Todd, *Safe and Sustainable World*, 100.

135. Quoted in Todd, *Safe and Sustainable World*, 113. "The New Alchemy Institute Ark, Opening Ceremony," WECR, M1045, box 7:4.

136. John Quinney, "Out of the Ark and into the World," *Whole Earth Review*, Spring 1989, no. 62:33–35; and John Todd, "Adventures of an Applied Ecologist, or Shit Happens," ibid.,36–39. The quote is from J. Baldwin on why "eco-righteous demonstration centers" were failing in the 1980s. Quinney was the New Alchemy director brought in to save the organization during financial crisis.

137. Alan Lithman to the Point Directors, July 24, 1973, HJP, Point Foundation Grants vol. 1.

138. Lithman, "Revisiting Auroville," 94.

139. Alan Lithman to Richard Austin, December 21, 1973, WECR, M1045, box 7:7; and Johnson, interview.

140. HJP, Point Foundation Grants vol. 1.

141. Mills, *In Service*, 172.

142. The Auroville Web site, http://www.auroville.org/.

143. Point Report for 1974, "Schedule of Investments," 7, PFR, M1441, box 1:2.

144. "The Plowboy Interview: Steve and Holly Baer," *Mother Earth News* 22 (July/August 1973): 8. The best contemporary analysis of Drop City's architecture is Bill Voyd, "Funk Architecture," in *Shelter and Society*, ed. Paul Oliver (New York: Frederick A. Praeger, 1969), 156–164.

145. "Plowboy Interview," 11.

146. Ibid., 8. Even forty-plus years later, Farrington Daniels's book feels fresh (*Direct Use of the Sun's Energy* [New Haven, CT: Yale University Press, 1964]).

147. Simon Sadler, "Drop City Revisited," *Journal of Architectural Education* 59, no. 3 (February 2006): 12.

148. "Plowboy Interview," 6.

149. Ibid, 12.

150. Zomeworks Technology Forum, http://www.zomeworks.com/tech/tech.html.

151. Stewart Brand, "Zomeworks Solar Energy Research," PFR, M1441, box 2:22.

152. Ibid.

153. Ibid.

154. Steve Baer to Stewart Brand, December 28, 1971, PFR, M1441, box 2:22. For a discussion of technology, science, and government, see Baer, *Sunspots* (Albuquerque, NM: Zomeworks, 1975), 21–22.

155. Ken Butti and John Perlin, "Solar Water Heaters in California, 1891–1930," reprinted in *Soft-Tech*, ed. J. Baldwin and Stewart Brand (New York: Penguin, 1978), 52–63.

156. Baer, *Sunspots,* 50–51.

157. Steve Baer, "Gravity Engines and the Diving Engine," *CoEvolution Quarterly* 2 (Summer 1974): 81–83, and "The Double Bubble Wheel Engine," in Baldwin and Brand, *Soft-Tech,* 30–32; and Baer, *Sunspots,* 77–84.

158. Brand to Mirikitani.

159. Brand, "History," 752.

160. Point 1974 Reports, Point Foundation Grant tables and summaries, PFR, M1441, box 1:3.

161. Brand, "History," 752.

162. Point Foundation business was regularly updated on the back pages of *CoEvolution Quarterly* and *Whole Earth Review.* Two particularly good summaries of Point in the late 1980s and into the 1990s may be found in Point Foundation, "The Future of Point: A Growing Dialog," *Whole Earth Review,* Winter 1990, no. 69:133–134, and "Point Foundation Report," *Whole Earth Review,* Summer 1991, no. 71:137–139.

5. The Final Frontier

1. Steven V. Roberts, "Mail Order Catalog of the Hip Becomes a National Best Seller," *New York Times,* April 12, 1970, 67.

2. Ernest Callenbach, *Ecotopia* (New York: Bantam, 1977).

3. For Ecotopia situated with the western tradition of "Denial and Dependence," see Patricia Nelson Limerick, *The Legacy of Conquest: The Unbroken Past of the American West* (New York: W. W. Norton, 1987), 94–95.

4. Edward Bellamy, *Looking Backward: 2000–1887* (New York: Signet Classics, 2000). Originally published in 1888, *Looking Backward* makes an excellent comparison with *Ecotopia.* Both books spend more time on the reconciliation of consumerism and politics than on any other single issue.

5. Michael Allen, "'I Just Want to Be a Cosmic Cowboy': Hippies, Cowboy Code, and the Culture of a Counterculture," *Western Historical Quarterly* 36 (Autumn 2005): 274–299.

6. For a much more complete analysis of the genre of science fiction within the context of western history, see Carl Abbott, *Frontiers Past and Future: Science Fiction and the American West* (Lawrence: University Press of Kansas, 2006).

7. I do not want to imply that Ernest Callenbach was a consumer sellout, but just that an important part of his philosophy included an attempt to move his readers toward a new type of socially and environmentally sensitive consumption. He goes to great lengths to discourage consumerism and present alternative ways to "live better" without being a slave to the market in his *Living Cheaply with Style: Live Better and Spend Less,* 2nd ed. (San Francisco: Ronin, 2000).

8. Samuel P. Hays, *Beauty, Health, and Permanence: Environmental Politics in the United States, 1955–1985* (Cambridge: Cambridge University Press, 1987). Even after twenty years, Hays's monumental study remains the most insightful analysis of environmental politics.

9. Here, too, are many similarities with other utopian writers, Bellamy and Upton Sinclair in particular. Both of these authors presented a case for socialism that was embraced by very few, while the consumer issues they discussed as a vehicle to present the socialism provided compelling food for thought and cause for action for millions.

10. Stewart Brand, "We Are As Gods," *Whole Earth Review*, Winter 1998, no. 95: 2. This thirtieth-anniversary celebration folio issue contains a complete reprint of the first edition of the 1968 catalog, with a large section of essays written in 1998 reflecting on issues and themes explored in the various *Whole Earth*s over the years.

11. David Wrobel, *Promised Lands: Promotion, Memory, and the Creation of the American West* (Lawrence: University Press of Kansas, 2002). The counterculture migrants to the West were, to use Hal Rothman's phrase, neonatives—new residents who become tenacious and protective boosters of their new home (Hal K. Rothman, *Devil's Bargains: Tourism in the Twentieth-Century American West* [Lawrence: University Press of Kansas, 1998]).

12. Stewart Brand is responding here to a series of letters complaining about his support for Governor Brown and the regional bias of the magazine's "Business and Letters" section (*CoEvolution Quarterly* 18 [Summer 1978]: 143). Although California might not be considered the West now, it definitely was in the 1960s and 1970s.

13. Peter Warshall, interview by the author, Tucson, Arizona, November 11, 2006; Peter Warshall, *Septic Tank Practices: A Guide to the Conservation and Re-use of Household Wastewater*, rept. (New York: Anchor, 1979), 174; and Andrew Brown, "The Hustler," *Guardian Unlimited*, April 30, 2005, 22–25.

14. Gurney Norman to Stewart Brand, June 18, 1976, WECR, M1045, box 7:1.

15. Paul Kleppner, "Politics without Parties: The Western States, 1900–1984," in *The Twentieth-Century West: Historical Interpretations*, ed. Gerald D. Nash and Richard W. Etulain (Albuquerque: University of New Mexico Press, 1989), 295–338; and Michael P. Malone and F. Ross Peterson, "Politics and Protest," in *The Oxford History of the American West*, ed. Clyde A. Milner, Carol A. O'Connor, and Martha A. Sandweiss (New York: Oxford University Press, 1994), 501–533. See also Richard D. Lamm and Michael McCarthy, *The Angry West: A Vulnerable Land and Its Future* (Boston: Houghton Mifflin, 1982).

16. Robert Graf, *Wilderness Preservation and the Sagebrush Rebellions* (Savage, MD: Rowman & Littlefield, 1990).

17. Robert Goldburg, "The Western Hero in Politics: Barry Goldwater, Ronald Reagan, and the Rise of the American Conservative Movement," in *The Political Legacies of the American West*, ed. Jeff Roche (Lawrence: University Press of Kansas, forthcoming).

18. Anne M. Butler, "Selling the Popular Myth," in *The Oxford History of the American West*, ed. Clyde A. Milner, Carol A. O'Connor, and Martha A. Sandweiss (New York: Oxford University Press, 1994), 784.

19. Stewart Brand, "History," in *The Whole Earth Epilog*, ed. Stewart Brand (New York: Point/Penguin, 1974), 753.

20. Ibid.

21. Resurrection of the *Whole Earth Catalog* with the *Epilog* is explored in the *Whole Earth Epilog* files, WECR, M1045, box 5:12. The contract between Stewart Brand and the Point Foundation, October 26, 1973, WECR, box 7:8.

22. Stewart Brand, *How Buildings Learn: What Happens after They're Built* (New York: Penguin, 1994). Brand's compelling arguments on behalf of low-road architecture are explored in a chapter "Nobody Cares What You Do in There," 24–33.

23. Andrew Fluegelman correspondence with Random House, WECR, M1045, box 5:12.

24. Steve Jobs, Stanford University commencement address, June 12, 2005; and text from "Stanford Report," June 14, 2005, http://news-service.stanford.edu/news/2005/june15/. Jobs mistakenly thought that the *Epilog* was the last catalog produced, but otherwise explained the appeal and significance of the catalog very well.

25. Lee Dembart, "'Whole Earth Catalog' Recycled as 'Epilog,'" *New York Times*, November 8, 1974, 41; and Andrew Fluegelman to Mr. James Silberman, November 27, 1974, WECR, M1045, box 5:12.

26. Stewart Brand, "Fanatic Life and Symbolic Death among the Computer Bums" and "Both Sides of the Necessary Paradox," in *II Cybernetic Frontiers*, ed. Stewart Brand (New York: Random House, 1974).

27. Ibid., 39.

28. Again, for a remarkable analysis of Brand and the computer revolution, see Frederick Turner, *From Counterculture to Cyberculture: Stewart Brand, the Whole Earth Network, and the Rise of Digital Utopianism* (Chicago: University of Chicago Press, 2006).

29. Gregory Bateson, *Steps toward an Ecology of Mind* (New York: Ballantine, 1972).

30. Stewart Brand, introductory notes to Gregory Bateson's "The Pattern Which Connects," *CoEvolution Quarterly* 18 (Summer 1978): 5.

31. The quote is from Brand, *II Cybernetic Frontiers*, 9.

32. The first quote is from Brand, "History," 453. The second two are from Brand, *II Cybernetic Frontiers*, 9.

33. Brand, "History," in *Whole Earth Epilog*, 753.

34. The relationship between Brand and Bateson is explored in David Lipset's *Gregory Bateson: The Legacy of a Scientist* (Englewood Cliffs, NJ: Prentice-Hall, 1980). This wonderfully detailed biography covers the relationship between aging, but still very vital, Bateson and the almost middle-aged Brand well.

35. Brand, "History," in *Whole Earth Epilog*, 453.

36. The best source on the history of *CoEvolution Quarterly* is Art Kleiner and Stewart Brand, eds., *News That Stayed News: Ten Years of CoEvolution Quarterly* (San Francisco: North Point, 1986). This thoughtfully selected collection of important articles includes a thorough appendix with an issue-by-issue history of *CoEvolution* and nice summaries of issues and trends by Brand and Kleiner. See also *CoEvolution/Whole Earth Review* files, WECR, M1045, boxes 24–26.

37. Transcript of Brand's talk at the 1976 World Game workshop held at the University of Pennsylvania. Particularly useful is the question-and-answer section, where Brand fields questions about his changing philosophy, *CoEvolution*, and his work with Governor Brown on environmental issues. Brand carefully outlines his evolving environmental philosophy and specifically answers questions about Spaceship Earth to Gaia shift (WECR, M1045, box 7:6).

38. Stewart Brand, ed., *Space Colonies* (New York: Penguin, 1977), 4.

39. Gregory Bateson appeared with John Todd at an environmental conference in Southampton, New York, in 1975, where Bateson "attacked contemporary man's relationship to

his environment, and condemned progress in the name of which addictive pollutants had been introduced" (quoted from Lipset, *Gregory Bateson*, 284–285).

40. The quote is from Brand's foreword to *Space Colonies*, 4.

41. Kleiner and Brand, *News That Stayed News*, 331.

42. Jaron Lanier, "Life at the Techno-humanist Interface," *Whole Earth Review*, Winter 1998, no. 95: 2. Lanier became one of the most consistently insightful critics of the cyberrevolution at the *Review*.

43. For more on *Limits to Growth* and the ideal of voluntary simplicity, see Art Kleiner, *The Age of Heretics: Heroes, Outlaws, and the Forerunners of Corporate Change* (New York: Doubleday, 1996), 294–297. D. H. Meadows, Jorgen Randers, and Dennis L. Meadows, *The Limits to Growth: The 30-Year Update* (London: Earthscan, 2004).

44. Kleiner and Brand, *News That Stayed News*, 331.

45. The end result of this convergence is most easily discernible in the *Whole Earth Review*, Winter 1998, no. 95, "Designing/Restoring" section, which features thoughtful discussions of issues of sustainability and quality of life from Jane Jacobs, William McDonough, Amory Lovins, Catherine Sneed, and John and Nancy Jack Todd. For a thoughtful overview of the politics of sustainability, see Gary C. Bryner, *Gaia's Wager: Environmental Movements and the Challenge of Sustainability* (New York: Rowman & Littlefield, 2001).

46. This was done explicitly in Stewart Brand's "Neighborhood Preservation Is an Ecological Issue," *CoEvolution Quarterly* 15 (Fall 1977): 40–45.

47. Stephanie Mills, "Salons and Their Keepers," *CoEvolution Quarterly* 2 (Summer 1974): 102–105. Brand's quote is from the preface to the *Ten Years of CoEvolution Quarterly* reprint of the article (10). Stephanie Mills, *In Service of the Wild: Restoring and Rehabiting Damaged Land* (Boston: Beacon, 1995).

48. Peter Warshall, "Natural History Comes to Whole Earth," *CoEvolution Quarterly* 2 (Summer 1974): 70–71.

49. Peter Warshall, guest editor of the "Watershed Consciousness" section with Stewart Brand's introduction of Warshall, *CoEvolution Quarterly* 12 (Winter 1976): 4–62.

50. Peter Warshall on Raymond Dasmann, in "Raymond Dasmann," *CoEvolution Quarterly* 12 (Fall 1976): 25. Warshall's comments preface two significant essays by Dasmann exploring the roots of emerging conflicts among environmental activists and scholars.

51. Stewart Brand, introduction to Peter Warshall's "Streaming Wisdom: Watershed Consciousness in the Twentieth Century," *CoEvolution Quarterly* 12 (Winter 1976/1977): 4–13 (quote, 4).

52. Henry Allen to Stewart Brand, June 24, 1975, WECR, M1045, box 12:8. Allen enclosed a copy of a review essay he had recently published in the *Washington Post* on Robert M. Pirsig's *Zen and the Art of Motorcycle Maintenance* (New York: Bantam, 1975). "I hope that most *CQ* readers, being by definition students of The Problem—have read it." For Allen and many readers, an obvious shared community of interest was captured by *CQ*.

53. World Game discussion, WECR, M1045, box 7:6–7.

54. Herman Kahn's best-known book is *On Thermonuclear War* (Princeton: Princeton University Press, 1960) and the most easily digestible is *Thinking the Unthinkable* (New York: Touchstone reprint edition, 2002). For his biography, see Sharon Ghamari-Tabrizi, *The Worlds of Herman Kahn: The Intuitive Science of Thermonuclear War* (Cambridge, MA: Harvard University Press, 2005).

55. Kleiner, *Age of Heretics*, 148–150.

56. Louis Menand, "FAT MAN: Herman Kahn and the Nuclear Age," *New Yorker* online, June 27, 2005, http://www.newyorker.com/archive/2005/06/27/050627crbo_books.

57. Fred M. Kaplan, *The Wizards of Armageddon* (Stanford, CA: Stanford University Press, 1991); and Kleiner, *Age of Heretics*, 148–150.

58. Kleiner, *Age of Heretics*, 297–298.

59. Ibid., 268. Kleiner draws his idea of the "meme . . . an intensely hardy" cultural element from geneticist Richard Dawkins.

60. World Game discussion.

61. Kleiner and Brand, *News That Stayed News*, outlines readership, distribution, and the economic realities of running a quarterly magazine without the massive appeal of the *Whole Earth*. Brand, "History," in *Whole Earth Epilog*, 331–337.

62. The first article on Gaia appears in the summer 1975 issue (Lynn Margulis and James Lovelock, "The Atmosphere as Circulatory System of the Biosphere—The Gaia Hypothesis," *CoEvolution Quarterly* 6 [Summer 1975]: 31–40). The best analyses of Gaia and its creators are by James Lovelock, "The Independent Practice of Science," *CoEvolution Quarterly* 25 (Spring 1980): 22–30; and Jeanne McDermott, "Lynn Margulis—Unlike Most Microbiologists," *CoEvolution Quarterly* 25 (Spring 1980): 31–38.

63. As quoted in Bryner, *Gaia's Wager*, xix.

64. Nice collections of essays discussing the Gaia hypothesis and surrounding controversies are in Lynn Margulis and Dorion Sagan, *Slanted Truths: Essays on Gaia, Symbiosis, and Evolution* (New York: Copernicus, 1997), and Lynn Margulis, "Another Four-Letter Word: Gaia," *Whole Earth Review*, Winter 1998, no. 95:4. A nice, concise, layperson's definition of Gaia theory is in Bryner, *Gaia's Wager*, xviii–xxi.

65. James Lovelock's *Gaia: A New Look at Life on Earth* (Oxford: Oxford University Press, 1979), *The Ages of Gaia* (New York: W. W. Norton, 1988), *Gaia: The Practical Science of Planetary Medicine* (London: Gaia, 1991), and *Homage to Gaia* (Oxford: Oxford University Press, 2000), Lovelock's autobiography.

66. Timothy Wessels to Stewart Brand, September 15, 1976, WECR, M1045, box 7:6. The quote is from Brand's introduction to Lovelock's "Independent Practice of Science," 22–30.

67. Stewart Brand, "Free Space," *CoEvolution Quarterly* 7 (Fall 1975): 4–29, covers Stewart Brand's introduction to Gerard O'Neill and prints O'Neill's testimony before Congress about space colonies, "Is the Surface of a Planet Really the Right Place for an Expanding Technological Civilization?" (10–19), an overview of his theories. Responses about space colonies are in "Is Balance Really Possible Where Even Gravity Is Manufactured?" *CoEvolution Quarterly* 9 (Spring 1976): 4–79; Stewart Brand, "'Spaceship Earth' Comes Home to Roost," *CoEvolution Quarterly* 10 (Summer 1976): 4–7; and Wendell Berry, "Wendell Berry Angry," in the same issue (8–11). Berry's article is the most angry response, as the title indicates. Brand, *Space Colonies;* and Gerard K. O'Neill, *The High Frontier: Human Colonies in Space* (Princeton, NJ: Space Studies Institute Press, 1989). O'Neill's vision is immortalized in science fiction author Larry Niven's Hugo Award–winning *Ring World* (New York: Del Ray, 1985). Freeman J. Dyson, "Pilgrim Fathers, Mormon Pioneers, and Space Colonists: An Economic Comparison," Proceedings of the American Philosophical Society 122 (April 24, 1978): 63–68. The best scholarly treatment

of the space-colonies debate is Peder Anker's excellent "The Ecological Colonization of Space," *Environmental History* 10 (April 2005): 239–268.

68. Stewart Brand's introduction to Gerard O'Neill's "High Frontier," in Brand, *Space Colonies*, 8–30.

69. O'Neill, *High Frontier*, 290–291; and Gerard K. O'Neill, "Space Colonies and Energy Supply to the Earth," *Science*, n.s., 190 (December 5, 1975): 943–947.

70. O'Neill, *High Frontier*, 291.

71. Ibid., 292.

72. Walter McDougall, *The Heavens and the Earth: A Political History of the Space Age* (New York: Basic, 1985).

73. Anker, "Ecological Colonization," 242.

74. Donald Worster, *Nature's Economy: A History of Ecological Ideas*, 2nd ed. (Cambridge: Cambridge University Press, 1994), 362, 371.

75. Ibid, 371.

76. Ibid, 370. Peter Taylor's quote is from his "Technocratic Optimism, H. T. Odum, and the Partial Transformation of Ecological Metaphor after World War II," *Journal of the History of Biology* 21 (Summer 1988): 213–244.

77. A remarkable volume of correspondence about space colonies along with very thoughtful letters from Stewart Brand's friends, public figures, and significant *CoEvolution Quarterly* contributors are contained in WECR, M1045, boxes 7:1 and 3:1–4. This debate dominated to such an extent that many other references are sprinkled throughout the *Whole Earth Catalog* collections.

78. Berry, "Wendell Berry Angry," 8.

79. Gurney Norman to Stewart Brand.

80. Stewart Brand discusses the influence of *Dune* on his thinking in his journals (1966–1969 Notebooks, 50, SBP, M1237).

81. Bruno Latour, *Science in Action: How to Follow Scientists and Engineers through Society* (Cambridge: Harvard University Press, 1987).

82. Berry, "Wendell Berry Angry," 9: "If we elect to live by such disruptions then we must resign ourselves to a life of desperate (and risky) solutions: the alternation of crisis and 'breakthrough' described by E. F. Schumacher."

83. R. [Robert] Crumb, "Space Day Symposium: Or What Ever the Hell It Was Called," *CoEvolution Quarterly* 15 (Fall 1977): 48–51 (quotes, 51).

84. Russell Schweickart, "Whose Earth?" in *The Next Whole Earth Catalog: Access to Tools*, ed. Stewart Brand (New York: Rand McNally, 1980), 19.

85. Huey Johnson, phone interview by the author, November 8, 2005. The best discussion of how the populist alternative sensibility translated into the halls of power and western politics of the 1970s is Sim Van Der Ryn's "Abstracted in Sacramento: The Late Great State Architect Tells What He Learned about Power," *CoEvolution Quarterly* 22 (Summer 1979): 16–21.

86. "Environmentalist, Huey Johnson, wins United Nations premier environmental prize," press release, November 2001, at United Nations Environment Programme Web site, http://www.unep.org.

87. Huey Johnson, *Green Plans: Greenprint for Sustainability* (Lincoln: University of Nebraska Press, 1995), 31–41.

88. "Environmentalist, Huey Johnson."

89. Claire T. Dedrick and Bill Press, "Perspective Article," WECR, M1045, box 4:5. The outline of Governor Brown's environmental policies and goals shows influence of Stewart Brand and Huey Johnson. A good sample *CoEvolution Quarterly* article demonstrating the work of J. Baldwin, Brand, and Sim Van Der Ryn during this time is Michael Phillips's "Don't Build a House till You've Looked at This," *CoEvolution Quarterly* 19 (Fall 1978): 100–102.

90. Brand, *Space Colonies*, 146.

91. Ibid.; and "California in the Space Age: An Era of Possibilities," WECR, M1045, box 25:4.

92. Brand, *Space Colonies*, 146.

93. Ibid.

94. Stewart Brand to Russell Schweickart, WECR, M1045, box 3:1; and Stewart Brand, "Jacques Cousteau at NASA Headquarters," *CoEvolution Quarterly* 10 (Summer 1976): 12–17.

95. Brand, "Jacques Cousteau," 12.

96. "California in the Space Age," the meeting moderated by Brand between Carl Sagan and Jacques Cousteau on the topic "Oceans in Need of Help from Space," which is an excellent discussion of the pro-space environmental view by its three leading spokespeople.

97. Stewart Brand, phone interview by the author, September 9, 2004.

98. William L. Kahrl, ed., *The California Water Atlas* (Sacramento: State of California, Governor's Office of Planning and Research, and California Department of Water Resources, 1978); and Warshall, interview.

99. As is always the case, the best source for the history of any of the *Whole Earth* ventures is the editors themselves. One of the most interesting and innovative aspects of the *Whole Earth* publications was the consistent effort to relate the process and history of creating the product that readers held. Nowhere is Brand's talent as a reader more obvious than in the introspective history sections and many introductions he penned respectfully with future readers in mind.

6. Free Minds, Free Markets

1. *Arcade: The Comics Review* makes its first appearance with, "The Garden," *CoEvolution Quarterly* 7 (Fall 1975): 67–70. Jay Kinney, "What's Left: Revisiting the Revolution," *CoEvolution Quarterly* 8 (Winter 1975): 19–30; Kinney, "Beyond Left and Right," *Whole Earth Review*, Summer 2000, no. 101: 22–29; and Kinney, phone interview with author, November 10, 2005.

2. Kinney, "Beyond Left and Right," 27. The "Third Way," Kinney wrote, comes from "the Democratic Leadership Council and its affiliated think tank, the Progressive Policy Institute (PPI), which were responsible for brainstorming Bill Clinton's center-left strategy" (ibid.).

3. Charlene Spretnak, "How About That Green Option?" and Stephanie Mills, "Hay Foot, Straw Foot," *Whole Earth Review*, Summer 2000, no. 101: 47–49 and 49, respectively.

4. Charlene Spretnak and Fritjof Capra, *Green Politics* (New York: Dutton, 1984); Charlene Spretnak, *The Resurgence of the Real: Body, Nature, and Place in a Hypermodern*

World (Reading, MA: Addison-Wesley, 1997); and Stephanie Mills, ed., *Turning Away from Technology* (San Francisco: Sierra Club, 1997).

5. Mills, "Hay Foot," 49.

6. Theodore Roszak, *From Satori to Silicon Valley: San Francisco and the American Counterculture* (San Francisco: Don't Call It Frisco, 1986), 21–24.

7. See the Electronic Frontier Foundation Web site at http://www.eff.org.

8. John Perry Barlow, "Crime and Puzzlement: In Advance of the Law on the Electronic Frontier," *Whole Earth Review*, Fall 1990, no. 68:44–57.

9. John Perry Barlow, "A Declaration of the Independence of Cyberspace," Electronic Frontier Foundation, http://homes.eff.org/~barlow/Declaration-Final.html.

10. Brian Doherty, "John Perry Barlow 2.0: The Thomas Jefferson of Cyberspace Reinvents His Body—and His Politics," *Reason Magazine* online, www.reason.com, August/September, 2004.

11. The quote is from Louis Rossetto, "Rebuttal of the Californian Ideology," Louis Rossetto's reply to Richard Barbrook and Andy Cameron's *The Californian Ideology*, on the *Alamut Bastion of Peace and Information* online community Web site, http://www.alamut.com/subj/ideologies/pessimism/califIdeo_II.html, with lots of online debate about California ideology and cyberlibertarians. For the most thorough critique, see Richard Barbrook and Andy Cameron, "California Ideology," on the Hypermedia Research Center Web site, http://www.hrc.wmin.ac.uk/theory-californianideology.html. Much of the discussion of the politics of cyberspace happens, no surprise, in Web discussions. For cyberlibertarians, see Langdon Winner, "Cyberlibertarian Myths and the Prospects for Community," in *Cyberethics: Social and Moral Issues in the Computer Age*, ed. Robert M. Baird, Reagan Mays Ramsower, and Stuart E. Rosenbaum (Amherst, NY: Prometheus, 2000), 319–331; and Thomas Streeter, "That Deep Romantic Chasm: Libertarianism, Neoliberalism, and the Computer Culture," in *Communication, Citizenship, and Social Policy*, ed. Andrew Calabrese and Jean-Claude Burgelman (New York: Rowman & Littlefield, 1999), 49–64.

12. Richard Barbrook and Andy Cameron, in particular (*California Ideology*).

13. Stewart Brand's libertarian leaning tendencies were first noticed by *Reason* magazine in 1975, when they began a correspondence that resulted in a profile that same year (Donna Rasnke to Stewart Brand, WECR, M1045, box 5:12).

14. John Perry Barlow, phone interview by the author, November 11, 2005. Garrett Hardin, "The Tragedy of the Commons," *Science* 162 (1968): 1243–1248. See also Robert H. Nelson, "Is 'Libertarian Environmentalism' an Oxymoron? The Crisis of Progressive Faith and the Environmental and Libertarian Search for a New Guiding Vision," and Thomas Michael Power, "Ideology, Wishful Thinking, and Pragmatic Reform: A Constructive Critique of Free-Market Environmentalism," both in *The Next West: Public Lands, Community, and Economy in the American West*, ed. John A. Baden and Donald Snow (Washington, DC: Island, 1997), 205–232 and 233–254, respectively.

15. Barlow was one of Dick Cheney's campaign managers during his first run for Congress. Both Barlow and Brand received awards from the leading libertarian journal, *Reason*, over the years for their contributions.

16. Barlow, phone interview; Barlow to author, May 16, 2007.

17. Michael Allen, "'I Just Want to Be a Cosmic Cowboy': Hippies, Cowboy Code, and the Culture of a Counterculture," *Western Historical Quarterly* 36 (Autumn 2005): 274–299.

18. Counterculture defined by Peter Braunstein and Michael William Doyle, "Introduction: Historicizing the American Counterculture of the 1960s and '70s," in their *Imagine Nation: The American Counterculture of the 1960s and '70s* (New York: Routledge, 2002), 7.

19. Thanks to Michael Doyle for this insight.

20. Again, thanks to Michael Doyle for this insight and discussions about counterculture politics.

21. R. Crumb to Stewart Brand, December 27, 1978, WECR, M1045, box 21:7. This is a wonderfully thoughtful and angry letter against a Michael Phillips's article on how to invest overseas.

22. World Game discussion, WECR, M1045, box 7:6, 9–10. Brand recounts a wonderful exchange between himself and J. Baldwin as they ate breakfast before the World Game workshop. They discussed the differences between liberals and conservatives and tried to place themselves within this always too restrictive framework.

23. Stewart Brand, "Money," in *The Last Whole Earth Catalog,* ed. Stewart Brand (New York: Portola Institute and Random House, 1971), 438.

24. On the complex relationship between the counterculture and the marketplace, see David Farber's "The Intoxicated State/Illegal Nation: Drugs in the Sixties Counterculture," in Braunstein and Doyle, *Imagine Nation,* 17–40. Art Kleiner, *The Age of Heretics: Heroes, Outlaws, and the Forerunners of Corporate Change* (New York: Doubleday, 1996), whose chapter on millenarians covers the counterculture business model.

25. Patricia Nelson Limerick, *The Legacy of Conquest: The Unbroken Past of the American West* (New York: Norton, 1987), 94–96.

26. See, for example, the first of the series, "Caring and Clarity: Conversation with Gregory Bateson and Edmund G. Brown, Jr., Governor of California," *CoEvolution Quarterly* 7 (Fall 1975): 32–47. For a different view on the Bateson/Brown encounter, see David Lipset, *Gregory Bateson: The Legacy of a Scientist* (Englewood Cliffs, NJ: Prentice-Hall, 1980), 287–289. Stewart Brand's work as Brown's adviser and the intellectuals and ideas he brought into the governor's office are discussed in the interview "Herman Kahn, Governor Jerry Brown, Amory Lovins: The New Class," in *News That Stayed News: Ten Years of CoEvolution Quarterly,* ed. Art Kleiner and Stewart Brand (San Francisco: North Point, 1986), 89–112, 332; and Kleiner, *Age of the Heretics.*

27. Stewart Brand, phone interview by the author, September 9, 2004.

28. Limerick, *Legacy of Conquest,* 78–96.

29. There is an impressive body of writing from the mid-1990s forward on the computer revolution and the politics it grew out of and spawned. Much of this scholarship examines the politics that lurked beneath the technology. Particularly useful for this essay were Stewart Brand, ed., *The Media Lab: Inventing the Future at MIT* (New York: Viking, 1987); Bruce Sterling, *The Hacker Crackdown: Law and Disorder on the Electronic Frontier* (New York: Bantam, 1993); Steven Levy, *Insanely Great: The Life and Time of the Macintosh, the Computer That Changed Everything* (New York: Viking, 1994); Stewart Brand, "WE OWE IT ALL TO THE HIPPIES: Forget antiwar protests, Woodstock, even long hair. The real legacy of the sixties generation is the computer revolution," *Time,* Spring 1995, no. 145:12; and Gregory Jordan, "Radical Nerds: The 60s Had Free Love; The 90s Have Free Information," *New York Times,* August 31, 1996, 7, 12.

30. Frank Herbert, *Dune* (New York: Ace, 2005), 40th anniversary edition; and Robert A. Heinlein, *The Moon Is a Harsh Mistress* (New York: Orb, 1997). First published in 1965 as a novella in the *Worlds of If* magazine, Heinlein's vision of a libertarian revolution on the new frontier of the Moon became an instant classic that influenced a generation.

31. Barlow, phone interview.

32. Historian Richard White provides the most captivating scholarly version of this tension in his essay "Are You an Environmentalist or Do You Work for a Living?" in *Uncommon Ground: Toward Reinventing Nature,* ed. William Cronon (New York: W. W. Norton, 1995), 171–185.

33. Richard White, *The Organic Machine: The Remaking of the Columbia River* (New York: Hill & Wang, 1995), 109. This remarkable little book uses the Columbia River as a metaphor for the contested territory where nature and culture, "the mechanical and the organic[,] meet" (110). For the most insightful proponents of a more historically grounded pragmatic environmental culture, see Patricia Nelson Limerick, "Mission to the Environmentalists," in her *Something in the Soil: Legacies and Reckonings in the New West* (New York: W. W. Norton, 2000): 171–185. William Cronon, "The Trouble with Wilderness; or, Getting Back to the Wrong Nature," in *Uncommon Ground: Toward Reinventing Nature,* ed. William Cronon (New York: W. W. Norton, 1995), 69–90.

34. Braden Allenby, *Reconstructing Earth: Technology and Environment in the Age of Humans* (Washington, DC: Island, 2005), 104, 108.

35. Hal K. Rothman, *The New Urban Park: Golden Gate National Recreation Area and Civic Environmentalism* (Lawrence: University Press of Kansas, 2004), 97.

36. Hal K. Rothman, *Devil's Bargains: Tourism in the Twentieth-Century American West* (Lawrence: University Press of Kansas, 1998), 23–24.

37. See ibid., 23–25.

38. The literature of consumer culture is too vast to list. Especially helpful for understanding how the counterculture blurred older distinctions between producer and consumer culture were the following: Thorstein Veblen, *The Theory of the Leisure Class: An Economic Study of Institutions* (London: George Allen & Unwin, 1957); Warren Susman, *Culture as History: The Transformation of American Society in the Twentieth Century* (New York: Pantheon, 1984); Lizabeth Cohen, *A Consumer's Republic: The Politics of Mass Consumption in Postwar America* (New York: Vintage, 2004); and Joseph Heath and Andrew Potter, *Nation of Rebels: Why Counterculture Became Consumer Culture* (New York: HarperBusiness, 2005).

39. John Markoff, *What the Dormouse Said: How the 60s Counterculture Shaped the Personal Computer Industry* (New York: Viking, 2005), 152–153.

40. A nice, concise discussion of 1960s leisure and the counterculture is in Braunstein and Doyle, "Historicizing the American Counterculture," 8–12.

41. Joseph E. Taylor, "On Climber, Granite, Sky," *Environmental History* 11 (January 2006): 130–133. Taylor points to the divisions within the Sierra Club between David Brower and a group of influential Yosemite climbers over the environmental versus recreation coverage in the *Sierra Club Bulletin.*

42. Ibid., 133.

43. Andrew Fluegelman, ed., *The New Games Book* (San Francisco: Headlands, 1976), and *More New Games* (New York: Doubleday, 1981); and Pat Farrington, "New Games Tournament," *CoEvolution Quarterly* 2 (Summer 1974): 94–101.

44. Fluegelman, *New Games Book*, 8.

45. Stewart Brand journals, 1973, SBP, M1237, 97–110. The idea of New Games explored here linked to a change of heart about another *Whole Earth* and work toward *Epilog*. Very comprehensive records on New Games, with emphasis on the environmental angle and the relationship between recreation, environmentalism, urban ecology, and economics, are in HJP, Resource Renewal Institute, Point Grants vol. 2, New Games file.

46. Fluegelman, *New Games Book*, 9.

47. George Leonard, *The Ultimate Athlete* (New York: Viking, 1974).

48. Fluegelman, *New Games Book*, 9.

49. Brand articulates his views on games and social change in the essay "Theory of Game Change," in Fluegelman, *New Games Book*, 137–140. Point Foundation veteran Bill Bryan also became a pioneer in significant changes in outdoor recreation in the early 1980s when he founded a pioneering ecotourism company, where he worked with industry leaders to create a responsible travel movement aimed at coming up with best practices for sustainable ecotravel (William L. Bryan Jr., "Appropriate Cultural Tourism: Can It Exist? Searching for an Answer," in *The Culture of Tourism, the Tourism of Culture: Selling the Past to the Present in the American Southwest*, ed. Hal K. Rothman [Albuquerque: University of New Mexico Press, 2003], 140–163). For information on Bryan's work in environmentally minded tourism, see Off the Beaten Path at http://www.offthebeatenpath.com, and Adventure Collection at http://www.adventurecollection.com for responsible tourism guidelines.

50. Huey Johnson, "New Games: Some Thoughts from an Environmentalist," HJP, Resource Renewal Institute, Point Grants, vol. 2.

51. Fluegelman, *New Games Book*, 93–135.

52. William deBuys, conversation with author, February 28, 2007.

53. Keith Power, "Searching for Brand New Earth Games," *Sports Illustrated*, January 7, 1974, 30–31; William O'Brien, "A Gamut of Games Scheduled for Marin," *San Francisco Examiner*, October 12, 1973, 4; and Susan Benson, Kathryn Meyer, and Pat Farrington, "The Energy Crisis and New Games: A Challenge and an Answer," *Outdoor Recreation Action*, Spring 1974, 44–46. Materials on New Games can be found in WECR, M 1045, 25:5–8.

54. Fluegelman, *New Games Book*, 13.

55. Ibid., 11.

56. *Outdoor Recreation for America: ORRRC Report, 1962* (Outdoor Recreation for America: A Report to the President and to the Congress by the Outdoor Recreation Resources Review Commission, Laurance S. Rockefeller, Chairman, January 1962.

57. Benson et al., "Energy Crisis and New Games."

58. Huey Johnson, "New Games: Some Thoughts from an Environmentalist," HJP, Resource Renewal Institute, Point Grants, vol. 2.

59. New Games Report, Point Foundation Report for 1974, PFR, M1441, 1:2.

60. Johnson, "New Games."

61. For the best statement about problems with too restrictive classifications of nature, see Cronon, "Trouble with Wilderness," 69. For a remarkably similar argument against elevating a mythically pristine wilderness at the expense of the rest of the environment, see Arthur Carhart, *Planning for America's Wildlands* (Harrisburg, PA: Telegraph, 1961). Carhart has often been criticized for his failure to support the wilderness bill at a time when

his influence and access to a national audience were at a high point. Carhart argues, convincingly, that wilderness as defined by the Wilderness Society did not really exist in any pure state: It was an "experience," a construct that lived "within your mind," rather than in a particular place. Carhart refused to support the wilderness bill in 1964 because he felt that arguing for wilderness purity would be a de facto concession to those who sought to develop lands not considered pristine. For information on the philosophy and activism of Edward Hobbs Hilliard, see Andrew Kirk, *Collecting Nature: The American Environmental Movement and the Conservation Library* (Lawrence: University Press of Kansas, 2001), 83–107.

62. Johnson, "New Games." An excellent source for Huey Johnson and TPL is the Carl Wimsen interview of Martin J. Rosen, "Trust for Public Land Founding Member and President, 1972–1997: The Ethics and Practice of Land Conservation," 1998–1999, Regional Oral History Office, University of California, Berkeley.

63. Yvon Chouinard, "On Corporate Responsibility for Planet Earth," in *Patagonia: The Edge Book Winter 2004 Catalog,* Winter 2004, 36–37. For specifics on catalog politics, see Yvon Chouinard, *Let My People Go Surfing: The Education of a Reluctant Businessman* (New York: Penguin, 2005), 129–133.

64. Galen Rowell points out the clear link between *Whole Earth* and Chouinard in his review of the *Chouinard Equipment Catalog,* by Yvon Chouinard and Tom Frost, in *American Alpine Journal,* 1973, 522–523. The fact that the catalog was reviewed as a book demonstrates the obvious significance of the volume even shortly after its release.

65. Drew Langsner, "Chouinard Equipment for Alpinists," in *The Last Whole Earth Catalog,* ed. Stewart Brand (New York: Portola Institute and Random House, 1971), 257.

66. Yvon Chouinard and Tom Frost, *Chouinard Equipment Catalog,* 1972.

67. Paul Hawken, *Growing a Business* (New York: Simon & Schuster, 1987), 61–63.

68. Andrew Kirk and Charles Palmer, "When Nature Becomes Culture: The National Register and Yosemite's Camp 4," *Western Historical Quarterly* 37 (Winter 2006): 497–506.

69. There is a lot of wonderful writing about this history of rock climbing and outdoor sports from the 1940s–1990s. Especially important for those interested in the critical period of innovation in Camp 4 is Steve Roper, *Camp 4: Recollections of a Yosemite Rockclimber* (Seattle: Mountaineers, 1994); Gary Arce, *Defying Gravity: High Adventure on Yosemite's Walls* (Berkeley, CA: Wilderness, 1996); Chris Jones, *Climbing in North America* (Berkeley: University of California Press, 1976); Paul Piana, *Big Walls: Breakthroughs on the Free-Climbing Frontier* (San Francisco: Sierra Club, 1997); Doug Scott, *Big Wall Climbing: Development, Techniques and Aids* (New York: Oxford University Press, 1981); Ed Bennett, "The Bay Chapter and the Birth of Modern Rock Climbing," Sierra Club Rock Climbing Section of San Francisco Bay *Yodeler,* June 1999; Jim Bridwell, "Brave New World," *Mountain* 31 (1973); Layton Kor, *Beyond the Vertical* (Boulder, CO: Alpine House, 1983); and Chouinard, *Let My People Go Surfing.*

70. Brand, in fact, featured one of Patagonia's San Francisco locations as the definition of a successfully adaptable low-road building (Stewart Brand, *How Buildings Learn: What Happens after They're Built* [New York: Penguin, 1994], 24–25).

71. Steve Roper to the author, January 13, 2004.

72. Yvon Chouinard and Tom Frost, "A Word," in their *Chouinard Equipment Catalog* (Ventura, CA: Great Pacific Iron Works, 1972).

73. Andrew G. Kirk, "Machines of Loving Grace: Appropriate Technology, Environment, and the Counterculture," in Doyle and Braunstein, *Imagine Nation,* 353–378; and

Paul Hawken, Amory Lovins, and L. Hunter Lovins, *Natural Capitalism: Creating the Next Industrial Revolution* (Boston: Little, Brown, 1999).

74. Tom Frost, phone interview by the author, February 4, 2005.

75. Outdoor Industry Foundation, "Outdoor Recreation Participation and Spending Study: A State-by-State Perspective," Outdoor Industry Foundation and Pew Charitable Trusts study, 2002.

76. Fred Turner, *From Counterculture to Cyberculture: Stewart Brand, the Whole Earth Network, and the Rise of Digital Utopianism* (Chicago: University of Chicago Press, 2006).

77. David Brooks, *Bobos in Paradise: The New Upper Class and How They Got There* (New York: Touchstone, 2000).

78. For a compelling look at the complexities of 1960s' political identity and particularly the intersections of liberalism and conservatism among the counterculture generation, see Rebecca E. Klatch, *A Generation Divided: The New Left, the New Right, and the 1960s* (Berkeley: University of California Press, 1999). Klatch argues against the "sellout" view of 1960s' political activists who moved away from political activism and toward traditional careers in business and industry. Mirroring the writings of those, like Stewart Brand, who lived through it, she argues that the politically active youth of the sixties took their ideologies with them to the marketplace.

79. Thomas Frank, *The Conquest of Cool: Business Culture, Counterculture, and the Rise of Hip Consumerism* (Chicago: University of Chicago Press, 1997). Also of interest is Heath and Potter, *Nation of Rebels;* Ken Goffman and Dan Joy, *Counterculture through the Ages: From Abraham to Acid House* (New York: Villard, 2004), which provides a very broad overview with thoughts on consumption along the way; and Sam Binkley, "Consuming Aquarius: Markets and the Moral Boundaries of the New Class, 1968–1980" (Ph.D. diss., New School University), November 2001.

80. Paul Hawken, "Surviving in Small Business," *CoEvolution Quarterly* 41 (Spring 1984): 14–17.

81. Brand, "WE OWE IT."

82. Ibid.

83. Hawken et al., *Natural Capitalism;* and Paul Hawken's *The Ecology of Commerce: A Declaration of Sustainability* (New York: Harper Collins, 1993) and his *Growing a Business,* which is the companion volume to Hawken's seventeen-part PBS series. See also William McDonough and Michael Braungart, *Cradle to Cradle: Remaking the Way We Make Thing* (New York: North Point, 2002).

84. Richard White, *The Organic Machine: The Remaking of the Columbia River* (New York: Hill & Wang, 1995), 35.

85. Leo Marx, *The Machine in the Garden: Technology and the Pastoral Ideal in America* (New York: Oxford University Press, 1964). For Marx on Emerson, see 229–242.

86. Daniel Bell, *The Cultural Contradictions of Capitalism* (New York: Basic, 1996). In this landmark book on the twisted relationship between capitalism and modernism, Bell dismisses the counterculture as a "children's crusade" that sought to "eliminate the line between fantasy and reality" (xxvi–xxvii).

87. The literature on business and the environment is vast. See Hawken et al., *Natural Capitalism;* Hawken, *Ecology of Commerce* and *Growing a Business; Harvard Business Review on Business and the Environment* (Cambridge: Harvard Business School Press, 2000); Hamish Pringle and Marjorie Thompson, *Brand Spirit: How Cause Related Marketing*

Builds Brands (Chichester, UK: John Wiley & Sons, 1999); Sue Adkins, *Cause Related Marketing: Who Cares Wins* (Oxford: Butterworth Heinemann, 1999); John Elkington, *Cannibals with Forks: The Triple Bottom Line of 21st Century Business* (Gabriola Island, Canada: New Society, 1998); Andrew Crane, *Marketing, Morality and the Natural Environment* (New York: Routledge, 2000); Alasdair Blair and David Hitchcock, *Environment and Business* (New York: Routledge, 2001); Jacquelyn A. Ottman, *Green Marketing: Opportunity for Innovation* (Chicago: NTC, 1998); and Carl Frankel, *In Earth's Company: Business, Environment and the Challenge of Sustainability* (Gabriola Island, Canada: New Society, 1998).

88. Bell, *Cultural Contradictions.*

89. Ibid.

90. Brooks, *Bobos in Paradise,* 41.

91. Stewart Brand, *The Whole Earth Catalog,* Spring 1970, 143. This version of the catalog is very useful for understanding Stewart Brand's views on the counterculture economy.

92. For information on the McToxics campaign, see the Center for Health, Environment and Justice Web site, http://www.chej.org/history.htm. For more on this center and Lois Gibbs, see Robert Gottlieb, *Forcing the Spring: The Transformation of the American Environmental Movement* (Washington, DC: Island, 1993), 167–169.

93. Aaron Davis, "John Perry Barlow: Wyoming's Estimated Prophet," Planet Jackson Hole.Com, htttp://www.planetjh.com (accessed July 28, 2005).

94. Doherty, "John Perry Barlow 2.0."

95. Ibid.

Epilogue: What Happened to Appropriate Technology?

1. Howard Rheingold, ed., *The Millennium Whole Earth Catalog* (San Francisco: Harper San Francisco, 1994). The quotes are from inside the front cover.

2. Statistics from 1970 to 2005 are from the National Association of Home Builders database, http://www.nahb.org.

3. Langdon Winner, *The Whale and the Reactor: A Search for Limits in an Age of High Technology* (Chicago: University of Chicago Press, 1986), 66. For another thoughtful critique, see Witold Rybczynski, "Appropriate Technology: The Upper Case Against," *Co-Evolution Quarterly* 13 (Spring 1977): 81–83.

4. The *Whole Earth Review* dedicated an issue to Biosphere II: Kevin Kelly, "Biosphere II: An Autonomous World, Ready to Go," *Whole Earth Review,* Summer 1990, no. 67: 2–13.

5. The quote is from ibid.

6. Ibid.

7. Peter Warshall, interview by the author, Tucson, Arizona, November 11, 2006.

8. Ibid.

9. For the Global Business Network, see Stewart Brand, *The Clock of the Long Now: Time and Responsibility* (New York: Basic, 1999); and http://www.gbn.com. Stewart Brand, "Environmental Heresies," *MIT Technology Review* online (May 2005). John Tierney, "An Early Environmentalist, Embracing New 'Heresies,'" *New York Times,* February 27, 2007, D1, 3.

10. The Long Now Foundation, www.longnow.org.

11. Carroll Pursell, *The Machine in America: A Social History of Technology* (Baltimore: Johns Hopkins University Press, 1995), 307.

12. Daren Fonda, "The Greening of Wal-Mart," *Time*, April 3, 2006, 48, 51. Asked if this was just another corporate "greenwash," Amory Lovins replied, "We don't go where we don't think there's a genuine interest in change." *Time* reported on global warming early and is a good source for the evolving presentation of science and environment in the popular media. Lovins was the first person to publish about global warming in 1968, when he was only twenty-one years old. For more on Lovins, see Art Kleiner, *The Age of Heretics: Heroes, Outlaws, and the Forerunners of Corporate Change* (New York: Doubleday, 1996), 303–306.

13. Stewart Brand, "Research Communities," in *Whole Earth Epilog*, ed. Stewart Brand (New York: Penguin/Point Foundation, 1974), 534.

14. Michael Phillips, *The Seven Laws of Money* (Boston: Shambhala, 1997); *The Briarpatch Book: Experiences in Right Livelihood and Simple Living from the Briarpatch Community* (San Francisco: New Glide/Reed, 1978).

Bibliography

The *Whole Earth Catalog* had many incarnations. The first edition appeared in November 1968 as *The Whole Earth Catalog: Access to Tools,* edited by Stewart Brand and published by the Portola Institute. Many revised versions followed, all with Brand as the lead editor but often with help from guest editors, between 1969 and 1971, when *The Last Whole Earth Catalog* (Portola Institute and Random House, 1971) was published. All of the *Whole Earth*s were reprinted many times, and often there were several supplemental editions in between. There were also several *Whole Earth* companion volumes, such as Brand's *Space Colonies* (New York: Penguin, 1977) and J. Baldwin and Stewart Brand, eds., *Soft-Tech* (New York: Penguin, 1978), that focused on particular themes or issues. For historians, four issues stand out in importance: *Last Whole Earth* has a wonderful history section that is picked up in *Whole Earth Epilog.* The winter 1988 *Whole Earth Review* features a 20th anniversary set of interviews with all of the significant contributors over the years, and the winter 1998 edition reprints the first edition of *Whole Earth* and provides thoughtful historical essays by major contributors. This bibliography is not a complete record of the voluminous publications of *Whole Earth.* Rather, I have listed only the works I found most useful for the study of the environmental thinking conveyed through *Whole Earth* in all its incarnations. For every environmentally themed article listed, there are at least several more on similar topics. The hundreds of environmental book reviews and concise section introductions are a vital source for anyone interested in the evolution of environmental thinking in *Whole Earth* over the years. Only a very few editions of the quarterlies and periodicals did not have significant environmental reviews too numerous to list.

Archival materials for this book came primarily from the Stanford Department of Special Collections at the Stanford University Green Library. I worked primarily with the *Whole Earth Catalog* Records (WECR), Point Foundation Records (PFR), Stewart Brand Papers (SBP), and Buckminster Fuller Collections (BFC). Other collections included Huey Johnson's Papers at the Resource Renewal Institute (RRI) and the New Alchemy Institute Collection, Department of Special Collections, Iowa State University.

Articles, Essays, and Chapters

Albright, Thomas, and Charles Perry. "The Last Twelve Hours of the Whole Earth." *Rolling Stone* 86 (July 8, 1971): 1, 6–8.
Allen, Michael. "'I Just Want to Be a Cosmic Cowboy': Hippies, Cowboy Code, and the Culture of a Counterculture." *Western Historical Quarterly* 36 (Autumn 2005): 274–299.
Alpine, Andy. "The Briarpatch Network." *CoEvolution Quarterly* 8 (Winter 1975): 8–9.
Anker, Peder. "The Ecological Colonization of Space." *Environmental History* 10 (April 2005): 239–268.

Baer, Steve. "The Clothesline Paradox." In *Soft-Tech,* ed. J. Baldwin and Stewart Brand, 50–51. New York: Penguin, 1978.

———. "The Double Bubble Wheel Engine." In *Soft-Tech,* ed. J. Baldwin and Stewart Brand, 30–32. New York: Penguin, 1978.

———. "Gravity Engines and the Diving Engine." *CoEvolution Quarterly* 2 (Summer 1974): 81–83.

———. "Zomes." In *Shelter,* ed. Lloyd Kahn, 134–135. Bolinas, CA: Shelter, 1973.

Bailey, Beth. "Sex as a Weapon: Underground Comix and the Paradox of Liberation." In *Imagine Nation: The American Counterculture of the 1960s and '70s,* ed. Peter Braunstein and Michael William Doyle, 305–324. New York: Routledge, 2002.

Baldwin, J. "The Autonomous Electronic Cottage." *Whole Earth Review,* Summer 1990, no. 67: 66.

———. "Catalytic Woodstoves." *CoEvolution Quarterly* 43 (Fall 1984): 89.

———. "The Case for Long Shots." *Strategy + Business* online, Autumn 2006. http://www.strategy-business.com.

———. "Designing Designers." *Whole Earth Review,* Summer 1995, no. 86: 14–16.

———. "Encounters with the Mentor." In *Design Outlaws on the Ecological Frontier,* ed. Chris Zelov and Phil Cousineau, 44–49. New York: Knossus, 1997.

———. "Making Things Right." *Whole Earth Review,* Fall 1995, no. 86: 6.

———. "The New Alchemists Are Neither Magicians nor Geniuses: They Are Hard Workers." *CoEvolution Quarterly* 12 (Winter 1976/1977): 104–111.

———. "One Highly Evolved Toolbox." *CoEvolution Quarterly* 5 (Spring 1975): 80–85.

———. "Stuff & Nonsense." *Whole Earth Review,* Winter 1995, no. 88: 106–109.

———. "Sun and Wind in New Mexico." In *Shelter,* ed. Lloyd Kahn, 164. Bolinas, CA: Shelter, 1973.

———. "Trails of an Ocean Ark Model." *CoEvolution Quarterly* 24 (Winter 1979): 56–58.

———. "The Ultimate Swiss Omni Knife." *Whole Earth Review,* 30th anniversary issue, Winter 1998, no. 95: 20.

———. "What's an Ecolog?" In *Whole Earth Ecolog: The Best of Environmental Tools and Ideas,* ed. J. Baldwin and Stewart Brand. New York: Harmony, 1990.

———. "World Game." In *Whole Earth Ecolog: The Best of Environmental Tools and Ideas,* ed. J. Baldwin and Stewart Brand, 7. New York: Harmony, 1990.

Barbrook, Richard, and Andy Cameron. "California Ideology." Hypermedia Research Center. http://www.hrc.wmin.ac.uk/theory-californianideology.html.

Barlow, John Perry. "Crime and Puzzlement: In Advance of the Law on the Electronic Frontier." *Whole Earth Review,* Fall 1990, no. 68: 44–57.

———. "A Declaration of the Independence of Cyberspace." Electronic Frontier Foundation. http://homes.eff.org/~barlow/Declaration-Final.html.

Bateson, Gregory. "The Pattern Which Connects." *CoEvolution Quarterly* 18 (Summer 1978): 5.

Bennett, Ed. "The Bay Chapter and the Birth of Modern Rock Climbing." Sierra Club Rock Climbing Section. San Francisco Bay *Yodeler,* June 1999.

Benson, Susan, Kathryn Meyer, and Pat Farrington. "The Energy Crisis and New Games: A Challenge and an Answer." *Outdoor Recreation Action,* Spring 1974, 44–46.

Berry, Wendell. "Wendell Berry Angry," *CoEvolution Quarterly* 10 (Summer 1976): 8–11.

Boulding, Kenneth E. "Plains of Science, Summits of Passion." *CoEvolution Quarterly* 5 (Spring 1975): 13.

Brand, Stewart. "Bigger, Finer, Slower." In *Whole Earth Ecolog: The Best of Environmental Tools and Ideas,* ed. J. Baldwin and Stewart Brand. New York: Harmony, 1990, 3.

———. "CoEvolution and the Biology of Communities." In *News That Stayed News: Ten Years of CoEvolution Quarterly,* ed. Art Kleiner and Stewart Brand, 3–9. San Francisco: North Point, 1986.

———. "The Earth from Space." *Rolling Stone,* May 15, 2003, 124.

———. "Environmental Heresies." *MIT Technology Review* online, May 2005, 1. http:// www.technologyreview.com/Energy/14406/.

———. "Environmentalists Stifle New Things." *Environmental Action Visions,* May/June 1985, 15.

———. "Free Space," *Coevolution Quarterly* 7 (Fall 1975): 4–29

———. "History." In *The Last Whole Earth Catalog,* ed. Stewart Brand, 439–441. New York: Portola Institute and Random House, 1971.

———. "Indians and the Counterculture." In *The Smithsonian Handbook of North American Indians.* Vol. 4, *History of Indian-White Relations,* ed. Wilcomb E. Washburn, 570–572. Washington, DC: Smithsonian Institution, 1988.

———. "Jacques Cousteau at NASA Headquarters." *CoEvolution Quarterly* 10 (Summer 1976): 12–17.

———. "Local Dependency." In *Soft-Tech,* ed. J. Baldwin and Stewart Brand, 50–51. New York: Penguin, 1978.

———. "Marlon Brando Plans a Whole Earth TV Special." A conversation with Stewart Brand and J. Baldwin. *CoEvolution Quarterly,* Winter 1975, 60–69.

———. "Neighborhood Preservation Is an Ecology Issue." *CoEvolution Quarterly* 15 (Fall 1977): 40–44.

———. "The New Class: Herman Kahn, Governor Jerry Brown, Amory Lovins." In *News That Stayed News: Ten Years of CoEvolution Quarterly,* ed. Art Kleiner and Stewart Brand, 89–112. San Francisco: North Point, 1986.

———. "Research Communities." In *The Whole Earth Epilog,* ed. Stewart Brand, 534. New York: Penguin/Point Foundation, 1974.

———. "Review of *The Resourceful Earth.*" *Whole Earth Review,* March 1985, no. 45: 15.

———. Resume. The Well. http://www.well.com/user/sbb/bio.html (accessed December 10, 2006).

———. "'Spaceship Earth' Comes Home to Roost." *CoEvolution Quarterly* 10 (Summer 1976): 4–7.

———. "Streaming Wisdom: Watershed Consciousness in the Twentieth Century." *CoEvolution Quarterly* 12 (Winter 1976/1977): 4–13.

———. "Theory of Game Change." In *The New Games Book,* ed. Andrew Fluegelman. Garden City, NY: 1976, 137–140.

———, ed. *II Cybernetic Frontiers.* New York: Random House, 1974.

———. "We Are as Gods." *Whole Earth Review,* Winter 1998, no. 95: 2.

———. "WE OWE IT ALL TO THE HIPPIES: Forget antiwar protests, Woodstock, even long hair. The real legacy of the sixties generation is the computer revolution." *Time,* special ed., Spring 1995, no. 145: 12.

————, ed. *Whole Earth Catalog*, Fall 1969, no. 129.

————. "Why Haven't We Seen the Whole Earth?" In *The Sixties: The Decade Remembered Now, by the People Who Lived It Then*, ed. Linda Obst, 168–169. New York: Random House/Rolling Stone, 1977.

Branwyn, Gareth. "Whole Earth Review." StreetTech. http://www.streettech.com/bcp/BCPgraf/CyberCulture/WholeEarthReview.html (accessed September 21, 1998).

Braun, Bruce, and Noel Castree. "The Construction of Nature and the Nature of Construction: Analytical and Political Tools for Building Survivable Futures." In *Remaking Reality: Nature at the Millennium*, ed. Bruce Braun and Noel Castree, 3–42. New York: Routledge, 1998.

Braunstein, Peter. "Forever Young: Insurgent Youth and the Sixties Culture of Rejuvenation." In *Imagine Nation: The American Counterculture of the 1960s and '70s*, ed. Peter Braunstein and Michael William Doyle, 243–274. New York: Routledge, 2002.

Braunstein, Peter, and Michael William Doyle. "Historicizing the American Counterculture of the 1960s and 1970s." In *Imagine Nation: The American Counterculture of the 1960s and 1970s*, ed. Peter Braunstein and Michael William Doyle, 5–14. New York: Routledge, 2002.

Bridwell, Jim. "Brave New World," *Mountain* 31 (1973).

Brown, Andrew. "The Hustler." *Guardian Unlimited*, April 30, 2005, 22–25.

————. "Whole Earth Visionary." *Guardian Online*, August 4, 2001.

Brown, Chip. "High Priest of the Low-Flow Shower Heads." *Outside Magazine*, November, 1991, 58.

Bryan, William L., Jr. "Appropriate Cultural Tourism: Can It Exist? Searching for an Answer." In *The Culture of Tourism, the Tourism of Culture: Selling the Past to the Present in the American Southwest*, ed. Hal K. Rothman, 140–163. Albuquerque: University of New Mexico Press, 2003.

Burns, Jennifer L. "O Libertarian, Where Is Thy Sting?" *Journal of Policy History* 19 (Fall 2007).

Butler, Anne M. "Selling the Popular Myth." In *The Oxford History of the American West*, ed. Clyde A. Milner, Carol A. O'Connor, and Martha A. Sandweiss, 771–801. New York: Oxford University Press, 1994.

Butti, Ken, and John Perlin. "Solar Water Heaters in California, 1891–1930." Reprinted in *Soft-Tech*, ed. J. Baldwin and Stewart Brand, 52–63. New York: Penguin, 1978.

————. "2,500 Years of Solar Architecture and Technology." *CoEvolution Quarterly* 24 (Winter 1979/80): 42–53.

Calthorpe, Peter. "Redefining Cities." *Whole Earth Review*, March 1985, no. 45: 1–2.

Carhart, Arthur. "Historical Development of Outdoor Recreation." In *Outdoor Recreation Literature: A Survey*. Report to the Outdoor Recreation Resources Review Commission. Study Report 27. Washington, DC: Government Printing Office, 1962.

————. "The Menaced Dinosaur National Monument." *National Parks Magazine* 26 (January/March 1952): 19–30.

————. "Recreation in the Forests." *American Forestry* 26 (May 1920): 268–272.

————. "State Must Not Toss Away Scenic Dinosaur Park." *Denver Post*, April 7, 1954, 41.

"Caring and Clarity: Conversation with Gregory Bateson and Edmund G. Brown, Jr., Governor of California." *CoEvolution Quarterly* 7 (Fall 1975): 32–47

Chouinard, Yvon. "On Corporate Responsibility for Planet Earth" In *Patagonia: The Edge Book Winter 2004 Catalog,* Winter 2004, 36–37.

Conway, James. "Stealing Beauty: The West's Common Lands Need Real Defenders." *Preservation* 58, no. 3 (May/June 2006): 28–33, 70–71.

Cronon, William. "A Place for Stories: Nature, History, and Narrative." *Journal of American History* 78 (March 1992): 1347–1376.

———. "The Trouble with Wilderness; or, Getting Back to the Wrong Nature." In *Uncommon Ground: Toward Reinventing Nature,* ed. William Cronon, 69–90. New York: W. W. Norton, 1995.

Crumb, R. [Robert]. "Space Day Symposium: Or What Ever the Hell It Was Called." *CoEvolution Quarterly* 15 (Fall 1977): 48–51.

Dasmann, Raymond. "Biogeographical Provinces." *CoEvolution Quarterly* 12 (Fall 1976): 32–37.

Davis, Aaron. "John Perry Barlow: Wyoming's Estimated Prophet." Planet Jackson Hole. Com. http://www.planetjh.com (accessed July 28, 2005).

Dembart, Lee. "'Whole Earth Catalog' Recycled as 'Epilog.'" *New York Times,* November 8, 1974, 41.

Doherty, Brian. "John Perry Barlow 2.0: The Thomas Jefferson of Cyberspace Reinvents His Body—and His Politics." *Reason Magazine* online, August/September 2004. http://www.reason.com/.

Doyle, Michael William. "Debating the Counterculture: Ecstasy and Anxiety over the Hip Alternative." In *The Columbia Guide to America in the 1960s,* ed. David Farber and Beth Bailey, 143–156. New York: Columbia University Press, 2001.

Dyson, Freeman J. "Pilgrim Fathers, Mormon Pioneers, and Space Colonists: An Economic Comparison." *Proceedings of the American Philosophical Society* 122 (April 24, 1978): 63–68.

Eckaus, R. S. "Appropriate Technology: The Movement Has Only a Few Clothes On." *Issues in Science and Technology* 3, no. 2 (Winter 1987): 62–71.

Ehrlich, Paul R. "Nuclear Winter: The Inside Story." *CoEvolution Quarterly* 42 (Summer 1984): 88–94.

Ehrlich, Paul R., and Anne H. Ehrlich. *One with Nineveh: Politics, Consumption, and the Human Future.* Washington, DC: Island, 2004.

Emshwiller, John R. "Amory Lovins Presses Radical Alternatives for Fueling the Nation." *Wall Street Journal,* March 16, 1981, 1.

Eric. "Report from Auroville." *CoEvolution Quarterly* 1 (Spring 1974): 72.

Farber, David. "The Intoxicated State/Illegal Nation: Drugs in the Sixties Counterculture." In *Imagine Nation: The American Counterculture of the 1960s and '70s,* ed. Peter Braunstein and Michael William Doyle, 17–40. New York: Routledge, 2002.

Farney, Dennis. "Land for Posterity: A Conservation Group Preserves Choice Sites by Aggressive Tactics." *Wall Street Journal,* March 11, 1970, 1.

Farrington, Pat. "New Games Tournament." *CoEvolution Quarterly* 2 (Summer 1974): 94–101.

Fonda, Daren. "The Greening of Wal-Mart." *Time,* April 3, 2006, 48, 51.

Friend, Gil. "Biological Agriculture in Europe." *CoEvolution Quarterly* 17 (Spring 1978): 60–63.

Frome, Michael. "Wilderness: 25 Years and Far from Finished." *National Parks Magazine* 63, nos. 7/8 (July 1989): 35–41.

Fuller, Buckminster. World Design Science Decade, 1965–1975, document 5. Quoted in *The Whole Earth Catalog*, ed. Stewart Brand. 1968 first edition reprint in *Whole Earth Review*, Winter 1999, no. 95: 4.

Goldburg, Robert. "The Western Hero in Politics: Barry Goldwater, Ronald Reagan, and the Rise of the American Conservative Movement." In *The Political Legacies of the American West*, ed. Jeff Roche. Lawrence: University Press of Kansas, forthcoming.

Goodell, Jeff. "The Rise and Fall of Apple Inc." *Rolling Stone*, April 4, 1996, 51–73, and April 18, 1996, 59–88.

Gross, Ronald. Review of the Big Rock Candy Mountain project, *New York Times*, May 21, 1972, BR44.

Hamilton, Cynthia. "Industrial Racism, the Environmental Crisis, and the Denial of Social Justice." In *Cultural Politics and Social Movements*, ed. Marcy Darnovsky, Barbara Epstein, and Richard Flacks, 189–196. Philadelphia: Temple University Press, 1995.

Hardin, Garrett. "Living on a Lifeboat" *BioScience* 24 (October 1974): 561–568.

———. "The Tragedy of the Commons." *Science* 162 (1968): 1243–1248.

Harvey, Mark. "Echo Park, Glen Canyon, and the Postwar Wilderness Movement." *Pacific Historical Review* 60 (February 1991): 43–67.

Hawken, Paul. "Surviving in Small Business." *CoEvolution Quarterly* 41 (Spring 1984): 14–17.

Hays, Samuel P. "From Conservation to Environment: Environmental Politics since World War Two." *Environmental Review* 6 (Fall 1982): 14–41.

Hess, John. "Farm-Grown Fish: A Triumph for the Ecologist and the Sensualist; Cabbage Problems Equally Effective." *New York Times*, September 6, 1973, 32.

Hickey, Dave. "Shooting the Land." In *The Altered Landscape*, ed. Peter E. Pool, 22–32. Reno: University of Nevada Press, 1999.

Hughes, Jonathan. "After Non-plan: Retrenchment and Reassertion." In *Non-plan: Essays on Freedom, Participation and Change in Modern Architecture and Urbanism*, ed. Jonathan Hughes and Simon Sadler, 178–180. Oxford: Architectural, 2000.

Illich, Ivan. "Beauty in the Junkyard." *Whole Earth Review*, Winter 1991, no. 73: 64–68.

Johnson, Huey. "Land Banking." *CoEvolution Quarterly* 2 (Summer 1974): 72–77.

Jones, Lisa. "The Biogladiator." *High Country News*, July 24, 1995.

Jordan, Gregory. "Radical Nerds: The 60s Had Free Love; The 90s Have Free Information." *New York Times*, August 31, 1996.

Kahn, Lloyd. "Pacific Domes." In *Domebook 2*, by Lloyd Kahn, 20–34. Bolinas, CA: Shelter, 1972.

Katz, Cindi. "Under a Falling Sky: Apocalyptic Environmentalism and the Production of Nature." In *Marxism in the Postmodern Age*, ed. Antonio Callari, Stephen Cullenberg, and Carole Biewener, 276–282. New York: Guilford, 1995.

Kelly, Kevin. "Biosphere II: An Autonomous World, Ready to Go," *Whole Earth Review*, Summer 1990, no. 67: 2–13.

———. "Tools Are the Revolution." *Whole Earth Review*, Winter 2000, no. 103: 4.

Kinney, Jay. "Beyond Left and Right." *Whole Earth Review*, Summer 2000, no. 101: 22–29.

———. "The Garden." *CoEvolution Quarterly* 7 (Fall 1975): 67–70.

———. "Libertarian Periodicals." *CoEvolution Quarterly* 25 (Spring 1980): 118–119.

———. "What's Left: Revisiting the Revolution." *CoEvolution Quarterly* 8 (Winter 1975): 19–30.

Kirk, Andrew G. "Machines of Loving Grace: Appropriate Technology, Environment, and the Counterculture." In *Imagine Nation: The American Counterculture of the 1960s and '70s*, ed. Peter Braunstein and Michael Doyle, 353–378. New York: Routledge, 2002.

Kirk, Andrew, and Charles Palmer. "When Nature Becomes Culture: The National Register and Yosemite's Camp 4." *Western Historical Quarterly* 37 (Winter 2006): 497–506.

Kleppner, Paul. "Politics without Parties: The Western States, 1900–1984," ed. Gerald D. Nash and Richard W. Etulain, 295–338. *The Twentieth-Century West: Historical Interpretations.* Albuquerque: University of New Mexico Press, 1989.

Krapfel, Paul. "Becoming Part of Gaia." *CoEvolution Quarterly* 43 (Fall 1984): 4–10.

Lambert, Cynthia F. "Huey D. Johnson: The Green Party." *Mother Earth News* 84 (November/December 1983). http://www.motherearthnews.com/Homesteading-and-Self-reliance/1983-11-01/Profiles.aspx (accessed April 30, 2007).

Langsner, Drew. "Chouinard Equipment for Alpinists." In *The Last Whole Earth Catalog*, ed. Stewart Brand, 257. San Francisco: Portola Institute and Random House, 1971.

Lanier, Jaron. "Life at the Techno-humanist Interface." *Whole Earth Review.* Winter 1998, no. 95: 2.

Lehmann-Haupt, Christopher. "Alas, It's Dead. Long Life to It!" *New York Times*, August 19, 1971, 33.

Leibhardt, Barbara. "Interpretation and Causal Analysis: Theories in Environmental History." *Environmental Review* 12 (Spring 1988): 23–36.

Leonard, Jennifer. "Stewart Brand on the Long View." Interview. In *Massive Change*, ed. Bruce Mau, 104–105. New York: Phaidon, 2004.

Limerick, Patricia Nelson. "Mission to the Environmentalists." In *Something in the Soil: Legacies and Reckonings in the New West*, 171–185. New York: W. W. Norton, 2000.

Lithman, Alan. "Revisiting Auroville: A Twenty-Year Project That's Still Growing." In *Helping Nature Heal: An Introduction to Environmental Restoration*, ed. Richard Nilsen, 94–96. Berkeley: Whole Earth Catalog/Ten Speed, 1991.

———. "The Subtlest of Catastrophes: Erosion vs. Reforestation at Auroville." *CoEvolution Quarterly* 25 (Spring 1980): 72–74.

Lovelock, James. "The Independent Practice of Science." *CoEvolution Quarterly* 25 (Spring 1980): 22–30.

Lovins, Amory. "Energy Lessons Learned and to be Learned: Verities That Will Astonish Some and Delight the Rest." *Whole Earth Review*, Winter 1998, no. 95: 31.

———. "Energy Strategy: The Road Not Taken." *Foreign Affairs* 55 (October 1976): 65–96.

Lovins, Amory, Hunter Lovins, and Paul Hawken. "A Road Map for Natural Capitalism." In *Harvard Business Review on Business and the Environment*, 7–10. Boston: Harvard Business School Press, 2000.

Maher, Neil. "Shooting the Moon." *Environmental History* 9 (July 2004): 526–531.

Malone, Michael P., and F. Ross Peterson. "Politics and Protest." In *The Oxford History of the American West*, ed. Clyde A. Milner, Carol A. O'Connor and Martha A. Sandweiss, 501–533. New York: Oxford University Press, 1994.

Maniaque, Caroline. "Searching for Energy." In *Ant Farm 1968–1978*, ed. Constance Lewallen and Steve Seid, 14–17. Berkeley: University of California Press, 2004.

Margulis, Lynn. "Another Four-Letter Word: Gaia." *Whole Earth Review*, Winter 1998, no. 95: 4.

Margulis Lynn, and James Lovelock. "The Atmosphere as Circulatory System of the Biosphere—The Gaia Hypothesis." *CoEvolution Quarterly* 6 (Summer 1975): 31–40.

McClanahan, Ed, and Gurney Norman. "The Whole Earth Catalog." *Esquire,* July 1970, 95–125.

McClosky, Michael. "Wilderness Movement at the Crossroads, 1945–1970." *Pacific Historical Review* 41 (August 1972): 346–361.

McDermott, Jeanne. "Lynn Margulis—Unlike Most Microbiologists." *CoEvolution Quarterly* 25 (Spring 1980): 31–38.

Meadows, Donella. "Seeing the Population Issue Whole." In *Beyond the Numbers,* ed. L. A. Mazur, 23–33. Washington, DC: Island, 1994.

Menand, Louis. "FAT MAN: Herman Kahn and the Nuclear Age." *New Yorker* online, June 27, 2005. http://www.newyorker.com/archive/2005/06/27/050627crbo_books.

Miller, Timothy. "The Sixties-Era Communes." In *Imagine Nation: The American Counterculture of the 1960s and '70s,* ed. Peter Braunstein and Michael Doyle, 327–352. New York: Routledge, 2002.

Mills, Stephanie. "Hay Foot, Straw Foot." *Whole Earth Review,* Summer 2000, no. 101: 47–49.

———. "Salons and Their Keepers." *CoEvolution Quarterly* 2 (Summer 1974): 102–105.

"Missal for Mammals." Review of *The Whole Earth Catalog,* ed. Stewart Brand. *Time,* November 21, 1969, 74–76.

"Moss the Tentmaker," *Time,* July 26, 1976.

Moy, Timothy. "The End of Enthusiasm: Science and Technology." In *The Columbia Guide to America in the 1960s,* ed. David Farber and Beth Bailey, 305–311. New York: Columbia University Press, 2001.

O'Brien, William. "A Gamut of Games Scheduled for Marin." *San Francisco Examiner,* October 12, 1973, 4.

O'Neill, Gerard K. "The High Frontier." In *Space Colonies,* ed. Stewart Brand, 8–30. New York: Penguin, 1977.

———. "Space Colonies and Energy Supply to the Earth." *Science,* n.s., 190 (December 5, 1975): 943–947.

On Growth and Form, review of by D'Arcy Wentworth Thompson. Reprinted in *The Whole Earth Catalog: Access to Tools,* edited by Stewart Brand. San Francisco: Portola Institute, 1968.

"Pangs and Prizes." *Time,* April 24, 1972.

Phillips, Michael. "Don't Build a House till You've Looked at This." *CoEvolution Quarterly* 18 (Summer 1978): 100–102.

"The Plowboy Interview: John Shuttleworth." *Mother Earth News* 31 (January/February 1975): 6–14.

"The Plowboy Interview: Steve and Holly Baer." *Mother Earth News* 22 (July/August 1973): 8.

Point Foundation. "The Future of Point: A Growing Dialog." *Whole Earth Review,* Winter 1990, 69: 133–134.

———. "Point Foundation Report." *Whole Earth Review,* Summer 1991, 71: 137–139.

Power, Keith. "Searching for Brand New Earth Games." *Sports Illustrated,* January 7, 1974, 30–31.

"Preservation and the Public Lands." *National Parks Magazine,* November 1965, 20–21.

"Public Land Conservation Grant Made." *Nevada State Journal,* June 1, 1972, 1.

Pursell, Carroll. "The Rise and Fall of the Appropriate Technology Movement in the United States, 1965–1985." *Technology and Culture* 34 (July 1993): 629–637.

Quinney, John. "Out of the Ark and into the World." *Whole Earth Review,* Spring 1989, no. 62: 33–35.

Rakestraw, Lawrence. "Conservation Historiography: An Assessment." *Pacific Historical Review* 41 (August 1972): 271–288.

Raymont, Henry. "Juror Quits Book Panel over 'Whole Earth Catalog.'" *New York Times,* April 5, 1972, 35.

"R. Buckminster Fuller." *Time,* January 10, 1964.

"Report on 1984 Hacker's Conference." *Whole Earth Review,* May 1985, 49.

Roberts, Steven V. "Mail Order Catalog of the Hip Becomes a National Best Seller." *New York Times,* April 12, 1970, 67.

Rossetto, Louis. "Rebuttal of the Californian Ideology." *Alamut Bastion of Peace and Information* online. http://www.alamut.com/subj/ideologies/pessimism/califIdeo_II.html.

Rowell, Galen. Review of the *Chouinard Equipment Catalog,* by Yvon Chouinard and Tom Frost. *American Alpine Journal,* 1973, 522–523.

Rybczynski, Witold. "Appropriate Technology: The Upper Case Against." *CoEvolution Quarterly* 13 (Spring 1977): 81–83.

Sadler, Simon. "Drop City Revisited." *Journal of Architectural Education* 59, no. 3 (February 2006): 5–16.

Scharff, Virginia. "Are Earth Girls Easy? Ecofeminism, Women's History, and Environmental History." *Journal of Women's History* 7, no. 2 (Summer 1995): 164–175.

Schweickart, Russell. "Whose Earth?" *The Next Whole Earth Catalog: Access to Tools,* ed. Stewart Brand, 19. New York: Rand McNally, 1980.

Secrest, Clark. "'No Right to be Poor': Colorado's Drop City." *Colorado Heritage,* Winter 1998, 14–21.

Sewell, William H., Jr. "A Theory of Structure: Duality, Agency, and Transformation." *American Journal of Sociology* 98 (July 1992): 1–29

Spretnak, Charlene. "How About That Green Option?" *Whole Earth Review,* Summer 2000, no. 101: 47–49.

Starr, Kevin. "*Sunset Magazine* and the Phenomenon of the Far West." *Sunset Magazine.* http://sunset-magazine.stanford.edu/html/influences_1.html.

Stewart, Francis. "The Case for Appropriate Technology." *Issues in Science and Technology* 3, no. 4 (Summer 1987): 101–109.

Stine, Jeffrey K., and Joel A. Tarr. "At the Intersection of Histories: Technology and the Environment." *Technology and Culture* 39 (October 1998): 601–640.

Streeter, Thomas. "That Deep Romantic Chasm: Libertarianism, Neolibertarianism, and the Computer Culture." In *Communication, Citizenship, and Social Policy,* ed. Andrew Calabrese and Jean-Claude Burgelman, 49–64. New York: Rowman & Littlefield, 1999.

Stumm, Jim. "Access to Libertarianism: 1990." *Whole Earth Review,* Summer 1990, no. 67: 54–55.

Taylor, Alan. "Unnatural Inequalities: Social and Environmental Histories." *Environmental History* 1, no. 4 (October 1996): 6–19.

Taylor, Joseph E. "On Climber, Granite, Sky." *Environmental History* 11 (January 2006): 130–133.

Taylor, Peter. "Technocratic Optimism, H. T. Odum, and the Partial Transformation of Ecological Metaphor after World War II." *Journal of the History of Biology* 21 (Summer 1988): 213–244.

Tibbs, Hardin B. C. "Industrial Ecology: An Environmental Agenda for Industry." *Whole Earth Review*, Winter 1992, no. 77: 4–19.

Tierney, John. "An Early Environmentalist, Embracing New 'Heresies.'" *New York Times*, February 27, 2007, D1, D3.

Todd, John. "Adventures of an Applied Ecologist, or Shit Happens." *Whole Earth Review*, Spring 1989, no. 62: 36–39.

———. "Ocean Arks." *CoEvolution Quarterly* 23 (Fall 1979): 46–55.

Turner, Fred. "Where the Counterculture Met the New Economy: The Well and the Origins of Virtual Community." *Technology and Culture* 46 (July 2005): 485–512.

Vaidhyanathan, Siva. "Rewiring the 'Nation': The Place of Technology in American Studies." *American Quarterly* 58 (September 2006): 555–567.

Van Der Ryn, Sim. "Abstracted in Sacramento: The Late Great State Architect Tells What He Learned about Power." *CoEvolution Quarterly* 22 (Summer 1979): 16–21.

———. "Hamilton Solar Village." *CoEvolution Quarterly* 23 (Fall 1979): 104–113.

Voyd, Bill. "Funk Architecture." In *Shelter and Society*, ed. Paul Oliver, 156–164. New York: Praeger, 1969.

Warshall, Peter. "The Ecosphere: Introducing an Einsteinian Ecology." *Whole Earth Review*, May 1985, no. 46: 28–31.

———. "Natural History Comes to Whole Earth," *CoEvolution Quarterly* 2 (Summer 1974): 70–71.

———. "Raymond Dasmann." *CoEvolution Quarterly* 12 (Fall 1976): 25.

———. "The Spiritual Labor of Whole Earth Healing." *Whole Earth* 91 (Winter 1997): 4–5.

———. "Watershed Consciousness." *CoEvolution Quarterly* 12 (Winter 1976): 4–62.

———. "A Whole Earth View of the Environmental Movement." Global Business Network online, March 2001. http://www.GBN.com.

Wells, Malcolm. "Those Ugly Solar Buildings." *CoEvolution Quarterly* 30 (Summer 1981): 71–73.

White, Richard. "Are You an Environmentalist or Do You Work for a Living?" In *Uncommon Ground: Toward Reinventing Nature*, ed. William Cronon, 171–185. New York: W. W. Norton, 1995.

———. "Discovering Nature in North America." *Journal of American History* 79 (December 1992): 874–891.

———. "Environmental History: The Development of a New Historical Field." *Pacific Historical Review* 54 (August 1985): 297–335.

———. "Environmental History, Ecology, and Meaning." *Journal of American History* 76 (March 1990): 1111–1116.

"Windmill Power." *Time*, December 2, 1974, 12.

Winner, Langdon. "Artifacts/Ideas and Political Culture. *Whole Earth Review*, Winter 1991, no. 73: 18–25.

———. "Building a Better Mousetrap: Appropriate Technology as a Social Movement." In *Appropriate Technology and Social Values: A Critical Appraisal*, ed. Franklin A. Long and Alexandra Oleson, 27–52. Cambridge, MA: Ballinger, 1980.

————. "Cyberlibertarian Myths and the Prospects for Community." In *Cyberethics: Social and Moral Issues in the Computer Age*, ed. Robert M. Baird, Reagan Mays Ramsower, and Stuart E. Rosenbaum, 319–331. Amherst, NY: Prometheus, 2000.

Worster, Donald. "The Ecology of Chaos and Order." *Environmental History Review* 14, nos. 1–2 (Spring/Summer 1990): 1–18.

————. "History as Natural History: An Essay on Theory and Method." *Pacific Historical Review* 53 (February 1984): 1–19.

————. "Transformations of the Earth: Toward an Agroecological Perspective in History." *Journal of American History* 76 (March 1990): 1087–1166

Wray, Matt. "A Blast from the Past: Preserving and Interpreting the Atomic Age." *American Quarterly* 58 (June 2006): 467–483.

Books and Dissertations

Abbott, Carl. *Frontiers Past and Future: Science Fiction and the American West*. Lawrence: University Press of Kansas, 2006.

Adams, Henry. *The Education of Henry Adams*. Boston: Houghton Mifflin, 1946.

Adkins, Sue. *Cause Related Marketing: Who Cares Wins*. Oxford: Butterworth Heinemann, 1999.

Adler, Jonathan H. *Ecology, Liberty and Property: A Free Market Environmental Reader*. Washington, DC: Competitive Enterprise Institute, 2000.

Alexander, Charles C. *Nationalism in American Thought, 1930–1945*. Chicago: Rand McNally, 1969.

Allen, Edward. *Stone Shelters*. Cambridge, MA: MIT Press, 1969.

Allenby, Braden, *Reconstructing Earth: Technology and Environment in the Age of Humans*. Washington, DC: Island, 2005.

Allin, Craig W. *The Politics of Wilderness Preservation*. Westport, CT: Greenwood, 1982.

Anderson, Terry L., and Donald R. Leal. *Free Market Environmentalism*. New York: Palgrave, 2001.

Applewhite, E. J. *Cosmic Fishing*. New York: Macmillan, 1985.

Arce, Gary. *Defying Gravity: High Adventure on Yosemite's Walls*. Berkeley, CA: Wilderness, 1996.

Armstrong, Susan J., and Richard G. Botzler. *Environmental Ethics: Divergence and Convergence*. New York: McGraw-Hill, 1993.

Auroville Community. *Auroville: The First Six Years*. Auroville, India: Auropublication, 1974.

————. *Auroville Perspective*. Pondicherry, India: Sri Aurobindo Ashram, 1973.

Baden, John A., and Donald Snow. *The Next West: Public Lands, Community, and Economy in the American West*. Washington, DC: Island, 1997.

Baer, Steve. *Dome Cookbook* (Corrales, NM: Lama Fund, 1968).

————. *Sunspots*. 2nd ed. Albuquerque, NM: Zomeworks, 1977.

Bailes, Kendall, ed. *Environmental History: Critical Issues in Comparative Perspective*. Lanham, MD: University Press of America, 1985.

Baird, Robert M., Reagan Mays Ramsower, and Stuart E. Rosenbaum, eds. *Cyberethics: Social and Moral Issues in the Computer Age*. Amherst, NY: Prometheus, 2000.

Baker, T. Lindsay. *A Field Guide to American Windmills*. Norman: University of Oklahoma Press, 1985.

Baldwin, J. *Bucky Works: Buckminster Fuller's Ideas for Today*. New York: John Wiley & Sons, 1996.

————, ed. *The Essential Whole Earth Catalog*. New York: Doubleday, 1986.

Baldwin, J., and Stewart Brand, eds. *Whole Earth Ecolog: The Best of Environmental Tools and Ideas*. New York: Harmony/Point Foundation, 1990.

————. *Soft-Tech*. New York: Penguin, 1978.

Barth, Gunther. *Fleeting Moments: Nature and Culture in American History*. New York: Oxford University Press, 1990.

Bateson, Gregory. *Steps toward an Ecology of Mind*. New York: Ballantine, 1972.

Bell, Daniel. *The Cultural Contradictions of Capitalism*. New York: Basic, 1996.

Bellamy, Edward. *Looking Backward: 2000–1887*. New York: Signet Classics, 2000.

Berger, Bennett M. *The Survival of a Counterculture: Ideological Work and Everyday Life among Rural Communards*. Berkeley: University of California Press, 1981.

Berman, Marshall. *All That Is Solid Melts into Air: The Experience of Modernity*. New York: Simon & Schuster, 1982.

————. *America in the Sixties: An Intellectual History*. New York: Free, 1968.

Berry, Wendell. *The Unsettling of America: Culture and Agriculture*. San Francisco: Sierra Club, 1977.

Betz, Mathew J., Pat McGowan, and Rolf T. Wigand. *Appropriate Technology: Choice and Development*. Durham, NC: Duke University Press, 1984.

Binkley, Sam. "Consuming Aquarius: Markets and the Moral Boundaries of the New Class, 1968–1980." Ph.D. diss., New School University, November 2001.

Birchard, Bill. *Nature's Keepers: The Remarkable Story of How the Nature Conservancy Became the Largest Environmental Organization in the World*. San Francisco: Jossey-Bass, 2005.

Blair, Alasdair, and David Hitchcock, *Environment and Business*. New York: Routledge, 2001.

Bookchin, Murray. *Post-scarcity Anarchism*. Berkeley: Ramparts, 1971.

————. *The Philosophy of Social Ecology: Essays on Dialectical Naturalism*. Montreal: Black Rose, 1990.

Borsook, Paulina. *Cyberselfish: A Critical Romp through the Terribly Libertarian Culture of High Tech*. New York: Public Affairs, 2000.

Botkin, Daniel. *Discordant Harmonies*. New York: Oxford University Press, 1990.

Boyer, Paul. *By the Bomb's Early Light: American Thought and Culture at the Dawn of the Atomic Age*. New York: Pantheon, 1985.

Boyle, T. C. *Drop City*. New York: Viking, 2003.

Brand, Stewart. *II Cybernetic Frontiers*. New York: Random House, 1974.

————. *The Clock of the Long Now: Time and Responsibility*. New York: Basic, 1999.

————. *How Buildings Learn: What Happens after They're Built*. New York: Penguin, 1994.

————, ed. *The Last Whole Earth Catalog*. San Francisco: Portola Institute & Random House, 1971.

————, ed. *The Media Lab: Inventing the Future at MIT*. New York: Viking, 1987.

————. *The Next Whole Earth Catalog*. New York: Rand McNally, 1980.

————, ed. *Space Colonies*. New York: Penguin, 1977.

————, ed. *The Whole Earth Catalog: Access to Tools*. San Francisco: Portola Institute, 1968.

————, ed., *Whole Earth Catalog: Find Your Place in Space.* Point Foundation, July 1970.

————. *The Whole Earth Epilog: Access to Tools.* New York: Point/Penguin, 1974.

Braun, Bruce, and Noel Castree, *Remaking Reality: Nature at the Millennium.* New York: Routledge, 1998.

Braunstein, Peter, and Michael Doyle, eds. *Imagine Nation: The American Counterculture of the 1960s and '70s.* New York: Routledge, 2002.

Brautigan, Richard. *The Pill versus the Spring Hill Mine Disaster.* San Francisco: Four Seasons, 1968.

Brooks, David. *Bobos in Paradise: The New Upper Class and How They Got There.* New York: Touchstone, 2000.

Brown, David E., Mindy Fox, and Leslie Hoffman, eds. *Sustainable Architecture White Papers.* New York: Earth Pledge, 2005 [2001].

Bryan, William. "An Identification and Analysis of Power-Coercive Change Strategies and Techniques Utilized by Selected Environmental Change Agents." Ph.D. diss., University of Michigan, 1971.

Bryner, Gary C. *Gaia's Wager: Environmental Movements and the Challenge of Sustainability.* New York: Rowman & Littlefield, 2001.

Buell, Lawrence. *The Environmental Imagination.* Cambridge, MA: Belknap Press of Harvard University Press, 1995.

Calabrese, Andrew, and Jean-Claude Burgelman, eds. *Communication, Citizenship, and Social Policy.* New York: Rowman & Littlefield, 1999.

Callenbach, Ernest. *Ecotopia.* New York: Bantam, 1977.

————. *Living Cheaply with Style: Live Better and Spend Less.* 2nd ed. San Francisco: Ronin, 2000.

Callicott, J. Baird. *In Defense of the Land Ethic: Essays in Environmental Philosophy.* Albany: State University of New York Press, 1989.

Capra, Fritjof. *The Web of Life: A New Scientific Understanding of Living Systems.* New York: Anchor, 1997.

Carhart, Arthur H. *How To Plan the Home Landscape.* New York: Doubleday, 1935.

————. *Planning for America's Wildlands.* Harrisburg, PA: Telegraph, 1961.

Carr, Marilyn. *The AT Reader: Theory and Practice in Appropriate Technology.* New York: Intermediate Technology Development Group of North America, 1985.

Carson, Rachel. *Silent Spring.* Boston: Houghton Mifflin, 1962.

Cheremisinoff, Nicholas P. *Fundamentals of Wind Energy.* Ann Arbor, MI: Ann Arbor Science, 1978.

Chouinard, Yvon. *Climbing Ice.* San Francisco: Sierra Club, 1978.

————. *Let My People Go Surfing: The Education of a Reluctant Businessman.* New York: Penguin, 2005.

Chouinard, Yvon, and Tom Frost. *Chouinard Equipment Catalog.* Ventura, CA: Great Pacific Iron Works, 1972.

Clark, Bruce, and Linda Henderson, eds. *From Energy to Information: Representation in Science and Technology, Art, and Literature.* Stanford, CA: Stanford University Press, 2002.

Clifford, James. *The Predicament of Culture: Twentieth-Century Ethnography, Literature, and Art.* Cambridge, MA: Harvard University Press, 1988.

Cohen, Lizabeth. *A Consumer's Republic: The Politics of Mass Consumption in Postwar America*. New York: Vintage, 2004.

Commoner, Barry. *The Closing Circle: Nature, Man, and Technology*. New York: Alfred A. Knopf, 1971.

Coonley, Douglas R. *Wind: Making It Work for You*. Philadelphia: Franklin Institute Press, 1979.

Corn, Joseph J., and Brian Horrigan. *Yesterday's Tomorrows: Past Visions of the American Future*. Baltimore: Johns Hopkins University Press, 1984.

Coyote, Peter. *Sleeping Where I Fall*. Washington, DC: Counterpoint, 1998.

Crane, Andrew. *Marketing, Morality and the Natural Environment*. New York: Routledge, 2000.

Cronon, William. ed. *Uncommon Ground: Toward Reinventing Nature*. New York: W. W. Norton, 1995.

Daniels, Farrington. *Direct Use of the Sun's Energy*. New Haven, CT: Yale University Press, 1964.

Darnovsky, Marcy, Barbara Epstein, Richard Flacks, eds. *Cultural Politics and Social Movements*. Philadelphia: Temple University Press, 1995.

Dickens, Peter. *Society and Nature: Towards a Green Social Theory*. Philadelphia: Temple University Press, 1992.

Dickson, David. *Alternative Technology and the Politics of Technical Change*. Glasgow: Fontana/Collins, 1974.

Didion, Joan. *Slouching towards Bethlehem*. New York: Farrar, Straus & Giroux, 1961.

Doherty, Brian. *This Is Burning Man: The Rise of a New American Underground*. New York: Little, Brown, 2004.

Douglas, Mary. *Purity and Danger: An Analysis of Concepts of Pollution and Taboo*. London: Routledge, 1980.

Douglas, Mary, and Aaron Wildavsky. *Risk and Culture: An Essay on the Selection of Technological and Environmental Dangers*. Berkeley: University of California Press, 1982.

Dowie, Mark. *American Foundations: An Investigative History*. Cambridge, MA: MIT Press, 2001.

Duerr, Hans Peter. *Dreamtime: Concerning the Boundary between Wilderness and Civilization*. New York: Basil Blackwell, 1985.

Ehrlich, Paul R. *The Population Bomb*. New York: Ballantine, 1968.

Ehrlich, Paul R., and Anne H. Ehrlich. *One with Nineveh: Politics, Consumption, and the Human Future*. Washington, DC: Island, 2005.

―――. *Population Resources Environment: Issues in Human Ecology*. San Francisco: W. H. Freeman, 1970.

Elkington, John. *Cannibals with Forks: The Triple Bottom Line of 21st Century Business*. Gabriola Island, Canada: New Society, 1998.

Elliot, Robert, and Arran Gare. *Environmental Philosophy*. University Park: Pennsylvania State University Press, 1983.

Ellul, Jacques. *The Technological Society*, translated by Joachim Neugroschel. New York: Continuum, 1980.

Entikin, J. Nicholas. *The Betweenness of Place: Towards a Geography of Modernity*. Baltimore: Johns Hopkins University Press, 1991.

Etulain, Richard. *Re-imagining the Modern American West: A Century of Fiction, History, and Art.* Tucson: University of Arizona Press, 1996.

Evernden, Neil. *The Natural Alien: Humankind and Environment.* Toronto: University of Toronto Press, 1985.

———. *The Social Creation of Nature.* Baltimore: Johns Hopkins University Press, 1992.

Farber, David R. *The Age of Great Dreams: America in the 1960s.* New York: Hill & Wang, 1994.

———. *The Sixties: From Memory to History.* Chapel Hill: University of North Carolina Press, 1994.

Farrell, James J. *The Spirit of the Sixties: Making Postwar Radicalism.* New York: Routledge, 1997.

Ferkiss, Victor. *Nature, Technology, and Society: Cultural Roots of the Current Environmental Crisis.* New York: New York University Press, 1993.

Feyerabend, Paul. *Against Method: Outline of an Anarchistic Theory of Knowledge.* London: Verso, 1979.

Findlay, John M. *Magic Lands: Western Cityscapes and American Culture after 1940.* Berkeley: University of California Press, 1992.

Fitch, James Marston. *Historic Preservation: Curatorial Management of the Built World.* Charlottesville: University Press of Virginia, 1990.

Fluegelman, Andrew, ed. *More New Games.* New York: Doubleday, 1981.

———. *The New Games Book.* San Francisco: Headlands, 1976.

Fox, Richard Wightman, and T. J. Jackson Lears. *The Culture of Consumption: Critical Essays in American History, 1880–1980.* New York: Pantheon, 1983.

Fox, Stephen. *The American Conservation Movement: John Muir and His Legacy.* Madison: University of Wisconsin Press, 1981.

Frank, Robert. *The Americans.* New York: Aperture, 1978.

Frank, Thomas. *The Conquest of Cool: Business Culture, Counterculture, and the Rise of Hip Consumerism.* Chicago: University of Chicago Press, 1997.

———. *One Market under God: Extreme Capitalism, Market Populism, and the End of Economic Democracy.* New York: Doubleday, 2000.

Frankel, Carl. *In Earth's Company: Business, Environment and the Challenge of Sustainability.* Gabriola Island, Canada: New Society, 1998.

Frykman, Jonas, and Orvar Löfgren. *Culture Builders: A Historical Anthropology of Middle-Class Life,* translated by Alan Crozier. New Brunswick, NJ: Rutgers University Press, 1987.

Fuller, R. Buckminster, and Robert Marks. *The Dymaxion World of Buckminster Fuller.* Garden City, NY: Anchor, 1973.

Geertz, Clifford. *The Interpretation of Cultures.* New York: Basic, 1973.

Ghamari-Tabrizi, Sharon. *The Worlds of Herman Kahn: The Intuitive Science of Thermonuclear War.* Cambridge, MA: Harvard University Press, 2005.

Gibbons, Euell. *Stalking the Wild Asparagus.* Chambersburg, PA: Alan C. Hood, 2005 [1962].

Giddens, Anthony. *A Contemporary Critique of Historical Materialism.* Berkeley: University of California Press, 1981.

Gipe, Paul. *Wind Energy Comes of Age.* New York: Wiley, 1995.

Gitlin, Todd. *The Sixties.* New York: Bantam, 1987.

Goffman, Ken, and Dan Joy, *Counterculture through the Ages: From Abraham to Acid House.* New York: Villard, 2004.

Gore, Al. *An Inconvenient Truth: The Planetary Emergency of Global Warming and What We Can Do about It*. New York: Rodale, 2006.

Gottlieb, Robert. *Forcing the Spring: The Transformation of the American Environmental Movement*. Washington, DC: Island, 1993.

Graf, Robert. *Wilderness Preservation and the Sagebrush Rebellions*. Savage, MD: Rowman & Littlefield, 1990.

Hackleman, Michael. *The Homebuilt, Wind-Generated Electricity Handbook*. Culver City, CA: Peace, 1975.

Haden, Dolores. *Building Suburbia: Green Fields and Urban Growth, 1820–2000*. New York: Vintage, 2003.

Haraway, Donna. *Primate Visions: Gender, Race, and Nature in the World of Modern Science*. New York: Routledge, 1989.

———. *Simians, Cyborgs, and Women: The Reinvention of Nature*. New York: Routledge, 1991.

Harding, Jim, ed. *Tools for the Soft Path*. San Francisco: Friends of the Earth, 1979.

Harvey, David. *The Condition of Postmodernity*. Cambridge, MA: Basil Blackwell, 1989.

Harvey, Mark W. T. *A Symbol of Wilderness: Echo Park and the American Conservation Movement*. Albuquerque: University of New Mexico Press, 1994.

Hawken, Paul. *The Ecology of Commerce: A Declaration of Sustainability*. New York: Harper-Collins, 1993.

———. *Growing a Business*. New York: Simon & Schuster, 1987.

Hawken, Paul, Amory Lovins, and L. Hunter Lovins, *Natural Capitalism: Creating the Next Industrial Revolution*. Boston: Little, Brown, 1999.

Hayles, Katherine N. *Chaos Bound: Orderly Disorder in Contemporary Literature and Science*. Ithaca, NY: Cornell University Press, 1990.

———. *The Cosmic Web: Scientific Field Models and Literary Strategies in the Twentieth Century*. Ithaca, NY: Cornell University Press, 1984.

Hays, Samuel P. *Beauty, Health, and Permanence: Environmental Politics in the United States, 1955–1985*. Cambridge: Cambridge University Press, 1987.

———. *Conservation and the Gospel of Efficiency: The Progressive Conservation Movement, 1890–1920*. Cambridge, MA: Harvard University Press, 1959.

———. *Explorations in Environmental History*. Pittsburgh: University of Pittsburgh Press, 1998.

Heath, Joseph, and Andrew Potter. *Nation of Rebels: Why Counterculture Became Consumer Culture*. New York: HarperBusiness, 2005.

Heffernan, Nick. *Projecting Post-Fordism: Capital, Class, and Technology in Contemporary American Culture*. London: Pluto, 2000.

Heider, Ulrike. *Anarchism: Left, Right, and Green*. San Francisco: City Lights, 1994.

Heinlein, Robert A. *The Moon Is a Harsh Mistress*. New York: Orb, 1997.

Helvarg, David. *The War against the Greens: The "Wise-Use" Movement, the New Right, and Anti-environmental Warfare*. San Francisco: Sierra Club, 1994.

Henderson, Linda Dalrymple. *The Fourth Dimension and Non-Euclidean Geometry in Modern Art*. Princeton, NJ: Princeton University Press, 1983.

Herber, Lewis [Murray Bookchin]. *Our Synthetic Environment*. New York: Alfred A. Knopf, 1962.

Herbert, Frank. *Dune.* New York: Ace, 2005, 40th anniversary edition.

Herring, Horace. *From Energy Dreams to Nuclear Nightmares: Lessons from the Anti–Nuclear Power Movement in the 1970s.* Charlbury, UK: Jon Carpenter, 2006.

Herron, John, and Andrew Kirk, eds. *Human Nature: Biology, Culture, and Environmental History.* Albuquerque: University of New Mexico Press, 1998.

Hills, Richard L. *Power from Wind: A History of Windmill Technology.* Cambridge: Cambridge University Press, 1994.

Hine, Robert V. *California Utopianism: Contemplations of Eden.* San Francisco: Boyd & Fraser, 1981.

Horowitz, Daniel. *Jimmy Carter and the Energy Crisis of the 1970s.* Boston: Bedford/St. Martin's, 2005.

Hoy, Suellen. *Chasing Dirt: The American Pursuit of Cleanliness.* New York: Oxford University Press, 1995.

Hughes, Jonathan, and Simon Sadler, eds. *Non-plan: Essays on Freedom, Participation and Change in Modern Architecture and Urbanism.* Oxford: Architectural, 2000.

Hughes, Thomas P. *American Genesis: A Century of Invention and Technological Enthusiasm, 1870–1970.* New York: Penguin, 1989.

Hunt, Lynn, ed. *The New Cultural History.* Berkeley: University of California Press, 1989.

Inglis, David Rittenhouse. *Wind Power and Other Energy Options.* Ann Arbor: University of Michigan Press, 1978.

Jackson, Kenneth. *Crabgrass Frontier: The Suburbanization of the United States.* New York: Oxford University Press, 1985.

Jacob, Jeffrey. *The New Pioneers: The Back-to-the-Land Movement and the Search for a Sustainable Future.* University Park: Pennsylvania State University Press, 1997.

Jacobs, Jane. *The Death and Life of Great American Cities.* New York: Vintage, 1992.

———. *Systems of Survival: A Dialogue on the Moral Foundations of Commerce and Politics.* New York: Random House, 1992.

Jéquier, Nicolas, ed. *Appropriate Technology: Problems and Promises.* Paris: Organisation for Economic Co-operation and Development, 1976.

Johnson, Huey D. *Green Plans: Greenprint for Sustainability.* Lincoln: University of Nebraska Press, 1995.

———. *No Deposit—No Return: Man and His Environment—A View toward Survival.* Reading, MA: Addison-Wesley, 1970.

Jones, Chris. *Climbing in North America.* Berkeley: University of California Press, 1976.

Kahn, Herman. *On Thermonuclear War.* Princeton, NJ: Princeton University Press, 1960.

———. *Thinking the Unthinkable.* New York: Touchstone, 2002.

Kahn, Lloyd. *Home Work: Handbuilt Shelter.* Bolinas, CA: Shelter, 2004.

———. *Refried Domes.* Bolinas, CA: Shelter, 1989.

———, ed. *Shelter.* Bolinas, CA: Shelter, 1973.

Kahn, Lloyd, and others. *Domebook 2.* Bolinas, CA: Shelter, 1972

Kahrl, William L, ed. *The California Water Atlas.* Sacramento: State of California, Governor's Office of Planning and Research, and California Department of Water Resources, 1978.

Kammen, Michael. *Mystic Cords of Memory: The Transformation of Tradition in American Culture.* New York: Vintage, 1991.

Kaplan, Fred M. *The Wizards of Armageddon*. Stanford, CA: Stanford University Press, 1991.

Keller, Evelyn Fox. *Reflections on Gender and Science*. New Haven, CT: Yale University Press, 1978.

Kelly, Kevin. *Out of Control: The Rise of the Neo-biological Civilization*. Reading, MA: Addison-Wesley, 1994.

———, ed. *Signal: Communication Tools for the Information Age: A Whole Earth Catalog*. New York: Harmony, 1988.

Kenner, Hugh. *Bucky: A Guided Tour of Buckminster Fuller*. New York: William Morrow, 1973.

Kidder, Tracy. *House*. Boston: Houghton Mifflin, 1985.

Kingsworth, Jimmie M., and Jacqueline S. Palmer. *Ecospeak: Rhetoric and Environmental Politics in America*. Carbondale: Southern Illinois University Press, 1992.

Kirk, Andrew G. *Collecting Nature: The American Environmental Movement and the Conservation Library*. Lawrence: University Press of Kansas, 2001.

Klatch, Rebecca E. *A Generation Divided: The New Left, the New Right, and the 1960s*. Berkeley: University of California Press, 1999.

Kleiman, Jordan Benson. "The Appropriate Technology Movement in American Political Culture." Ph.D. diss., University of Rochester, 2000.

Kleiner, Art. *The Age of Heretics: Heroes, Outlaws, and the Forerunners of Corporate Change*. New York: Doubleday, 1996.

Kleiner, Art, and Stewart Brand, eds., *News That Stayed News, 1974–1984: Ten Years of Co-Evolution Quarterly*. San Francisco: North Point, 1986.

Kor, Layton. *Beyond the Vertical*. Boulder, CO: Alpine House, 1983.

Kornbluth, Jesse, ed. *Notes from the New Underground*. New York: Viking, 1968.

Krausse, Joachim, and Claude Lichtenstein, eds. *Your Private Sky, R. Buckminster Fuller: The Art of Design Science*. Baden, Switzerland: Lars Muller, 1999.

Kuhn, Thomas S. *The Structure of Scientific Revolutions*. 2nd ed. Chicago: University of Chicago Press, 1970.

Lamm, Richard, and Michael McCarthy. *The Angry West: A Vulnerable Land and Its Future*. Boston: Houghton Mifflin, 1982.

Landis, Scott. *The Workshop Book: A Craftsman's Guide to Making the Most of Any Work Space*. Newton, CT: Taunton, 1987.

Lasch, Christopher. *The True and Only Heaven: Progress and Its Critics*. New York: W. W. Norton, 1991.

Latour, Bruno. *Science in Action: How to Follow Scientists and Engineers through Society*. Cambridge, MA: Harvard University Press, 1987.

Lears, T. J. Jackson. *Fables of Abundance: A Cultural History of Advertising in America*. New York: Basic, 1994.

———. *No Place of Grace: Antimodernism and the Transformation of American Culture, 1880–1920*. New York: Pantheon, 1981.

Leonard, George. *The Ultimate Athlete*. New York: Viking, 1974.

Leopold, Aldo. *A Sand County Almanac: And Sketches Here and There*. New York: Oxford University Press, 1949.

Levy, Steven. *Hackers: Heroes of the Computer Revolution*. New York: Penguin, 1994.

———. *Insanely Great: The Life and Times of the Macintosh, the Computer That Changed Everything*. New York: Viking, 1994.

Lewallen, Constance M., and Steve Seid. *Ant Farm 1968–1978*. Berkeley: University of California Press, 2004.

Lewis, Martin. *Green Delusions: An Environmentalist Critique of Radical Environmentalism*. Durham, NC: Duke University Press, 1992.

Limerick, Patricia Nelson. *The Legacy of Conquest: The Unbroken Past of the American West*. New York: W. W. Norton, 1987.

———. *Something in the Soil: Legacies and Reckonings in the New West*. New York: W. W. Norton, 2000.

Lipset, David. *Gregory Bateson: The Legacy of a Scientist*. Englewood Cliffs, NJ: Prentice-Hall, 1980.

Livingston, James. *Pragmatism and the Political Economy of Cultural Revolution, 1850–1940*. Chapel Hill: University of North Carolina Press, 1997.

Long, Franklin A., and Alexandra Oleson, eds. *Appropriate Technology and Social Values: A Critical Appraisal*. Cambridge, MA: Ballinger, 1980.

Lothstein, Arthur, ed. *"All We Are Saying . . .": The Philosophy of the New Left*. New York: Capricorn, 1970.

Louter, David. *Windshield Wilderness: Cars, Roads, and Nature in Washington's National Parks*. Seattle: University of Washington Press, 2006.

Lovelock, James. *The Ages of Gaia*. New York: W. W. Norton, 1988.

———. *Gaia: A New Look at Life on Earth*. Oxford: Oxford University Press, 1979.

———. *Gaia: The Practical Science of Planetary Medicine*. London: Gaia, 1991.

———. *Homage to Gaia*. Oxford: Oxford University Press, 2000.

Lovins, Amory. *The Energy Controversy: Soft Path Questions and Answers*. San Francisco: Friends of the Earth, 1979.

Lowenthal, David. *The Heritage Crusade and the Spoils of History*. Cambridge. MA: Cambridge University Press, 1998.

Macy, Christine, and Sarah Bonnemaison, *Architecture and Nature: Creating the American Landscape*. New York: Routledge, 2003.

Mander, Jerry. *In the Absence of the Sacred: The Failure of Technology and the Survival of the Indian Nations*. San Francisco: Sierra Club, 1991.

Marcuse, Herbert. *One-Dimensional Man: Studies in the Ideology of Advanced Industrial Society*. Boston: Beacon, 1964.

Margulis, Lynn, and Dorion Sagan. *Slanted Truths: Essays on Gaia, Symbiosis, and Evolution*. New York: Copernicus, 1997.

Markoff, John. *What the Dormouse Said: How the 60s Counterculture Shaped the Personal Computer Industry*. New York: Viking, 2005.

Marks, Robert, ed. *Buckminster Fuller: Ideas and Integrities*. Englewood Cliffs, N.J.: Prentice-Hall, 1963.

Marx, Leo. *The Machine in the Garden: Technology and the Pastoral Ideal in America*. New York: Oxford University Press, 1964.

Mau, Bruce. *Massive Change*. New York: Phaidon, 2004.

McDonough, William, and Michael Braungart. *Cradle to Cradle: Remaking the Way We Make Things*. New York: North Point, 2002.

McDougall, Walter. *The Heavens and the Earth: A Political History of the Space Age*. New York: Basic, 1985.

McLennan, Jason F. *The Philosophy of Sustainable Design: The Future of Architecture.* Kansas City, MO: Ecotone, 2004.

Meadows, D. H., Jorgen Randers, and Dennis L. Meadows. *The Limits to Growth: The 30-Year Update.* London: Earthscan, 2004.

Melville, Keith. *Communes in the Counterculture: Origins, Theories, Styles of Life.* New York: William Morrow, 1972.

Menand, Louis. *The Metaphysical Club: A Story of Ideas in America.* New York: Farrar, Straus & Giroux, 2001.

Merchant, Carolyn. *The Death of Nature: Women, Ecology, and the Scientific Revolution.* San Francisco: Harper & Row, 1980.

————. *Ecological Revolutions: Nature, Gender, and Science in New England.* Chapel Hill: University of North Carolina Press, 1989.

Merrill, Karen R. *The Oil Crisis of 1973–1974.* Boston: Bedford/St. Martin's, 2007.

Miles, Barry. *Hippie.* New York: Sterling, 2004.

Miller, Char. *Gifford Pinchot and the Making of Modern Environmentalism.* Washington, DC: Island, 2001.

————. *Ground Work: Conservation in American Culture.* Durham, NC: Forest History Society, 2007.

Miller, Perry. *The Life of the Mind in America.* New York: Harvest, 1965.

Miller, Timothy. *The Hippies and American Values.* Knoxville: University of Tennessee Press, 1991.

————. *The 60s Communes: Hippies and Beyond.* Syracuse, NY: Syracuse University Press, 1999.

Mills, Stephanie. *In Service of the Wild: Restoring and Reinhabiting Damaged Land.* Boston: Beacon, 1995.

————, ed. *Turning Away from Technology: A New Vision for the 21st Century.* San Francisco: Sierra Club, 1997.

Mitman, Gregg. *The State of Nature: Ecology, Community, and American Social Thought, 1900–1950.* Chicago: University of Chicago Press, 1992.

Mukerji, Chandra, and Michael Schudson. *Rethinking Popular Culture: Contemporary Perspectives in Cultural Studies.* Berkeley: University of California Press, 1991.

Mumford, Lewis. *The Myth of the Machine: The Pentagon of Power.* New York: Harcourt Brace Jovanovich, 1970.

————. *Technics and Civilization.* New York: Harcourt, Brace & World, 1963.

Myers, Norman, and Jennifer Kent. *The New Consumers: The Influence of Affluence on the Environment.* Washington, DC: Island, 2004.

Nash, Gerald D. *The American West in the Twentieth Century: A Short History of an Urban Oasis.* Albuquerque: University of New Mexico Press, 1973.

Nash, Hugh, ed. *The Energy Controversy: Soft Path Questions and Answers.* San Francisco: Friends of the Earth, 1979.

Nash, Roderick. *Wilderness and the American Mind.* 4th ed. New Haven, CT: Yale University Press, 2001.

Nearing, Helen, and Scott Nearing. *Living the Good Life: How to Live Sanely and Simply in a Troubled World.* 2nd ed. New York: Galahad, 1970.

Newton, Norman T. *Design on the Land: The Development of Landscape Architecture.* Cambridge, MA: Harvard University Press, 1971.

Nicholas, Liza, Elaine M. Bapis, and Thomas J. Harvey, eds. *Imagining the Big Open: Nature, Identity, and Play in the New West.* Salt Lake City: University of Utah Press, 2003.

Nielsen, Waldemar A. *The Big Foundations: A Twentieth Century Fund Study.* New York: Columbia University Press, 1972.

Nilsen, Richard, ed. *Helping Nature Heal: An Introduction to Environmental Restoration.* Berkeley: Whole Earth Catalog/Ten Speed, 1991.

Niven, Larry. *Ring World.* New York: Del Ray, 1985.

Norwood, Vera. *Made from This Earth: American Women and Nature.* Chapel Hill: University of North Carolina Press, 1993.

Nye, David. *America as Second Creation: Technology and Narratives of New Beginnings.* Cambridge, MA: MIT Press, 2004.

———. *American Technological Sublime.* Cambridge, MA: MIT Press, 1996.

Obst, Linda. *The Sixties: The Decade Remembered Now, by the People Who Lived It Then.* New York: Random House/Rolling Stone, 1977.

Oelschlaeger, Max. *The Idea of Wilderness: From Prehistory to the Age of Ecology.* New Haven, CT: Yale University Press, 1991.

Oliver, Paul, ed. *Shelter and Society.* New York: Frederick A. Praeger, 1969.

Olkowski, Helga, Bill Olkowski, Tom Javits, and the Farallones Institute staff. *The Integral Urban House: Self-reliant Living in the City.* San Francisco: Sierra Club, 1979.

O'Neill, Gerard. *The High Frontier: Human Colonies in Space.* Princeton, NJ: Space Studies Institute Press, 1989.

Orr, David. *Earth in Mind.* Washington, DC: Island, 1994.

———. *The Nature of Design: Ecology, Culture, and Human Intention.* New York: Oxford University Press, 2002.

Osborn, Fairfield. *Our Plundered Planet.* Boston: Little, Brown, 1948.

Ottman, Jacquelyn A. *Green Marketing: Opportunity for Innovation.* Chicago: NTC, 1998.

Paehlke, Robert C. *Environmentalism and the Future of Progressive Politics.* New Haven, CT: Yale University Press, 1989.

Parks, Dennis. *Living in the Country Growing Weird: A Deep Rural Adventure.* Reno: University of Nevada Press, 2001.

Perry, Charles. *A History of the Haight-Ashbury.* New York: Vintage, 1985.

Petroski, Henry. *The Evolution of Useful Things.* New York: Alfred A. Knopf, 1992.

Petulla, Joseph M. *American Environmental History: The Exploitation and Conservation of Natural Resources.* San Francisco: Boyd and Fraser, 1977.

———. *American Environmentalism: Values, Tactics, Priorities.* College Station: Texas A&M University Press, 1980.

Phillips, Michael. *The Briarpatch Book: Experiences in Right Livelihood and Simple Living from the Briarpatch Community.* San Francisco: New Glide/Reed, 1978.

———. *The Seven Laws of Money.* Boston: Shambhala, 1997.

Piana, Paul. *Big Walls: Breakthroughs on the Free-Climbing Frontier.* San Francisco: Sierra Club, 1997.

Pirsig, Robert M. *Zen and the Art of Motorcycle Maintenance.* New York: Bantam Books, 1975.

Pollan, Michael. *Second Nature: A Gardener's Education.* New York: Atlantic Monthly, 1991.

Potter, David M. *People of Plenty: Economic Abundance and the American Character.* Chicago: University of Chicago Press, 1954.

Price, Roberta. *Huerfano: A Memoir of Life in the Counterculture.* Amherst: University of Massachusetts Press, 2004.

Pringle, Hamish and Marjorie Thompson. *Brand Spirit: How Cause Related Marketing Builds Brands.* Chichester, UK: John Wiley, 1999.

Pursell, Carroll. *The Machine in America: A Social History of Technology.* Baltimore: Johns Hopkins University Press, 1995.

Rabbit, Peter. *Drop City.* New York: Olympia Press, 1971.

Rainbook: Resources for Appropriate Technology. New York: Schocken, 1977.

Reich, Charles A. *The Greening of America: How the Youth Revolution Is Trying to Make America Livable.* New York: Random House, 1970.

Rheingold, Howard, ed. *The Millennium Whole Earth Catalog.* San Francisco: Harper San Francisco, 1994.

Righter, Robert W. *Wind Energy in America: A History.* Norman: University of Oklahoma Press, 1996.

Rolston, Holmes. *Conserving Natural Value.* New York: Columbia University Press, 1994.

———. *Environmental Ethics: Duties to and Values in the Natural World.* Philadelphia: Temple University Press, 1988.

Rome, Adam. *The Bulldozer in the Countryside: Suburban Sprawl and the Rise of American Environmentalism.* Cambridge: Cambridge University Press, 2001.

Roper, Steve. *Camp 4: Recollections of a Yosemite Rockclimber.* Seattle: Mountaineers, 1994.

Roszak, Theodore. *The Cult of Information: A Neo-Luddite Treatise on High-Tech, Artificial Intelligence, and the True Art of Thinking.* Berkeley: University of California Press, 1994.

———. *From Satori to Silicon Valley: San Francisco and the American Counterculture.* San Francisco: Don't Call It Frisco, 1986.

———. *The Making of a Counter Culture: Reflections on the Technocratic Society and Its Youthful Opposition.* Berkeley: University of California Press, 1969.

———. *The Voice of the Earth.* New York: Simon & Schuster, 1992.

———. *Where the Wasteland Ends: Politics and Transcendence in Postindustrial Society.* Garden City, NY: Anchor, 1973.

Rothman, Hal K. *Devil's Bargains: Tourism in the Twentieth-Century American West.* Lawrence: University Press of Kansas, 1998.

———. *The Greening of a Nation? Environmentalism in the United States since 1945.* New York: Harbrace, 1997.

———. *The New Urban Park: Golden Gate National Recreation Area and Civic Environmentalism.* Lawrence: University Press of Kansas, 2004.

———. *Saving the Planet: The American Response to the Environment in the Twentieth Century.* Chicago: Ivan R. Dee, 2000.

Rubin, Charles T. *Conservation Reconsidered: Nature, Virtue, and American Liberal Democracy.* Lanham, MD: Rowman & Littlefield, 2000.

Rybczynski, Witold. *Home: A Short History of an Idea.* New York: Penguin, 1986.

———. *Paper Heroes: A Review of Appropriate Technology.* Garden City, NY: Anchor, 1980.

————. *Stop the 5-Gallon Flush.* Montreal: Minimum Cost Housing Group, 1975.

————. *Taming the Tiger: The Struggle to Control Technology.* New York: Penguin, 1985.

Sadler, Simon, *The Situationist City.* Cambridge, MA: MIT Press, 1998.

Safdie, Moshe. *Beyond Habitat.* Cambridge, MA: MIT Press, 1970.

Sale, Kirkpatrick. *The Green Revolution: The American Environmental Movement, 1962–1992.* New York: Hill & Wang, 1993.

Schell, Orville. *The Town That Fought to Save Itself.* New York: Pantheon, 1976.

Schumacher, E. F. *Small Is Beautiful: Economics as if People Mattered.* New York: Harper & Row, 1973.

Scott, Doug. *Big Wall Climbing: Development, Techniques and Aids.* New York: Oxford University Press, 1981.

Seager, Joni. *Earth Follies: Coming to Feminist Terms with the Global Environmental Crisis.* New York: Routledge, 1993.

Shabekoff, Philip. *A Fierce Green Fire.* New York: Hill & Wang, 1993.

Shi, David E. *The Simple Life: Plain Living and High Thinking in American Culture.* New York: Oxford University Press, 1985.

Simon, Julian L., and Herman Kahn, eds. *The Resourceful Earth: A Response to Global 2000.* Hagerstown, MD: Basil Blackwell, 1984.

Skyttner, Lars. *General Systems Theory: Ideas and Applications.* River Edge, NJ: World Scientific, 2001.

Smith, Henry Nash. *Virgin Land: The American West as Symbol and Myth.* Cambridge, MA: Harvard University Press, 1950.

Smith, Michael L. *Pacific Visions: California Scientists and the Environment, 1850–1915.* New Haven, CT: Yale University Press, 1987.

Snyder, Robert, ed. *Buckminster Fuller: Autobiographical Monologue / Scenario.* New York: St. Martin's, 1980.

Soja, Edward W. *Postmodern Geographies: The Reassertion of Space in Critical Social Theory.* London: Verso, 1989.

Soleri, Paolo. *Arcology: The City in the Image of Man.* Cambridge, MA: MIT Press, 1969.

Spretnak, Charlene. *The Resurgence of the Real: Body, Nature, and Place in a Hypermodern World.* Reading, MA: Addison-Wesley, 1997.

Spretnak, Charlene, and Fritjof Capra. *Green Politics.* New York: Dutton, 1984.

Sprin, Anne Whiston. *The Granite Garden: Urban Nature and Human Design.* New York: Basic, 1984.

Starr, Kevin. *Material Dreams: Southern California through the 1920s.* New York: Oxford University Press, 1990.

Steele, James. *Ecological Architecture: A Critical History.* New York: Thames & Hudson, 2005.

Steffen, Alex, ed. *World Changing: A User's Guide for the 21st Century.* New York: Abrams, 2006.

Sterling, Bruce. *The Hacker Crackdown: Law and Disorder on the Electronic Frontier.* New York: Bantam, 1993.

Stevens, Jay. *Storming Heaven: LSD and the American Dream.* New York: Grove, 1987.

Strong, Douglas Hillman. *Dreamers and Defenders: American Conservationists.* Lincoln: University of Nebraska Press, 1988.

Susman, Warren. *Culture as History: The Transformation of American Society in the Twentieth Century.* New York: Pantheon, 1984.

Sutter, Paul. *Driven Wild: How the Fight against Automobiles Launched the Modern Wilderness Movement.* Seattle: University of Washington Press, 2002.

Szasz, Ferenc Morton. *The Day the Sun Rose Twice: The Story of the Trinity Site Nuclear Explosion, July 16, 1945.* Albuquerque: University of New Mexico Press, 1995.

Taylor, Bob Pepperman. *Our Limits Transgressed: Environmental Political Thought in America.* Lawrence: University Press of Kansas, 1992.

Theobald, Robert, and Stephanie Mills, eds. *The Failure of Success: Ecological Values vs Economic Myths.* Indianapolis, IN: Bobbs-Merrill, 1973.

Todd, Nancy Jack. *A Safe and Sustainable World: The Promise of Ecological Design.* Washington, DC: Island, 2005.

Todd, Nancy Jack, and John Todd. *From Eco-cities to Living Machines: Principles of Ecological Design.* Berkeley, CA: North Atlantic, 1993.

Toffler, Alvin. *The Third Wave.* New York: Bantam, 1982.

Tuan, Yi-Fu. *Space and Place: The Perspective of Experience.* Minneapolis: University of Minnesota Press, 1977.

Turner, Fred. *From Counterculture to Cyberculture: Stewart Brand, the Whole Earth Network, and the Rise of Digital Utopianism.* Chicago: University of Chicago Press, 2006.

Udall, Stewart. *The Quiet Crisis.* New York: Holt, Rinehart and Winston, 1963.

Van Der Ryn, Sim. *The Toilet Papers: Recycling Waste and Conserving Water,* rept. ed. Sausalito, CA: Ecological Design Press, 1999.

———, and Stuart Cowan. *Ecological Design.* Washington, DC: Island, 1996.

Van Dresser, Peter. *Development on a Human Scale: Potentials for Ecologically Guided Growth in Northern New Mexico.* New York: Praeger, 1972.

———. *Homegrown Sundwellings.* Santa Fe, NM: Lightning Tree, 1977.

Veblen, Thorstein. *The Theory of the Leisure Class: An Economic Study of Institutions.* London: George Allen & Unwin, 1957.

Vogt, William. *Road to Survival.* New York: William Sloane, 1948.

Wagstaff, J. M., ed. *Landscape and Culture: Geographical and Archaeological Perspective.* New York: Basil Blackwell, 1987.

Wallace, Mike. *Mickey Mouse History and Other Essays on American Memory.* Philadelphia: Temple University Press, 1996.

Ward, Barbara. *Spaceship Earth: The Impact of Science on Society.* New York: Columbia University Press, 1966.

———, and Rene Dubos. *Only One Earth: The Care and Maintenance of a Small Planet.* New York: W. W. Norton, 1972.

Warshall, Peter. *Septic Tank Practices: A Guide to the Conservation and Re-use of Household Wastewater,* rept. ed. New York: Anchor, 1979.

———. "Sociodemography of Free-Ranging Male Rhesus Macaques." Ph.D. diss., Harvard University, 1973.

———, ed. *30th Anniversary Celebration: Whole Earth Catalog.* San Rafael, CA: Point Foundation, 1999.

Watson, Charles, and T. H. Watkins, *The Lands No One Knows: America and the Public Domain.* San Francisco: Sierra Club, 1975.

Weiner, Douglas R. *Models of Nature: Conservation, Ecology, and Cultural Revolution.* Bloomington: Indiana University Press, 1988.

Westrum, Ron. *Technologies and Society: The Shaping of People and Things.* Belmont, CA: Wadsworth, 1991.

White, Richard. *The Organic Machine: The Remaking of the Columbia River.* New York: Hill & Wang, 1995.

Wiener, Norbert. *Cybernetics: Or Control and Communications in the Animal and the Machine.* 2nd ed. Boston: MIT Press, 1965.

Williams, Raymond. *Problems in Materialism and Culture.* London: Verso, 1980.

Wilson, Chris. *The Myth of Santa Fe: Creating a Modern Regional Tradition.* Albuquerque: University of New Mexico Press, 1997.

Winner, Langdon. *The Whale and the Reactor: A Search for Limits in an Age of High Technology.* Chicago: University of Chicago Press, 1986.

Wolfe, Tom. *The Electric Kool-Aid Acid Test.* New York: Bantam, 1997.

Worster, Donald, ed. *The Ends of the Earth: Perspectives on Modern Environmental History.* Cambridge: Cambridge University Press, 1988.

———. *Nature's Economy: A History of Ecological Ideas.* 2nd ed. Cambridge: Cambridge University Press, 1994.

———. *Rivers of Empire: Water, Aridity, and the Growth of the American West.* New York: Oxford University Press, 1985.

———. *Under Western Skies: Nature and History in the American West.* New York: Oxford University Press, 1992.

———. *The Wealth of Nature: Environmental History and the Ecological Imagination.* New York: Oxford University Press, 1993.

Wrede, Stuart, and William Howard Adams. *Denatured Visions: Landscape and Culture in the Twentieth Century.* New York: Museum of Modern Art, 1991.

Wrobel, David. *The End of American Exceptionalism: Frontier Anxiety from the Old West to the New Deal.* Lawrence: University Press of Kansas, 1993.

———. *Promised Lands: Promotion, Memory, and the Creation of the American West.* Lawrence: University Press of Kansas, 2002.

Wyant, William K. *Westward in Eden: The Public Lands and the Conservation Movement.* Los Angeles: University of California Press, 1982.

Yergin, Daniel. *The Prize: The Epic Quest for Oil, Money and Power.* New York: Free Press, 1991.

Zelov, Chris, and Phil Cousineau, eds. *Design Outlaws on the Ecological Frontier.* Easton, PA: Knossus, 1997.

Zimmerman, Michael E. *Contesting Earth's Future: Radical Ecology and Postmodernity.* Berkeley: University of California Press, 1994.

Zung, Thomas T. K., ed. *Buckminster Fuller: Anthology for the New Millennium.* New York: St. Martin's, 2001.